In Search
of Swampland

In Search of Swampland

A Wetland Sourcebook and Field Guide

Second Edition

Ralph W. Tiner

Rutgers University Press
New Brunswick, New Jersey, and London

Library of Congress Cataloging-in-Publication Data

Tiner, Ralph W.
   In search of swampland : a wetland sourcebook
and field guide / Ralph W. Tiner. — 2nd ed.
       p.   cm.
   Includes bibliographical references (p.) and
index.
   ISBN-13: 978-0-8135-3681-1 (pbk: alk. paper)
   1. Wetlands — Northeastern States.
2. Wetlands — Lake States.   I. Title.

QH104.5.N58T56   2005
578.768'0974—dc22   2005042857

A British Cataloging-in-Publication record for this
book is available from the British Library.

Manufactured in the United States of America

In Search
of Swampland

In Search of Swampland

A Wetland Sourcebook and Field Guide

Second Edition

Ralph W. Tiner

Rutgers University Press

New Brunswick, New Jersey, and London

Library of Congress Cataloging-in-Publication Data

Tiner, Ralph W.
  In search of swampland : a wetland sourcebook
and field guide / Ralph W. Tiner. — 2nd ed.
       p.   cm.
  Includes bibliographical references (p.) and
index.
    ISBN-13: 978-0-8135-3681-1 (pbk: alk. paper)
    1. Wetlands — Northeastern States.
2. Wetlands — Lake States.   I. Title.

QH104.5.N58T56   2005
578.768′0974—dc22   2005042857

A British Cataloging-in-Publication record for this
book is available from the British Library.

Manufactured in the United States of America

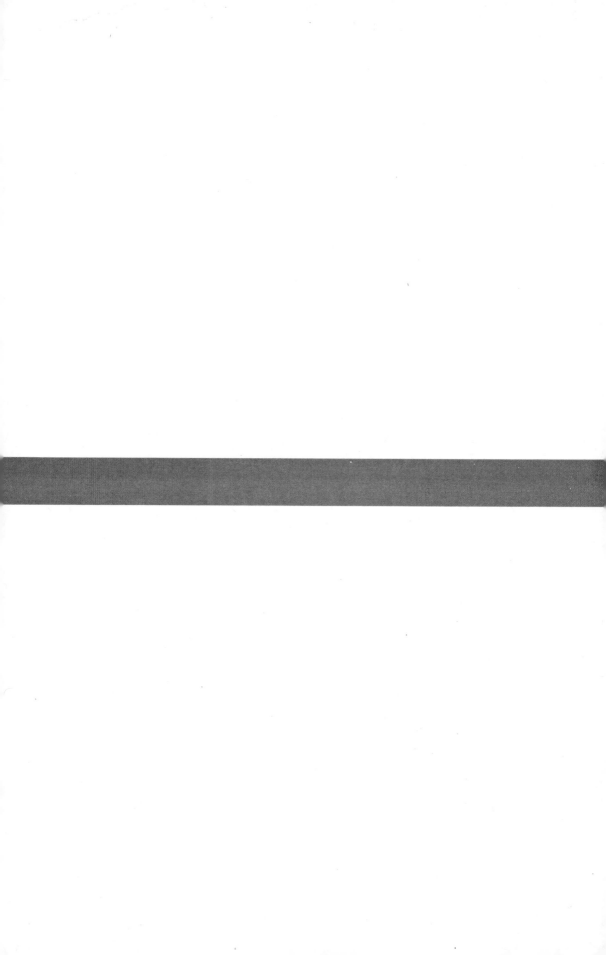

In Search
of Swampland

In Search of Swampland

A Wetland Sourcebook and Field Guide

Second Edition

Ralph W. Tiner

Rutgers University Press

New Brunswick, New Jersey, and London

Library of Congress Cataloging-in-Publication Data

Tiner, Ralph W.
   In search of swampland : a wetland sourcebook
and field guide / Ralph W. Tiner. — 2nd ed.
        p.   cm.
   Includes bibliographical references (p.) and
index.
   ISBN-13: 978-0-8135-3681-1 (pbk: alk. paper)
   1. Wetlands — Northeastern States.
2. Wetlands — Lake States.   I. Title.

QH104.5.N58T56   2005
578.768'0974—dc22   2005042857

A British Cataloging-in-Publication record for this
book is available from the British Library.

Manufactured in the United States of America

Dedicated to my wife, Barbara,
and our children: Andrew, Avery, and Dillon

Contents

List of Illustrations and Tables

Figures

Plates

Tables

Preface

Today there is great public sentiment in favor of wetlands, as evidenced by the number of government regulations and programs dealing with them. Once viewed by most people as wastelands, wetlands—marshes, swamps, and bogs—are now recognized as one of the world's most valuable natural resources. They are the temperate-zone equivalent of rain forests, serving vital life-sustaining functions in water-quality renovation, aquatic-ecosystem productivity, and biodiversity, as well as providing important socioeconomic benefits such as flood-damage protection, shoreline stabilization, and commercial and recreational fisheries.

Like their tropical counterparts, wetlands have been and are still being heavily used by human society, often with little regard to the ecological consequences. The early view of wetlands as wastelands led most people to believe that converting swampland to agricultural land or cities and suburbs was a desirable, worthwhile, and even necessary endeavor. For over two centuries, wetlands of all kinds in the United States have been filled for residential, commercial, and industrial development or drained for cropland. To date, over half of the country's wetlands that existed when the Pilgrims landed in Plymouth, Massachusetts, have been destroyed.

With the discovery of important functions of wetlands and the recognition of the adverse consequences resulting from wetland alteration, the government began protecting wetlands through a variety of mechanisms. Since the 1960s, a number of laws, regulations, ordinances, and bylaws have been established by federal, state, and local governments to help control and slow the destruction of wetlands. Many wetlands have been acquired by government agencies and private organizations and added to the nation's natural-resource trust as protected and managed fish and wildlife habitats. In the 1990s, programs have been established by the federal government and some states to repair and restore damaged wetlands.

Heightened public awareness and concern for increasing wetland protection have stimulated much scientific research. During the past three decades, there has been an explosion of scientific studies about all aspects of wetlands. With a better understanding of wetlands and a need to identify the limits of wetlands for regulatory purposes, the federal government began developing wetland identification and delineation techniques. In the 1960s and 1970s, wetlands were mainly identified by analyzing vegetation. A quick look at the plants was often all that was needed to recognize a wetland. Upon initiating the National Wetlands Inventory in the mid-1970s, the U.S. Fish and Wildlife Service (FWS) soon realized that this approach was inadequate for wetland identification since many seasonally wet areas did not possess vegetation that was distinctly different from that of adjacent upland (dry land), especially in areas of low relief. An examination of the soils then became an important part of identifying wetlands. With support from other federal agencies, the FWS developed lists of plants and soils associated with wetlands. At the same time, two federal regulatory agencies, the U.S. Army Corps of Engineers and the Environmental Protection Agency, began developing standardized techniques to identify regulated wetlands. From the mid-1970s to 1990, formal methods to identify and delineate wetlands were established and refined. Today, such methods are required to determine the extent of government jurisdiction on both public and private lands across the nation. Similar methods are also used by many states to implement wetland protection laws.

Purpose of the Book

This book serves as an introduction to wetlands and the identification of them. It is a primer rather than an exhaustive treatment of the subject, designed to give people with little or no training a better understanding and appreciation of wetlands and the basic tools for identifying wetlands, their plant and animal life, and hydric soils. As such, the book is intended for a wide audience, including science teachers and students (college and secondary school levels), biologists, ecologists, soil scientists, and other natural and physical scientists, personnel of government agencies, environmental consultants, natural-resource planners, consulting engineers, public-health officials, park naturalists, real-estate agents, environmental lawyers, farmers, developers, landowners, bankers, and anyone affected by wetland regulations who might be interested in learning more about wetlands. The book is written in a nontechnical style, yet it is impossible to convey some of the concepts without using some technical terms. It may be used as a teaching tool—a textbook for introductory courses in wetlands or environmental science—or simply as a guidebook to make the reader more familiar with wetlands.

This book is a revision of the original edition published in 1998. The book has been reprinted a few times and is used as a basic wetland textbook at some universities. Given this widespread usage, I felt the need to add some new material that should be of interest to students as well as to the average reader. The book retains the basic format of the original, but has been expanded to include more information on a variety of topics including hydrology, wetland formation, wetlands of the Great Lakes states, hydric soils, invasive species, wetlands at risk, and wetland management and conservation. Some more recent publications have been added to the Additional Readings at the end of each chapter. A timeline of significant events affecting wetlands has been added to Chapter 9. A new chapter on wetland classification and assessment (Chapter 14) has been added to the wetland identification guide since this information is crucial for interpreting wetland maps and understanding the relationship between wetland characteristics and functions. I have also expanded information on poisonous plants in the preface. A recent experience with a species related to water hemlock made it imperative that I include more information on this topic.

The book is divided into two parts, a Wetland Primer and a Wetland Identification Guide. Part I (the primer) is an overview of wetland ecology, status, trends, and conservation.

It contains nine chapters covering wetland concepts, formation, wetland characteristics (hydrology, hydric soils, hydrophytic vegetation, and wetland wildlife), functions and values, status and trends (including causes of wetland loss and degradation), and wetland protection and management. At the end of each chapter is a list of selected references that can provide more information about each subject. Other materials used in preparing this book are listed in the references section near the end of the book. Part II is a field guide to wetland plants, soils, and animals and to wetland identification, delineation, classification, and evaluation. These five chapters include descriptions and illustrations of common wetland plants and animals (with simple identification keys), information on how to distinguish typical hydric soils from better-drained soils, general procedures for identifying wetlands in the field, and an introduction to wetland classification and functional assessment. The book also contains three appendices: Appendix A, an action agenda (offering some ideas on how people can help improve wetland protection), Appendix B, a listing of some wetlands in the Northeast worth visiting, and Appendix C, a listing of some poisonous plants.

By design, the book focuses on wetlands of the northeastern United States (from Maine through Maryland and west to Ohio and Kentucky). These wetlands are used to discuss wetland ecology and related wetland identification issues. For this revised edition, the discussion has been expanded to include numerous references to wetlands of the Great Lakes region. References to wetlands in other parts of the country are also made to illustrate some key points in Part I. More than 160 plants and 120 animals characteristic of Northeast wetlands are illustrated and described in Part II, with additional references to more than 120 other plants and over 100 other animals. While the emphasis is on the more common species, several species of special interest or concern are also referred to and illustrated in Chapter 5. Since the plants and animals described in the book are common throughout much of the eastern United States, the field guide may be useful over a wider region, allowing for the omission of strictly southern or midwestern species. Much of the primer material is more generic, so the book should have wide application as an introductory textbook on wetlands.

I hope that this book provides readers with a broad understanding of wetlands and that this increased knowledge leads to a greater appreciation of the important roles wetlands play. I would be most gratified if it stimulates an interest in wetland conservation and motivates readers to make a personal commitment to preserve, restore, and enhance wetlands for people and wildlife.

Note of Caution

Before doing any field work, you should take two precautions: 1) learn to recognize poisonous plants, and 2) dress appropriately to minimize tick contact. Also, if biting insects (mosquitoes, greenheads, and deer flies) are a problem, insect repellents, available at drugstores and supermarkets, are recommended.

Numerous poisonous plants have colonized wetlands of the Northeast and Great Lakes states. Perhaps the most familiar are poison ivy (species 128 in Chapter 10) and poison sumac (species 105) which produce oils that cause serious skin irritations (e.g., blistering, oozing sores in more extreme cases). Poison ivy is easily recognized by its compound leaves composed of three leaflets and when growing as a climbing vine, by its "hairy" or wiry stems. It grows in both wetlands and uplands, especially in disturbed sites. Poison sumac, a tall shrub or short tree (to 20 feet tall) restricted to wetlands (usually on organic soils), may be identified by its large compound leaves composed of seven to thirteen leaflets. It is easiest to recognize in the fall when its leaves turn a brilliant orange color. Both species bear whitish berries that may be present in winter. Two wetland plants cause skin to sting upon contact: wood nettle (species 71) and stinging nettle (species 71b). The remaining plants considered poisonous usually must be eaten (see Appendix C), although contact with the juices of water hemlock (species 91b) can be lethal. In fact, I've read reports where children made pea-shooters from the stems of this plant and died shortly afterwards. A single bite of its roots may be fatal, so be very careful when dealing with this species and related plants. It is interesting to note that its relatives include edible plants such as carrot. Water parsnip (species 91a) has been listed as both an edible plant and poisonous in various references. The Secwepemc people of British Co-

lumbia ate water parsnip roots raw, fried, or steamed and didn't appear to suffer ill effects; but they considered the flowering inflorescence to be poisonous (http://www.secwepemc.org). Another relative, stiff cowbane (*Oxypolis rigidior*), is listed as poisonous by some sources. A word of caution: beware of herbs with compound leaves bearing many small white flowers in umbrella-like clusters (umbels) growing in wet places. Some other notable plants (not listed in Appendix C) with leaves that are poisonous when ingested include: rosebay rhododendron (species 101), sheep laurel (species 96), willows (species 113 and 153), and buttonbush (species 109), while others with toxic berries are Virginia creeper (species 129) and two invasive shrubs to small trees—common buckthorn (*Rhamnus cathartica*) and glossy buckthorn (*Frangula alnus*, formerly *Rhamnus frangula*) (Figure 8.7).

Ticks and biting insects are present in wetlands as they are elsewhere. People are alarmed at the increasing incidence of Lyme disease, an illness caused by the bacterium *Borrelia burgdorferi*, which is transmitted by infected deer ticks. Deer ticks are about half the size of the more common wood or dog tick (about 1/8 inch long v. 1/4 inch). The deer tick is reddish to orangish brown, with a dark spot behind the head. The wood tick is brown, with a large whitish spot behind the head. Deer ticks and Lyme disease are associated with any open space—woods and fields, wetlands and nonwetlands—frequented by deer.

To reduce tick risk, dress appropriately: 1) tuck your pant legs into your socks, 2) spray your boots and clothing with a DEET-type insect repellent (don't spray this on skin, especially that of children), and 3) wear a hat (this also keeps deer flies out of your hair!). If you find a tick imbedded in your skin, remove it with tweezers. Be aware that it takes a few hours for an infected tick to begin feeding (and thereby transferring the bacterium to its host), so if you discover a deer tick on you and remove it before the end of the day, you have lowered the risk of infection. Prompt removal of ticks (within twenty-four hours) is important. Wash thoroughly and save the tick for identification. Also, note that not all deer ticks are infected with the bacterium.

The following symptoms may indicate Lyme disease. The first signs usually appear within a few weeks of the bite: a reddish skin rash of raised red bumps that may increase in size to 20 inches or more in diameter. The bumps may have clear centers. Later, open sores may develop in the rash. Also, flulike symptoms including headache, fever, fatigue, soreness, and nausea may occur. Within several weeks to months, neurological and cardiac complications may develop. Finally, in the most advanced stages, chronic arthritis and more serious neurological problems may result. Should any of these symptoms develop, consult your physician or local health department. The disease is readily treated with antibiotics in the early stages, so act promptly. For more information on this disease and the ecology of deer ticks and the bacterium, see Ginsberg (1993) or contact your physician or the Gundersen Medical Foundation, La Crosse, Wisconsin.

Acknowledgments

This type of book could have been completed only with the help of many other people. The existence of many fine books on a variety of topics (listed as additional readings at the end of each chapter or in the reference section) was a great aid in preparing the book. Not surprisingly, the works of others inspired me to choose wetland conservation as a career and to help broaden people's awareness of wetlands through my own work. I'd like to pay special tribute to three scientists whose writings gave me a solid foundation to build upon: Neil Hotchkiss of the U.S. Fish and Wildlife Service, whose illustrated field guides to wetland and aquatic plants helped me learn the more common marsh species; Bill Niering of Connecticut College, whose writings conveyed the ecological and social significance of wetlands and the need to protect them; and John Teal of the Woods Hole Biological Institution, whose work on coastal wetlands (especially *Life and Death of the Salt Marsh*) heightened my awareness and gave me a reason to pursue a career in wetland conservation.

Several colleagues and friends reviewed all or parts of the draft manuscript. Their kind yet critical review helped improve this book. Special thanks go to Peter Veneman, Bill Sipple, Robert Brooks, Becky Brooks, Christy Foote-Smith, Stephen Brown, Amanda Lindley Stone, Andrew Stone, Ken Metzler, and Pam Dansereau, and especially to Aram Calhoun for reviewing the revised text and new materials included in this expanded edition.

After a quick glance through the book, readers will see that it is extensively illustrated. Most of these fine illustrations came from other field guides and government publications. In particular, nearly all of the plant drawings were prepared by Abigail Rorer (some courtesy of Dennis Magee). Abbie prepared the drawing of the northern diamondback terrapin especially for this book, while the other amphibian and reptile illustrations are also hers; they appeared in a USDA Forest Service publication, *New England Wildlife: Habitat, Natural History, and Distribution*. Most of the bird drawings, which also come from this source, were prepared by Charles Joslin. Other bird illustrations were done especially for this book by Julien Beauregard (some courtesy of the U.S. Fish and Wildlife Service) and Nancy Haver. Mammal drawings by Roslyn A. Alexander, originally prepared for *Mammals of Ontario*, were used with permission from Anne Innis Dagg (copyright 1974). Julien Beauregard also prepared the drawing of the Delmarva fox squirrel. Other copyrighted illustrations appearing in the text have been acknowledged in the captions, and I appreciate the assistance provided by the following persons: Andrew Stone of the American Ground Water Trust, L. Harold Stevenson of the Estuarine Research Federation, Heather Lengyel of the University of Michigan Press, and M. J. Crowley of the *Star-Ledger*.

Most of the photos are mine, and those of others are credited to the photographer. I'd like to thank Geoff Knapp, Fred Knapp, Robin Burr, Al Dole, Bill French, Susie von Oettingen, Hank Tyler, Barbara Tiner, Bill Zinni, the U.S. Fish and Wildlife Service, and the U.S. Army Corps of Engineers for use of their photos. The Kollmorgen Corporation, Munsell Color, New Windsor, New York, graciously granted permission to print copies of two Munsell soil-color charts (Plate 23).

Many individuals provided information that was used in preparing this book, including Dennis Magee, Ken Metzler, Bryan Windmiller, Scott Jackson, George Nicholas, Jim Fortner, Rick Kanaski, Lyda Craig, Wade Wander, Russ Cohen, Russ Cole, Patricia Riexinger, Joseph Dowhan, Bob Hosking, Jr., Ken Thoman, Sue Marcha, Wright Hitt, Joseph Winnicki, John Hansen, Joseph Bracale, Hollace Hoffman, John Keator, Paul Sedor, William Foley, William Cerynik, Robert Goodman, Ron Taglairino, Helen Maurella, Lisa

Schmidt, Dave Davis, Daniel Belknap, and Bart Wallin. I also wish to acknowledge the assistance provided by budding naturalists Matthew Rice and Andrew and Avery Tiner, and express thanks to the Guenther family of Sherborn, for collecting various amphibians and reptiles and bringing them to me for examination.

A personal note of thanks to Simon (aka "Rugrat"), our beloved Westie, who survived after ingesting a seed of stiff cowbane (*Oxypolis rigidior*), a potentially poisonous relative of the deadly water hemlock (*Cicuta maculata*). He experienced five days of lethargy, uncoordinated movement, and temporary blindness before passing the seed and eventually coming out of the trauma. This utterly regrettable incident moved me to learn more about toxic plants and pass on some of this information in the Note of Caution section of the Preface and Appendix C. Hopefully, this added material will broaden readers' awareness of this and other toxic wetland species.

I'd also like to thank Art Schnure, former chairman of the Sherborn Conservation Commission, for his computer help with the original edition of this book and to the Rutgers University Press team that prepared the original book and this revised edition for publication, especially Karen Reeds and Doreen Valentine for the first edition and Audra Wolfe who encouraged me to write this revision.

Finally, a great note of thanks is extended to the Sweet Water Trust, which provided funding for the color plates. Their contribution has made it possible for readers to better experience the beauty and variety of wetlands and to visualize the differences between hydric and non-hydric soils.

PART I

Wetland Primer

CHAPTER 1

Swampland, Marshland, Wetland
Wetland Concept and Definitions

What Is a Wetland?

For centuries, the terms "swampland" and "marshland," or simply "swamps" and "marshes," have been used interchangeably by most people to refer to wet, soggy ground. Many other terms, some regional, have also been applied to such wet areas, including "bog," "fen," "mire," "peatland," "pocosin," "pothole," "slough," "playa," "wet meadow," and "muskeg." These terms usually describe a particular type of wetland and are less generic than "marsh" and "swamp."

The average citizen's view of wetlands over the course of human history is probably dominated by the limitations of human uses of land caused by soil wetness or periodic flooding, or the unique plants and animals that occur in wet places. From the human-use standpoint, wetlands are lands that are too wet to do most things normally done on dry land, such as farming or constructing buildings, without drainage "improvement" or filling. The Swamp and Overflowed Lands Acts passed by the U.S. Congress in the mid-1800s exemplified this notion of wetland by giving certain states the authority to claim federal-owned swamps and overflowed lands for the purposes of draining and converting them to productive uses, namely agriculture. "Swamp" was defined as "land, whether open or timbered, above tide water that is too wet for cultivation." "Overflowed land" was "bottom land along streams that can not be cultivated safely because of overflow." Flooding or prolonged soil saturation places significant limitations on human uses, imposes serious stress on plants and animals, and initiates biochemical processes that promote the development of wetland soil. Recurrent prolonged wetness was then and remains today the fundamental element—the master variable—in the concept of wetland.

Wetland may be wet all year long or simply long enough to stress vegetation or affect land use. The area may be wet every year, or less often at some recurring frequency. The variations in wetness are limitless. It is clear, however, that the term is not restricted to land that is always covered by water. If all wetlands were this wet, they would include only ponds, lakes, rivers, or other bodies of water—that is, submerged or underwater lands.

Since the 1950s, "wetland" has taken on a more formal meaning as a general descriptor for all types of wet habitats: marshes, swamps, bogs, pocosins, fens, water-lily beds, and similar areas. In 1956, the U.S. Fish and Wildlife Service published *Wetlands of the United States*, containing the results of the first nationwide inventory of wet habitats. This report and others that followed identified a continuing loss of these valuable habitats. Public concern for the well-being of wetlands and alarm at the loss of valued functions such as flood storage and water renovation eventually led to the passage of state and federal laws to protect wetlands. Since the 1960s, "wetlands" has been applied almost universally to government laws and regulations, such as New Jersey's Wetlands Act of 1970, New York's Freshwater Wetlands Act, Massachusetts's Wetlands Protection Act, Maryland's Nontidal Wetlands Act, and the federal Clean Water Act, and to government programs like the National Wetlands Inventory and the Massachusetts Wetlands Restoration Program.

Wetlands are a diverse collection of areas where the presence of water for extended periods exerts a controlling influence on the plant community, soil properties, and animals living in or using them. Wetlands exist in various settings in the natural landscape, including 1) low-lying areas adjacent to rivers and other bodies of water subject to flooding, 2) isolated depressions that collect surface water, and 3) areas of groundwater discharge (both flats and slopes) (see Chapter 3). Wetlands include permanently

a. Cypress swamp (North Carolina).

b. Hardwood swamp (New Jersey).

c. Northern bog (Maine).

d. Inland marsh (Nevada).

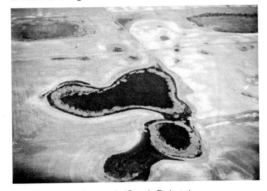

e. Prairie-pothole marsh (South Dakota).

f. Muskeg (Alaska). (Source: U.S. Fish and Wildlife Service)

Figure 1.1. Examples of wetlands in the United States.

inundated shallow marshes and water-lily beds, seasonally flooded wooded swamps, temporarily inundated floodplains, saturated bogs, seasonally saturated woods (e.g., pitch-pine lowlands), seasonal seeps, and tidally flooded coastal marshes. Most wetlands are characterized by certain types of plants that give them more distinctive names, such as red-maple swamp, cattail marsh, sedge meadow, leatherleaf bog, pitch-pine lowland, and cordgrass marsh (Figure 1.1). Although most wetlands are vegetated, many are not, such as tidal mud flats and sandbars, intertidal rocky shores, periodically exposed shores of rivers, reservoirs, and lakes, and shallow bottoms of bodies of water.

Key elements in the concept of wetland include the following: 1) wetlands are not necessarily wet all the time; 2) they may be vegetated or not; 3) they occur along bodies of water or in isolation (surrounded by upland); 4) they form in depressions, in flat areas, and even on slopes—wherever water is present in the soil for extended periods on a recurring basis; 5) they may be naturally formed or artificially created;

and 6) they occur in saltwater and freshwater environments. Implicit in this list is the variability among wetlands.

Definitions have been drafted for scientific purposes as the foundation for wetland inventories and for regulatory purposes to indicate the scope of government jurisdiction. These definitions are of vital importance since they serve as the basis for identifying and protecting lands that perform certain valued functions.

Scientific Definitions

Although some people have claimed that "wetland" is a relatively new and strictly regulatory term with meaning only in a legal context, it was used by scientists well before any wetland law or regulation was conceived. "Wetland" was among several terms ("muck," "overflowed land," and "swamp land") used to describe wet areas at the beginning of the twentieth century. In 1920 the scientist A. P. Dachnowski complained that the widespread and interchangeable use of these terms was hindering progress in peat studies. Despite his urging, the term "wetland" continued to be used by both the public and scientists. In 1928 the ecologist P. Viosca's study of Louisiana wetlands, which was published in *Ecology,* a scientific journal devoted to explaining relationships between organisms and the environment, was perhaps the first paper to extoll the values of wetlands as natural resources and to attempt to dispel the ill-conceived public notion of wetlands as wastelands.

1953 Fish and Wildlife Service Definition

In 1953, the U.S. Fish and Wildlife Service (FWS) published its first wetland classification system: "Classification of Wetlands of the United States" served as the official national classification scheme until the late 1970s and was the foundation for the FWS's nationwide inventory of wetland habitats important to waterfowl. It was also adopted by the Department of Agriculture's Soil Conservation Service (now the Natural Resources Conservation Service). Numerous inventories of wetlands conducted by the FWS from the 1950s to the 1970s identified threats to the nation's important

waterfowl wetlands. The FWS's first national report on wetlands, *Wetlands of the United States,* was published in 1956.

The general definition of wetlands—"lowlands covered with shallow and sometimes temporary or intermittent waters"—gives the impression that all wetlands are in low topographic positions and must be flooded at some time. Permanently flooded lands are excluded from this definition, except for the vegetated shallows of lakes and ponds. Also excluded are areas that are not flooded long enough to promote the growth of wetland vegetation. In addition to the general definition, the wetland classification system contains definitions of many wetland types, such as inland fresh meadows, inland deep marshes, shrub swamps, bogs, wooded swamps, and coastal salt meadows. These definitions make reference to water depth and duration during the growing season and to typical vegetation. The 1956 report included a discussion of wetland soils (hydromorphic, halomorphic, and alluvial).

1979 Fish and Wildlife Service Definition

With increasing national concern about losses of wetlands and a growing public awareness and appreciation of wetland functions and values, the FWS established the National Wetlands Inventory project (NWI) in the mid-1970s to collect, analyze, and disseminate scientific information on the characteristics and extent of U.S. wetlands. One of the first tasks of the NWI was to decide on a wetland definition and classification system to form the basis for mapping. The earlier FWS system used to survey wetlands important for waterfowl was not sufficient and had been inconsistently applied across the country.

This was the first ecologically based wetland definition with extensive scientific peer review and public input. The definition and accompanying wetland classification system went through three major drafts and extensive field testing prior to final publication in 1979 as *Classification of Wetlands and Deepwater Habitats of the United States.* Since then, the document has been widely used throughout the United States and in other countries for identifying, classifying, and inventorying wetlands.

According to the FWS definition, "*Wetlands are lands transitional between terrestrial and aquatic systems where the water table is usually at or near the surface or the land is covered by shallow water. For purposes of this classification wetlands must have one or more of the following three attributes: 1) at least periodically, the land supports predominantly hydrophytes; 2) the substrate is predominantly undrained hydric soil; and 3) the substrate is nonsoil and is saturated with water or covered by shallow water at some time during the growing season of each year."*

This definition recognizes that wetlands may or may not be vegetated. It does not specify the duration of flooding or the height of the water table, but suggests that inundation and saturation should be sufficient to be reflected in either 1) the plant community by the predominance of hydrophytes (plants growing in water or saturated soils) or 2) the soils by the prevalence of undrained hydric soils (waterlogged or frequently flooded soils), except where vegetation and hydric soils do not occur. All wetlands must have enough water at some time of year to stress plants and animals not adapted for life in water or hydric soils. Wetlands can be identified by any of the three attributes in the definition. National and regional lists of hydrophytes and national and state lists of hydric soils have been prepared to aid in wetland recognition.

1995 National Research Council Definition

From 1993 to 1995, the National Research Council (NRC) conducted a thorough review of the scientific basis for wetland identification and delineation in the United States. In *Wetlands: Characteristics and Boundaries*, the NRC proposed the following conceptual definition to provide a reference point for its examination of wetland characteristics: *"A wetland is an ecosystem that depends on constant and recurrent, shallow inundation or saturation at or near the surface of the substrate. The minimum essential characteristics of a wetland are recurrent, sustained inundation or saturation at or near the surface and the presence of physical, chemical, and biological features reflective of recurrent, sustained inundation or saturation. Common diagnostic features of wetlands are hydric soils and hydrophytic vegetation. These features will be present except where specific physiochemical, biotic, or anthropogenic factors have removed them or prevented their development."*

Like the FWS definition, this definition recognizes that, although hydrophytes and hydric soils are typically present, some wetlands may lack such features. Moreover, the absence of such vegetation or soils does not prevent classification of an area as wetland as long as it has the appropriate hydrology. This definition effectively serves as the current scientific foundation for the term "wetland" and should provide the scientific rationale for regulatory agencies to improve wetland protection in the future.

Regulatory Definitions

Federal, state, and local governments have developed laws to control uses of wetlands and to protect them from development. Regulations are promulgated by agencies to implement these laws. Each regulation includes a definition of wetland to designate the scope of government jurisdiction. Regulatory definitions must have a technical foundation, but they also incorporate other considerations such as geographic exclusions and applications in the field (e.g., does the area satisfy certain criteria; is it isolated or too small to regulate; is it a created wetland?). Consequently, regulated wetlands are often only a portion of the wetlands defined by scientists, and each law, bylaw, or regulation may define wetland somewhat differently to suit its administrative purposes.

From a legal standpoint, a wetland is whatever the law or government regulation or local zoning ordinance says it is. Many laws recognize the existence of artificially created wetlands, such as those formed from irrigation seepage or earthen-dam seepage, or in manmade ponds or roadside ditches. Several state laws include all deepwater areas (rivers, lakes, and reservoirs) and the entire hundred-year floodplain as wetlands. These additions were presumably made for convenience, since regulating activities in wetlands without controlling similar activities in adjacent bodies of water makes little sense.

Federal Regulatory Definition

Since 1975, the U.S. Army Corps of Engineers has been regulating most of the nation's wetlands under authority of the Clean Water Act (see Chapter 9). Wetlands are regarded as a component of the "waters of the United States" for regulatory purposes and are defined as *"those areas that are inundated or saturated by surface or ground water at a frequency and duration sufficient to support, and that under normal cirumstances do support, a prevalence of vegetation typically adapted for life in saturated soil conditions. Wetlands generally include swamps, marshes, bogs, and similar areas."*

This definition emphasizes frequent flooding and saturation that make the soil wet enough to favor the growth of marsh, swamp, bog, and comparable plants. Nonvegetated wetlands (e.g., tidal mud flats) are not considered regulated wetlands. Interpreting this definition in the field requires finding positive evidence of hydrophytic vegetation, hydric soils, and wetland hydrology, with limited exceptions. The application of this three-parameter approach (via the Corps's wetlands delineation manual) has resulted in the exclusion of certain vegetated wetlands from government jurisdiction (e.g., seasonally saturated and temporarily flooded wetlands).

Nonvegetated wetlands and aquatic beds are included with coral reefs and riffle-pool complexes as other "waters of the United States" for regulatory purposes. Water lily and other aquatic beds are not "wetlands" since they do not grow on soils; rather, they colonize flooded substrates. Aquatic beds are considered "vegetated shallows," one of several special aquatic sites recognized in the regulations.

Other Regulatory Definitions

Definitions have been developed by states for wetland protection statutes and by other federal agencies for quasi-regulatory programs. Table 1.1 presents some of these definitions used in the Northeast and the Great Lakes region.

While the more recent state laws tend to use the federal regulatory definition, many state laws, especially tidal-wetland protection laws, passed prior to active federal involvement in wetland protection have different definitions, with most emphasizing the presence of certain plants for designating wetlands. Some definitions include exemptions for particular geographic areas, like the exclusion of the Hackensack Meadowlands in New Jersey's Coastal Wetland Protection Act. Such areas are omitted because they are covered by other laws or for political or economic considerations.

The 1985 Food Security Act was the first federal law to define wetlands in the act itself. Other laws (e.g., the Clean Water Act) have relied on regulations promulgated by government agencies to define wetlands. The Food Security Act gave the Natural Resources Conservation Service (NRCS) responsibility for implementing the "swampbuster" provision, an attempt to reduce wetland drainage on agricultural lands. A geographic exception was made for Alaskan permafrost lands with high agricultural potential. The federal Emergency Wetland Resources Act of 1986 also uses this definition (without the reference to permafrost lands).

Field Interpretation of the Federal Regulatory Definition

The ultimate test for any wetland definition is its application to identify wetlands. A useful and practical definition should produce accurate and repeatable results in the field when used by different individuals. A scientific definition that considers the complex interrelationships among plants, soils, and hydrology requires scientific techniques consistently applied in the field by persons with training in botany and soil science. A science-based definition should be changed only as warranted by new information on wetland ecology.

In contrast, regulatory definitions based, in part, on policy considerations may be changed to meet the needs of society or the moods of legislators. The interpretation of such definitions tends to be more fluid. Although the federal regulatory definition of wetland has not changed in thirty years, its interpretation and application in the field has greatly changed, in response first to increased demands for stronger protection, and more recently to demands for less protection (regulatory relief) from property-rights advocates.

Wetland delineation manuals have been developed to aid in consistent application of the federal regulatory definition. Regulatory guidance memoranda have also been issued to further clarify points of interpretation or to change certain requirements. The following sections briefly summarize the evolution of the current interpretation of the regulatory definition. They should provide some insight into the difficulties of applying scientific concepts in a regulatory and political environment.

Table 1.1. Selected regulatory wetland definitions.

State or Federal Department	Definition
Maine	Freshwater wetlands are defined as "freshwater swamps, marshes, bogs, and similar areas which are: A) Of 10 or more contiguous acres, or of less than 10 contiguous acres and adjacent to a surface water body, excluding any river, stream or brook, such that in a natural state, the combined surface area is in excess of 10 acres; and B) Inundated or saturated by surface or ground water at a frequency and for a duration sufficient to support, and which under normal circumstances do support, a prevalence of wetland vegetation typically adapted for life in saturated soils." (Note: This is the state's inland-wetland regulatory definition.)
Vermont	"Wetlands means those areas of the state that are inundated by surface or ground water with a frequency sufficient to support significant vegetation or aquatic life that depend on saturated or seasonally saturated soil conditions for growth and reproduction. Such areas include but are not limited to marshes, swamps, sloughs, potholes, fens, river and lake overflows, mud flats, bogs and ponds, but excluding such areas as grow food or crops in connection with farming activities."
Massachusetts	"Salt Marsh means a coastal wetland that extends landward up to the highest high tide line, that is, the highest spring tide of the year, and is characterized by plants that are well adapted or prefer living in saline soils. Dominant plants within salt marshes are salt meadow cord grass (*Spartina patens*) and/or salt marsh cord grass (*Spartina alterniflora*). A salt marsh may contain tidal creeks, ditches and pools." (Note: This is the state's coastal-wetland regulatory definition; it was published in the first state wetland law in the nation in 1962.)
Massachusetts	"Bordering Vegetated Wetlands are freshwater wetlands which border on creeks, rivers, streams, ponds and lakes. The types of freshwater wetlands are wet meadows, marshes, swamps and bogs. They are areas where the topography is low and flat, and where the soils are annually saturated. The ground and surface water regime and the vegetational community which occur in each type of freshwater wetland are specified in the Act." (Note: This is the state's inland-wetland regulatory definition; it pertains only to wetlands that border bodies of water.)
Rhode Island	"Coastal wetlands include salt marshes and freshwater or brackish wetlands contiguous to salt marshes. Areas of open water within coastal wetlands are considered a part of the wetland. Salt marshes are areas regularly inundated by salt water through either natural or artificial water courses and where one or more of the following species predominate: [8 plants listed]. Contiguous and associated freshwater or brackish marshes are those where one or more of the following species predominate: [9 plants listed]." (Note: This is the state's official coastal-wetland regulatory definition.)
Rhode Island	Freshwater wetlands are defined to include "but not be limited to marshes; swamps; bogs; ponds; river and stream flood plains and banks; areas subject to flooding or storm flowage; emergent and submergent plant communities in any body of fresh water including rivers and streams and that area of land within fifty feet (50') of the edge of any bog, marsh, swamp, or pond." (Note: This is the state's official inland-wetland regulatory definition. Various wetland types are further defined, based on hydrology and indicator plants.)
Connecticut	"Wetlands are those areas which border on or lie beneath tidal waters, such as, but not limited to banks, bogs, salt marshes, swamps, meadows, flats or other low lands subject to tidal action, including those areas now or formerly connected to tidal waters, and whose surface is at or below an elevation of one foot above local extreme high water." (Note: This is the state's tidal-wetland regulatory definition. The definition also includes a list of indicator plants.)

Table 1.1. (continued)

State or Federal Department	Definition
Connecticut	"Wetlands mean land, including submerged land, which consists of any of the soil types designated as poorly drained, very poorly drained, alluvial, and floodplain by the National Cooperative Soils Survey, as may be amended from time to time, of the Soil Conservation Service of the United States Department of Agriculture." (Note: This is the state's inland-wetland regulatory definition; it focuses on soil types.)
New Jersey	Coastal wetlands are "any bank, marsh, swamp, meadow, flat or other low land subject to tidal action in the Delaware Bay and Delaware River, Raritan Bay, Sandy Hook Bay, Shrewsbury River including Navesink River, Shark River, and the coastal inland waterways extending southerly from Manasquan Inlet to Cape May Harbor, or any inlet, estuary or those areas now or formerly connected to tidal waters whose surface is at or below an elevation of 1 foot above local extreme high water, and upon which may grow or is capable of growing some, but not necessarily all, of the following:" (19 plants listed). Coastal wetlands exclude "any land or real property subject to the jurisdiction of the Hackensack Meadowlands Development Commission." (Note: This is the state's tidal-wetland regulatory definition; it contains a geographic exclusion.)
New Jersey	"Freshwater wetland means an area that is inundated or saturated by surface water or groundwater at a frequency and duration sufficient to support, and that under normal circumstances does support, a prevalence of vegetation typically adapted for life in saturated soil conditions, commonly known as hydrophytic vegetation; provided, however, that the department, in designating a wetland, shall use the 3-parameter approach . . . developed by the U.S. Environmental Protection Agency, and any subsequent amendments thereto." (Note: This is the state's official inland- or freshwater-wetland regulatory definition.)
New York	"Freshwater wetlands means lands and waters of the state as shown on the freshwater wetlands map which contain any or all of the following: (a) lands and submerged lands commonly called marshes, swamps, sloughs, bogs, and flats supporting aquatic or semi-aquatic vegetation of the following types: [lists indicator trees, shrubs, herbs, and aquatic species]; (b) lands and submerged lands containing remnants of any vegetation that is not aquatic or semi-aquatic that has died because of wet conditions over a sufficiently long period . . . provided further that such conditions can be expected to persist indefinitely, barring human intervention; (c) lands and waters substantially enclosed by aquatic or semi-aquatic vegetation . . . the regulation of which is necessary to protect and preserve the aquatic and semi-aquatic vegetation; and (d) the waters overlying the areas set forth in (a) and (b) and the lands underlying (c)." (Note: This is the state's inland-wetland regulatory definition.)
Maryland	Tidal wetlands are "all State and private tidal wetlands, marshes, submerged aquatic vegetation, lands, and open water affected by the daily and periodic rise and fall of the tide within the Chesapeake Bay and its tributaries, the coastal bays adjacent to Maryland's coastal barrier islands, and the Atlantic Ocean to a distance of 3 miles offshore of the low water mark." (Note: This is the state's tidal-wetland regulatory definition; it includes deepwater areas.)
Michigan	Wetland is defined as "land characterized by the presence of water at a frequency and duration sufficient to support, and that under normal circumstances does support, wetland vegetation or aquatic life, and is commonly referred to as a bog, swamp, or marsh." (Note: This is the state's regulatory definition.)
Minnesota	Wetlands are defined as "lands transitional between terrestrial and aquatic systems where the water table is normally at or near the surface or the land

(continued)

Table 1.1. (continued)

State or Federal Department	Definition
	is covered by shallow water. For purposes of this definition, wetlands must have the following three attributes: (1) have a predominance of hydric soils; (2) are inundated or saturated by surface or ground water at a frequency and duration sufficient to support a prevalence of hydrophytic vegetation typically adapted for life in saturated soil conditions; and (3) under normal circumstances support a prevalence of such vegetation. For the purpose of regulation . . . the term wetland does not include public water wetlands." (Note: This is the state's regulatory definition and it excludes marshes, ponds, and reservoirs of a certain size in specific geographic areas. Public water wetlands are shallow fresh marshes, deep fresh marshes, and open freshwater shallow ponds and reservoirs not included in the state's definition of "public waters" that are 10 acres or greater in unincorporated areas or 2.5 acres or more in incorporated areas.)
Illinois	Wetland is defined as "land that has a predominance of hydric soils and that is inundated or saturated by surface or groundwater at a frequency and duration sufficient to support, and that under normal circumstances does support, a prevalence of hydrophytic vegetation typically adapted for life in saturated soil conditions. (Note: This is the state's definition for administering environmental policy related to state-initiated or -sponsored activities by signatory agencies; this definition was adapted from the U.S. Department of Agriculture's definition that is presented below.)
U.S. Department of Agriculture	"Wetlands are defined as areas that have a predominance of hydric soils and that are inundated or saturated by surface or ground water at a frequency and duration sufficient to support, and under normal circumstances do support, a prevalence of hydrophytic vegetation typically adapted for life in saturated soil conditions, except lands in Alaska identified as having a high potential for agricultural development and a predominance of permafrost soils." (Note: This definition is used by the Natural Resources Conservation Service in administering the swampbuster provision of the Food Security Act of 1988 and amendments.)

Varied Interpretations of the Federal Regulatory Definition

The federal government, through the Army Corps of Engineers, began regulating wetlands on a large scale in the 1970s. Until the late 1980s, only the wettest wetlands were regulated across much of the country, with few exceptions (e.g., New England). Each Corps district was responsible for deciding what it would regulate as wetland. There was no consistency across the country in applying the federal regulatory definition, despite publication of the *Corps of Engineers Wetlands Delineation Manual* in 1987. The use of this manual by Corps districts was discretionary—they could use it, modify it, or use their own locally derived methods to identify wetlands subject to the Clean Water Act. At the same time, the Environmental Protection Agency (EPA), which administers the federal wetland regulatory program jointly with the Corps, developed its own manual for identifying regulatory wetlands. Although both manuals were using the same definition, they applied different rules, often leading to different determinations and boundaries in the ground. This caused both interagency disputes over limits of regulated wetlands affected by proposed projects and confusion for individuals requesting permits to alter wetlands on their properties.

Developing a National Standard

After one year of testing their respective manuals, the Corps and the EPA met to discuss how to resolve conflicts. They invited the FWS and the Soil Conservation Service (now, as noted above, the NRCS) into the discussion. Because there was much agreement among the agencies regarding technical aspects of identifying wetlands, it was recommended that a single technically based wetland delineation manual be written.

To compile this interagency manual, a committee of wetland delineation and mapping experts and policy analysts drawn from the Corps, the EPA, the FWS, and the NRCS—the Federal Interagency Committee for Wetland Delineation—was established. From August 1988 to January 1989, the committee worked on what was published as *Federal Manual for Identifying and Delineating Jurisdictional Wetlands*. Despite the term "jurisdictional," the manual was developed strictly as a technical document for identifying vegetated wetlands. In fact, early drafts were entitled *Federal Manual for Identifying and Delineating Vegetated Wetlands*. The title was changed in the final draft at the request of the Corps members of the committee.

When, on January 10, 1989, heads of the four agencies officially adopted "this Federal Manual as the technical basis for identifying and delineating jurisdictional wetlands in the United States," it was the first time that these rival federal agencies agreed to a single set of standards for wetland identification. They were able to reach this consensus, in large part, because policy was intentionally kept out of the manual. Each agency would decide how to use the manual for its own programs.

The Corps and the EPA planned to use the manual for implementing Clean Water Act regulations, and on January 19, 1989, they issued a memorandum of agreement stating that the new manual would be used to identify the geographic limits of wetlands subject to this act. At that moment, the federal government finally instituted a technical standard for identifying regulated wetlands, putting an end to many interagency disputes.

States began to adopt this standard, and the country appeared to be on its way to a unified concept of wetland, ending over a decade of debate on this issue. Instead of a set of different jurisdictional boundaries for a given wetland (a Corps boundary, an EPA boundary, a state boundary, and, in some cases, a local-government boundary), a single boundary could be drawn based on a repeatable and scientifically defensible method. The various government agencies could then focus attention on how best to evaluate the impacts of proposed development on specific wetlands.

The effect of adopting a national standard was enormous in regions where only the wetter wetlands had been regulated, such as the Southeast. Under the new procedures, much more area was now classified as federally regulated wetland. For some regions, there was little change, as in New England, where technically based methods consistent with the 1989 manual were already being employed by Corps personnel. Not surprisingly, dissenting voices were soon raised. Many farmers and land developers felt that the government had significantly changed the rules and expanded the scope of federal jurisdiction without giving proper notice and opportunity for public comment. On the other hand, the government argued that it was only extending throughout the country what was already being done in some parts of the country. Moreover, a wetland determination using the 1989 interagency manual was quite similar to a delineation performed using the EPA manual, so there was little change in the EPA's policy.

Changing the Standard

In 1990, the interagency committee began revising the manual to clarify points that had been unintentionally or willfully misapplied during the past year. Mounting pressure from groups opposed to the federal government's strengthened wetland regulations forced political intervention in the revision process. The Bush administration and the EPA rejected proposed technical revisions, disbanded the interagency committee, and proposed more stringent hydrology requirements for identifying regulated wetlands. In order to qualify as wetland, an area would have to be flooded for fifteen or more consecutive days or saturated to the surface for twenty-one or more consecutive days during the growing season in most years, or be periodically flooded by tidal water in most years. Using this measure, about half of the wetlands in the lower forty-eight states would not be considered regulated wetlands. This proposal was soundly rejected based on a hundred thousand comments received from scientists, state wetland regulatory agencies, governors, nonprofit environmental organizations, and the public at large.

Undaunted, various groups representing farmers, developers, and private-property advocates effectively lobbied their representatives

Figure 1.2. Pressures are constantly being applied to change wetland definitions and reduce wetland protection, as illustrated in this political cartoon. (Copyright 1995 *Star-Ledger*. All rights reserved. Reprinted with permission.)

and senators to get Congress to provide relief from federal wetland regulations. In the Water Resources Development Act of 1990 (which authorizes the Corps operating budget), Congress included a statement saying in effect that the Corps must not use any wetland delineation manual that had not undergone administrative review, and if it did, the agency would be denied its operating budget. Consequently, the Corps had to abandon use of the interagency manual.

In October 1991, the Corps officially adopted its 1987 manual as the national standard. This did not, however, reestablish pre-1989 conditions, because districts were now required to use the manual. The Corps later provided specific guidance on how to interpret several key provisions of the manual that had been sources of gross misapplications. The clarifications were scientifically based, resulting in wetland determinations similar to many of those derived from the 1989 interagency manual, although many wetlands (e.g., seasonally satu-

rated and temporarily flooded types) still did not meet the requirements for regulation.

Changes in delineation methods can have significant impacts on wetland protection and on development. For example, an article in the newsletter of the New York Homebuilders Association in January 1994 described a proposed subdivision that in accordance with the 1989 manual had 12.4 acres of wetland, whereas, when the 1987 Corps manual was used, only 1.9 acres were identified as wetland. Another thirty-two lots could be developed.

The debate continues. At the request of Congress, the EPA in 1993 commissioned the National Research Council (NRC) to study the scientific basis for wetland delineation and characterization. The report published in 1995 and quoted from above, *Wetlands: Characteristics and Boundaries*, serves as the scientific foundation for wetland identification and delineation, as well as related topics. The report concludes that the federal regulatory program is scientifically sound in most cases, but could be made technically more accurate, noting, for example, that it is "often scientifically defensible to infer information about one factor from another in the absence of alterations or mixed evidence" and that hydrology is "adequately character-ized by hydric soils or hydrophytic vegetation if there is no evidence of alteration of hydro-logic conditions" (p. 8). This view contradicts the Corps manual's requirement that positive evidence of hydrophytic vegetation, hydric soils, and wetland hydrology be found to certify a wetland. A major recommendation of the re-port is that the federal agencies prepare a new federal wetland delineation manual incorporat-ing improvements in scientific knowledge since 1987. Although no new federal manual has been developed, the Corps is working on re-gional supplements to its manual to clarify its use and expand the suite of allowable indicators for certain parameters. As of January 2005, no such supplement has been published, but sup-plements for Alaska and the arid West should be forthcoming.

While the struggle among diverse and com-peting interests to protect or deregulate wet-lands seems to be inevitable, the regulatory defi-nition should be based on science for several reasons. As we have seen, wetlands are varied, complex, and valuable resources that require an understanding of the interrelationships among plants, soils, and hydrology for identification and evaluation. Determinations and boundary delineations performed by trained individuals using technically sound methods produce more accurate and reproducible results than arbitrary and capricious procedures. The limits of wet-lands can be identified following a consistent and scientifically defensible set of guidelines. Fi-nally, a unified, standard, scientifically based concept of wetland is good government: there should be only one wetland line established on the ground rather than multiple lines based on different definitions. Government agencies (fed-eral, state, and local) have the ultimate author-ity to regulate uses of wetlands and will develop necessary criteria to exercise their authority. Af-ter all, the regulations, not the definition of wet-land, are what dictate acceptable and unaccept-able activities.

Although science may form a foundation for wetland policy, it is politics that shapes it.

For an interesting perspective on the politics influencing wetland definitions and regula-tions, see William Lewis' book *Wetlands Ex-plained: Wetland Science, Policy, and Politics in America*.

Wetland Definition for This Book

For the purpose of this book, the FWS defini-tion of wetlands is used, since it is more eco-logically sound than the regulatory definitions and is consistent with the National Research Council's definition. The FWS definition has been the national standard for wetland mapping for nearly two decades. It also has been exten-sively peer reviewed and open to public review and comment. It is a standard that is inter-preted strictly on a technical basis and has not been subjected to varied interpretations due to the politics of wetland regulation. Finally, the FWS wetland classification system was officially adopted in 1996 as the federal geographic-data standard for wetlands, meaning that it is the sys-tem that the federal government (all agencies) will use to collect information on the location, characteristics, extent, and condition of the na-tion's wetlands.

Additional Readings

Lewis, W. M. 2001. *Wetlands Explained: Wetland Science, Policy, and Politics in America*. Ox-ford University Press, New York.

National Research Council, Committee on Char-acterization of Wetlands. 1995. *Wetlands: Characteristics and Boundaries*. National Academy Press, Washington, DC.

Tiner, R. W. 1993. Wetlands are ecotones: reality or myth? In *Wetlands and Ecotones: Studies on Land-Water Interactions*, ed. B. Gopal, A. Hillbricht, and R. G. Wetzel, pp. 1–15. National Institute of Ecology, New Delhi, India. (Available from U.S. Fish and Wildlife Service, Hadley, MA 01035)

World Wildlife Fund and Environmental Defense Fund. 1991. *How Wet is a Wetland? The Im-pacts of the Proposed Revisions to the Federal Wetlands Delineation Manual*. World Wild-life Fund and Environmental Defense Fund, Washington, DC, and New York.

CHAPTER 2

Water, the Lifeblood of Swampland
Wetland Hydrology

Hydrology is the engine behind wetland formation and maintenance; it is the master variable. The defining characteristic of all wetlands is that they are wet for at least significant periods at recurring intervals sufficient to create conditions that stress most plants and animals. Since many wetlands are dry for extended periods, the challenge for wetland identification and delineation is to be able to recognize and interpret indirect indicators—soil, plant, and other indicators of prolonged seasonal wetness—at times when water is not present. To better appreciate the current debate over wetland definitions and understand the essence of wetland, a basic knowledge of wetland hydrology is needed.

What Is Hydrology?

Hydrology is the science or study of properties, distribution, and movement (circulation) of water on land, in bodies of water, underground, and in the atmosphere. The hydrologic cycle is the movement of water from Earth's atmosphere to the land surface, underground, back to the surface or to bodies of water, and finally back to the atmosphere (Figure 2.1). Understanding the hydrology of forests, estuaries, wetlands, deserts, rivers, and lakes requires knowledge of the sources of water and the movement of water into, through, and out of them. Important factors affecting the hydrologic dynamics of a given area include climate (especially temperature and precipitation patterns), soils and underlying geologic formations, vegetation, and landscape or topographic position. Wetland hydrology describes the characteristics and patterns of the movement of water in wetlands and is the essence of their being.

Water Budget

Before addressing the issue of what constitutes sufficient wetness to be considered wetland hydrology, we should first consider why an area is wet. Wetlands form in places where there is frequently water in or on the soil for a prolonged period (see Chapter 3). Prolonged wetness is simply a matter of more water coming into a system than is moving out. This wetness may occur year-round, seasonally, or intermittently. The timing, frequency, and duration are critical factors affecting soil development and plant growth.

The water budget of an area, wetland or otherwise, is an accounting of water inflows (gains or inputs) and outflows (losses or outputs) (Figure 2.2). The water-budget equation is used by hydrologists and other scientists to evaluate the net change of the volume of water in a defined area over time (Table 2.1). Change in volume equals inputs minus outputs, as expressed by the formula

$$\Delta V = [P + Si + Gi + Ti] - [ET + So + Go + To].$$

Wetlands develop at locations where there is at least a recurring seasonal increase in water inputs versus outputs. In the Northeast, where precipitation is fairly constant throughout the year, this increase usually occurs from winter to early spring, since evaporation is low and plant activity is minimal then. The nature of this change in water volume affects soil development, plant communities, and animal life at such locations.

Water Sources (Inputs)

In the water-budget equation, four possible sources of water are identified: 1) precipitation (P), 2) surface-water inflow (Si), 3) groundwater inflow (Gi), and 4) tidal inflow (flood tides; Ti). Precipitation is water falling from the atmosphere—rain, snow, sleet, hail, or fog. Surface-water inflow comes from river and stream overflows or elevated lake levels. It also comes from water running off the surface of the ground during storms. Groundwater inflow is water

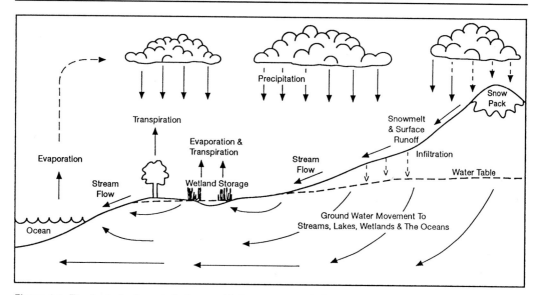

Figure 2.1. The hydrologic cycle is the earth's interconnected circulatory system of water movement. Water is constantly moving among the land, the oceans and other bodies of water, and the atmosphere. Evaporation by the sun and plant transpiration cause water to change into a gas (water vapor), which condenses to form clouds, and eventually much returns to the earth as rain, snow, sleet, or fog. On the land surface, water moves by gravity, which causes it to flow downhill or percolate into the soil, where groundwater supplies are replenished. Approximately 75 percent of the earth's fresh water is frozen in polar ice and glaciers, 25 percent is groundwater, 0.55 percent can be found in rivers, lakes, soils, and wetlands, and 0.035 percent is vapor or liquid in the atmosphere. (Source: Stone and Stone 1994; copyright 1994 the American Ground Water Trust. Reprinted with permission.)

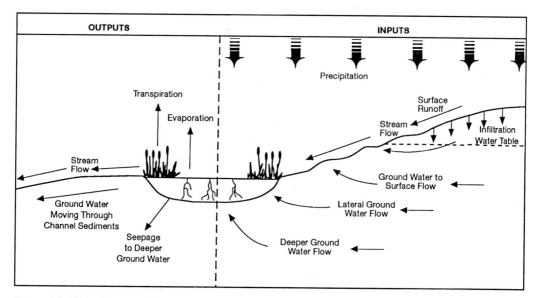

Figure 2.2. Water inputs and outputs for a nontidal wetland. Not all wetlands receive water from all sources; some wetlands are strictly surface-water driven. The water budget for an individual wetland is not easily determined since it is difficult to measure some of the inputs and outputs. (Source: Stone and Stone 1994; copyright 1994 the American Ground Water Trust. Reprinted with permission.)

Table 2.1. Examples of annual water budgets for selected wetlands in North America.

Wetland (Location)	Inputs				Outputs				Net Storage
	P	Si	Gi	Ti	ET	So	Go	To	
Mangrove Swamp (Florida)	8	—	—	88	7	6	2	85	−4
Bog (Massachusetts)	100	—	—	—	70	8	8	—	14
Nevin Wetland (Wisconsin)	7	4	89	—	8	92	0	—	0
Cypress River Swamp (Illinois)	23	70	7	—	22	71	6	—	1
Okefenokee Swamp (Georgia)	76	23	0	—	55	43	2	—	−1
Hidden Valley Marsh (Ontario, Canada)	11	53	36	—	12	35	50	—	3
Arctic Fen (Northwest Territories, Canada)	22	52	26	—	27	40	<1	—	>32
Experimental Marsh (Ohio)	3	97	—	—	2	72	26	—	0

Note: Units are percentages of inputs and outputs. P = precipitation, Si = surface water inflow, Gi = groundwater inflow, Ti = tidal inflow, ET = evapotranspiration, So = surface water outflow, Go = groundwater outflow, and To = tidal outflow. (Sources: Carter 1996; Twilley and Chen 1998; Zhang and Mitsch 2002 for experimental marsh)

discharged to the land surface from underground, with seeps and springs being prime examples of discharges. Many, if not most, Northeast wetlands are groundwater-discharge sites. Flood (rising) tides bring seawater into coastal waters and inundate low-lying tidal marshes and swamps. While all wetlands receive water through precipitation, the other sources are site-dependent. For example, most wetlands are beyond tidal influence and therefore lack tidal water input.

Water Losses (Outputs)

Water losses occur through: 1) evapotranspiration (ET), 2) surface water outflow (So), 3) percolation or groundwater outflow (Go), and 4) tidal outflow (ebb tides; To). Evapotranspiration is the combination of evaporation and plant transpiration. Evaporation increases with increasing air temperature and exposure of the land or water surface to the sun. Plant transpiration is the natural uptake of water from the soil by plants; most of this is lost to the atmosphere in the form of water vapor. Surface-water outflow is due to water draining off the land surface, usually into rivers, streams, and other

drainageways. Groundwater outflow is water exiting an area through percolation (downward movement in the soil), which may replenish local groundwater supplies (aquifers). (Most groundwater recharge, however, occurs in uplands and along the edges of wetlands.) Water also leaves tidal wetlands during the ebb (falling) tides. At low tide in tidal wetlands, any incoming fresh water drains freely from the wetland into the adjacent body of water.

How Wet Is a Wetland?

Anyone involved in creating or affected by wetland regulations and policy knows that this is not a trivial question. The answer defines the geographic extent of public and private lands subject to government land-use regulations and therefore identifies which lands may ultimately receive some form of protection from or restriction of use.

Considerable effort has been devoted to measuring the hydrology of rivers, streams, and forests, yet very few studies have examined the hydrology of wetlands. The hydrology of a given wetland is changeable, varying annually, seasonally, and daily. Moreover, no two wetlands have exactly the same hydrology, although there are undoubtedly many similarities among specific wetland types. Hydrologic assessments require long-term studies to document the fluctuations of the surface-water levels and the position of the water table. To date, scientific research has not focused on wetlands for several reasons: 1) the interest in this topic is only recent, spurred on by the regulatory implications; 2) wetland identification has traditionally relied on plants and soils, with little cause to evaluate hydrology; 3) the long-term commitment of resources (dollars and time) required to undertake such a task was not available; and 4) there simply were not many hydrologists interested or trained in wetlands.

The lack of hydrologic data for most wetlands probably is the main reason that definitions of wetland have not included a specific duration and frequency of wetness. Vegetation and soil properties are strongly influenced by site wetness. These two parameters are usually present, are readily evaluated, and are reasonable indicators of site wetness provided an area has not been significantly drained. In contrast, the presence of water in most wetlands is variable and more dependent on the season, and the mere presence of water at a given moment does little to explain the long-term hydrology.

This lack of extensive knowledge about wetland hydrology, however, does not prevent scientists from making reasonable assessments about its scope. In *Wetlands: Characteristics and Boundaries*, the National Research Council (NRC) concluded that *wetland hydrology should be considered to be saturation within one foot of the soil surface for two weeks or more during the growing season in most years* (about every other year on average). The upper foot contains most of the plant roots that would be adversely affected by anaerobic conditions resulting from prolonged saturation. The NRC acknowledged that there may be regional variations due to climate, vegetation, soils, and geologic differences, but found that no such data currently exist. Until information to the contrary is produced, this threshold should be considered the minimum time necessary to create conditions that support the growth of hydrophytic vegetation and to define wetlands hydrologically.

Saturation within one foot of the ground surface is related to the position of the water table and the soil characteristics as well as other factors such as evapotranspiration rates. In all soils, there is a zone above the water table (the level of freestanding water in an unlined hole) where nearly all the soil pores are filled with water due to surface tension (Figure 2.3). This zone is called the "capillary fringe" or "saturated zone." In sandy soils, the capillary fringe is thin, typically less than 1 inch for coarse sands, because the pores are large and water molecules do not adhere to one another easily in pores of this size (Table 2.2). In clay soils, pores are extremely small, and water can be held under tension with little effort, so the capillary fringe is thick, up to 24 inches in some soils. Loamy soils fall between these two extremes. Soil texture (through soil porosity), therefore, plays a major role in determining the depth of saturation in soils. This relationship between capillarity and soil characteristics explains the need for establishing water-table depth requirements to define hydric soils and the zone in which soil properties must be examined to verify the presence of hydric soil.

While setting a minimum standard appears sensible and necessary, it is impractical to require the collection of long-term hydrologic data (i.e., flooding frequency and duration and

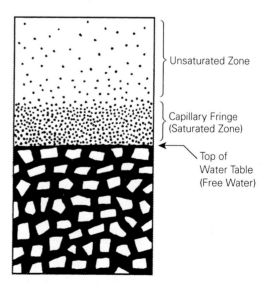

Figure 2.3. Above the water table lies a zone called the "capillary fringe," where nearly all the pores of the soil are filled with water. This water is held under tension in pores between the soil particles, causing saturation above the water table that affects plant growth, soil development, and animal life.

Table 2.2. Thickness of the capillary fringe is in large part a function of soil texture.

Soil Texture	Range in Thickness of the Fringe (inches)*
Coarse sand	0.4–2.8
Very fine sand	1.6–4.8
Loamy coarse sand	2.0–5.6
Loamy very fine sand	4.0–8.0
Coarse sandy loam	3.2–7.2
Very fine sandy loam	6.2–10.2
Loam	8.0–12.0
Silt loam	10.0–16.0
Silt	14.0–20.0
Sandy clay	8.0–12.0
Silty clay	16.0–24.0
Clay	10.0–16.0

*based on lab studies; estimates probably higher than in the field.
(Source: Natural Resources Conservation Service)

Table 2.3. Some indicators of wetland hydrology used for wetland delineation purposes.

Direct Observations of Water*
    Presence of surface water
    Free water within 1 foot of the surface in an open pit
    Saturation within 1 foot of the surface

Indirect Observations
    Water marks (sign of prolonged inundation)
    Sediment deposits (evidence of flooding)
    Drift lines (sign of water-carried debris deposited in wetland and floodplain)
    Bare, nonvegetated areas on floodplains (sign of scouring by floodwaters)
    Drainage patterns (evidence of surface water flowage)
    Oxidized rhizospheres along living roots within one-foot of the soil surface (evidence of plant growing under anaerobic conditions)
    Water-stained leaves (sign of prolonged inundation)
    Algae (evidence of prolonged flooding)
    Peat moss (sign of long periods of saturation at the soil surface; even periodic inundation in some wetlands)
    Aquatic invertebrates (evidence of life dependent on wetlands and aquatic habitats)
    Buttressed trunks, hypertrophied lenticels, aerenchyma tissue, and other plant morphological adaptations (responses of plants to prolonged inundation and soil saturation)

*observations during the growing season are preferred over those made during the non-growing season
Note: These indicators are considered after verifying the presence of wetland plants (hydrophytes) and hydric soils to help insure that these wetland features reflect current conditions. Some of the indirect indicators can be found in nonwetlands following extreme flooding events (e. g., twenty-year frequency flood); peat moss can be found occasionally in evergreen forests in northern climates.

weekly changes in the water table) to identify wetlands at most sites. Practical considerations dictate the use of direct and indirect indicators of wetland hydrology (see Table 2.3 and Plate 38 for examples) at sites where the hydrology has not been significantly diminished. The threshold is critical for evaluating sites where the hydrology has been significantly altered. These sites demand a more detailed assessment of water-table fluctuations.

Hydrology and the Growing Season

The growing season has been considered to be the critical period for determining wetland hydrology. This term is usually applied to farming activities: it is the time when the danger of frost or freezing is minimal, so that planting of crops will not be risky. The federal government has used the frost-free period as the growing season for assessing wetland hydrology (see the *Corps of Engineers Wetlands Delineation Manual*), yet this concept has little relevance to the growth of nonagricultural species.

The growing season is plant-specific—it is different for each type of plant and for plants of the same species in different locations within a state and even on one side of a mountain versus the other. Some native species are actively growing roots, moving nutrients and water through their cells and even flowering before the beginning of the frost-free period. Skunk cabbage, the dominant herb in most seasonally flooded red-maple swamps throughout the Northeast, flowers as early as February, when snow is still on the ground. The fact that skunk cabbage flowers and leafs before other species emerge undoubtedly gives it a significant edge over would-be competitors. Other early bloomers are pussy willow (flowers open in March), red maple, marsh marigold, leatherleaf, spring beauty, and trout lily. These and other plants typically bloom before the trees around them are fully leafed. Many shrubs and trees also bloom before their leaves appear, including highbush blueberry, alders, American elm, ashes, oaks, and sweet gum.

Autumn is a time of much root growth, as evidenced by the reseeding of residential lawns. The absence of the tree canopy (no shading) and the relatively high soil moisture (due to lower air temperatures and reduced evapotranspiration) create favorable conditions for root growth. Many woody plants also exhibit significant root growth at this time. Since trees are no longer putting energy into leaf growth, nutrients taken up by them are concentrated in the root system.

Evergreen trees and shrubs, cranberry vines, cool-season grasses, and certain sedges grow year-round or virtually so in northern climes. Any plant that is green in winter is probably active, although at much reduced levels. These circumstances suggest that whenever the soil is not frozen, there is activity going on in certain wetland plants. Some wetland soils in northern areas do not freeze—remember the brittle white ice along the edges of the skating pond, where cattails grow. Microbial activity is likely to occur in these soils in winter.

Winter wetness is critically important to some species. Certain evergreens are highly susceptible to the effects of drying or drought due to evaporation from cold, dry winds. Evergreen rhododendron, bog plants, and cranberries (members of the heath family, or Ericaceae) all might significantly benefit from winter wetness. Winter flooding of cranberry bogs is a standard management practice. Studies of loblolly pine in the Southeast demonstrated that winter wetness is important for increasing productivity. If wetness during the growing season is critical to defining wetland, then the season should reflect growth of any kind in wetlands, including microbial activity in the soil.

The NRC reached similar conclusions, recommending regional studies to investigate the relevance of the growing season as a condition for defining the critical period for wetland hydrology and suggesting that it may be more appropriate to use a time-temperature concept. This concept would establish the duration of wetland hydrology on a regional basis related to the effect of temperature on plants, animals, soil microbes, and soil formation.

Wetland Hydrology

Wetlands, in concept, occur along the soil-moisture continuum or gradient between permanent deep water and dry land (Figure 2.4). The conceptual nature of this statement is emphasized, since many wetlands do not border a permanent body of water but rather exist in areas having high seasonal water tables where surface water is absent or rarely present. At the lower end of the soil-moisture gradient, wetlands tend to be obviously wet for very long periods, with clear signs of distinctive vegetation and soils. Along the upper end of the gradient, where the soils are seasonally saturated for various intervals, plants are often less indicative of the wetland boundary, and soils usually offer better clues for wetland identification and delineation. These soils remain wet near the surface for sufficient periods to significantly affect plant growth and soil development. This area has been called the "transition zone" by some scientists. Most of the political debate over wetland regulations is focused on this zone and upper-end wetlands with similar characteristics.

The variation in site wetness over the course of time is called the "hydroperiod." Some variation is short-term (days or weeks), other variation is seasonal (e.g., wetter in spring in the Northeast) or interannual (e.g., between years or decades). The greatest differences in water levels are usually interannual (i.e., wet years vs. dry years). Although many wetlands exhibit a similar pattern, each wetland has its own unique pattern related to local precipitation (rainfall and snowfall), groundwater flows, topography (landscape position), and underlying soils and geology. Examples of hydroperiods for different wetland types are shown in Figure 2.5.

Wetlands can be divided into two general types based on hydrology, tidal and nontidal. Tidal wetlands are subjected to periodic flooding from ocean-generated tides. Nontidal wetlands occur beyond tidal influence and are affected by freshwater inflows from precipitation, surface waters (rivers and streams), and groundwater (underground aquifers).

Tidal Wetland Hydrology

In coastal areas, ocean-driven tides are the dominant hydrologic feature of many wetlands. Along the Atlantic coast, tides are semidiurnal

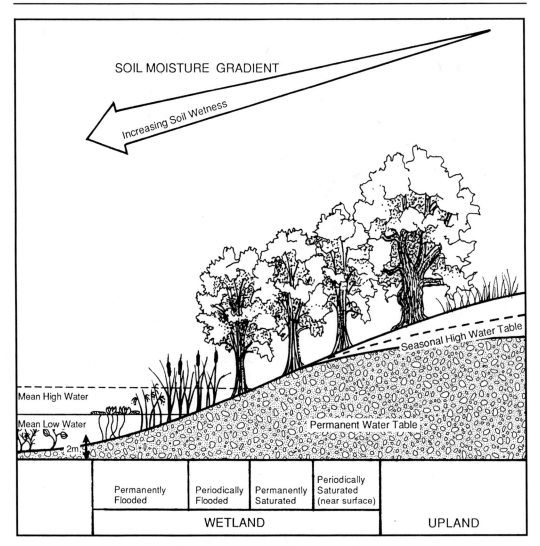

SOIL MOISTURE GRADIENT

Increasing Soil Wetness

Seasonal High Water Table

Mean High Water

Mean Low Water

2m

Permanent Water Table

	Permanently Flooded	Periodically Flooded	Permanently Saturated	Periodically Saturated (near surface)	
	WETLAND				UPLAND

Figure 2.4. Wetlands often lie between dry land and water, where they are subjected to an array of hydrologies. Conceptually, wetlands exist along the soil-moisture gradient between areas that are never flooded or saturated near the surface and permanent bodies of water. The seasonal high water table represents the upper level attained by the water table for a significant period in an average year; in the Northeast, it usually occurs from midwinter to mid-spring.

and symmetrical, with a period of 12 hours and 25 minutes. This means that there are roughly two high tides and two low tides each day (Figure 2.5). Since the tides are largely controlled by the position of the moon relative to the sun, the highest and lowest astronomic tides— spring tides—usually occur during full and new moons. (It should be noted that there is no relation to the season of spring.) Coastal storms such as northeasters cause the most extreme high and low tides during an otherwise average year. Such tides may even on occasion inundate adjacent low-lying uplands. Of course,

hurricane-driven tides can cause record flooding. Wind direction and strength may also influence tide heights. Extremely low tides can be created in embayments by strong winds blowing offshore.

Ocean-driven tides can sometimes move inland a hundred miles or more, as evidenced by the tides in the Hudson and Delaware Rivers in the Northeast. As the tides move upstream, there is a gradual mixing of salt water with the fresh water being carried seaward by the rivers. This creates a salty-to-brackish body of water called an "estuary" (Figure 2.6). With variable

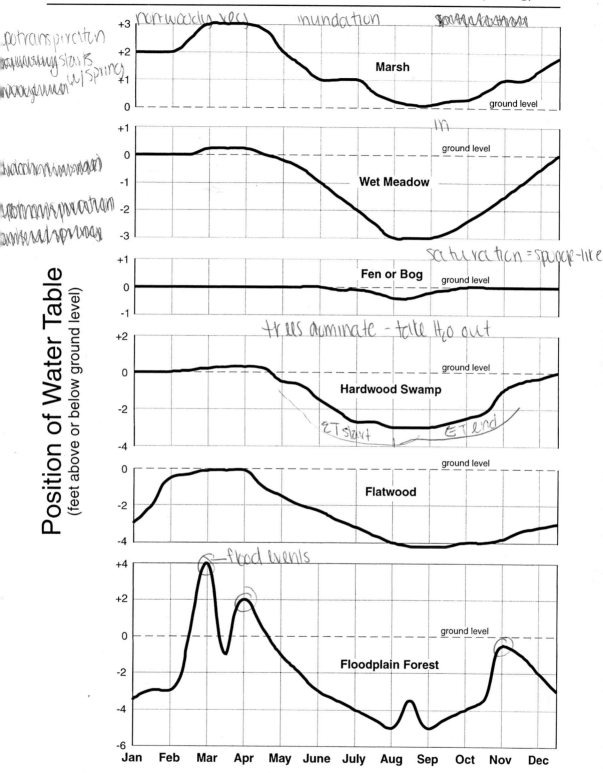

Figure 2.5. Examples of hydroperiods for nontidal wetlands. (Note: These are examples for the referenced type; there is variability in water table fluctuations within types also.)

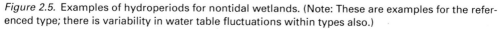

SUMMER

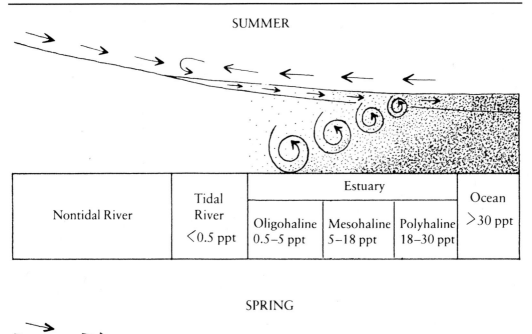

Nontidal River	Tidal River <0.5 ppt	Estuary			Ocean >30 ppt
		Oligohaline 0.5–5 ppt	Mesohaline 5–18 ppt	Polyhaline 18–30 ppt	

SPRING

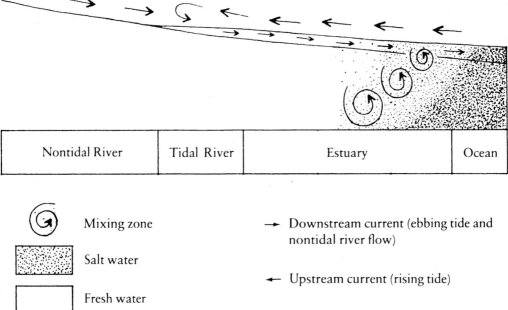

Nontidal River	Tidal River	Estuary	Ocean

⟳ Mixing zone   → Downstream current (ebbing tide and nontidal river flow)

▨ Salt water

▢ Fresh water   ← Upstream current (rising tide)

Figure 2.6. Generalized salinity and current patterns in coastal rivers. Note: 1) the upstream limit of the estuary is defined by maximum saltwater penetration (in summer); 2) the position of the salt-fresh mixing zone changes seasonally; 3) salinities decrease upstream; and 4) tides cause bidirectional currents in the estuary and ocean, as opposed to the one-directional (downstream) flow of the nontidal river. Heavy discharges in spring may eliminate or otherwise dampen tidal fluctuations, especially in rivers with enormous watersheds. (Source: Tiner 1993; copyright 1993 Ralph W. Tiner. Reprinted with permission.)

salinities and differences in tidal flooding, a wide range of wet environments develop. Above this mixing zone, the influence of tides is still felt as water levels in freshwater tidal rivers move up and down with the tides.

Tidal ranges vary due to many factors, including the shape of an estuary and its basin, the volume of water entering the estuary from the sea, and the discharge volume of the river. In the Northeast, mean tidal ranges vary from

20 feet in Maine's St. Croix River to as little as ½ foot in New Jersey's Barnegat Bay to less than an inch in the upper reaches of tidal rivers (Table 2.4). The world's highest tides (50 feet or more) occur in the Bay of Fundy in Atlantic Canada between New Brunswick and Nova Scotia. Rivers with a vast drainage basin and great discharges of fresh water—like the Mississippi River, which drains most of the eastern United States—may have no tide at peak discharges and only a little tidal effect near the mouth at low-flow periods.

Table 2.4. Examples of tidal ranges in the Northeast.

Location	Tidal Range (feet)	
	Mean	Spring
St. Croix River, Calais, ME	20.0	22.8
Frenchman Bay, Bar Harbor, ME	10.5	12.1
Hampton Harbor, Hampton, NH	8.3	9.5
Merrimack River, Newburyport, MA	7.8	9.0
Cape Cod Bay, Wellfleet, MA	10.6	11.6
Cape Cod Bay, Plymouth, MA	9.5	11.0
Narragansett Bay, Newport, RI	3.5	4.4
Connecticut River, Saybrook, CT	3.5	4.2
Connecticut River, Hartford, CT	1.9	2.3
Long Island Sound, Stonington, CT	2.7	3.2
Long Island Sound, Mamaroneck, NY	7.3	8.6
Hudson River, Jersey City, NJ	4.4	5.3
Hudson River, West Point, NY	2.7	3.1
Hudson River, Albany, NY	4.6	5.0
Barnegat Bay, Mantoloking, NJ	0.5	0.6
Barnegat Bay, Barnegat Inlet, NJ	3.1	3.8
Delaware River, New Castle, DE	5.6	6.0
Delaware River, Philadelphia, PA	6.0	6.3
Delaware River, Trenton, NJ	6.8	7.1
Chesapeake Bay, Kent Island, MD	1.2	1.4
Chincoteague Bay, Public Landing, MD	0.4	0.5
Potomac River, Piney Point, MD	1.4	1.6
Potomac River, Colonial Beach, VA	1.7	1.9
Potomac River, Washington, DC	2.8	3.2

Differences in the frequency of tidal flooding create two readily identifiable zones in tidal wetlands, regularly and irregularly flooded zones (Figure 2.7). The former, also called a "low marsh," is flooded at least once a day by the tides, while the latter ("high marsh") is flooded less often and exposed to air most of the time. The uppermost marsh may be flooded only during storm tides, which are more frequent in winter. Even when not flooded, the soils remain saturated near the surface. Plants that have adapted to differences in tidal flooding (and to the differing soil salinities) often serve as useful indicators of these two zones (Table 2.5).

Nontidal Wetland Hydrology

Nontidal wetlands are more abundant and diverse than their tidal counterparts. They receive their water mainly from surface-water runoff, groundwater discharge, and direct precipitation. Surface water running off the land from snowmelt or heavy rainfall either collects in isolated basins surrounded by upland or enters rivers and lakes, which may eventually overflow their banks or shores. In many areas, water percolates into the soil, replenishing soil moisture and ultimately groundwater supplies. This water may be discharged into depressional wetlands or into sloping wetlands in spring or seepage areas (Figure 2.8). In other cases, water inputs from direct precipitation or shallow lateral groundwater flow may become perched above a confining layer (e.g., hardpan, dense basal till, or compact clay) saturating the soil above it and sometimes causing rainwater to pond on the surface for extended periods. Some wetlands receive only surface water, while others are groundwater fed. Most wetlands in the Northeast probably receive water from both sources.

Rivers in the Northeast usually discharge their greatest volumes in winter and spring. Flooding is heaviest during April and May, despite relatively uniform rainfall throughout the year. Snowmelt, spring rains on frozen soil, low evaporation, and low plant uptake of soil water have a great combined effect on surface-water runoff. In summer, less water is available to runoff due to high evaporation, active uptake by plants, and interception of rainfall by plant leaves.

Water-table fluctuations follow a similar and generally predictable pattern and are largely

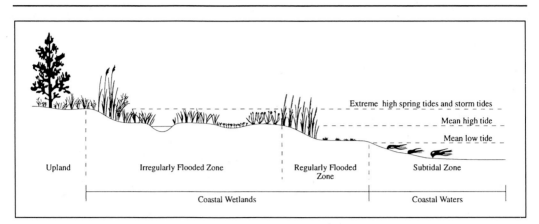

Figure 2.7. Tidal wetlands can be divided into two zones based on the frequency of tidal flooding: 1) regularly flooded and 2) irregularly flooded. The highest moon-driven tides ("spring tides") occur during full and new moons, while coastal storms can produce even higher tides that occasionally inundate low-lying uplands. (Source: Tiner and Burke 1995)

Table 2.5. Examples of plant indicators of wetland hydrology for selected Northeast wetland types.

Water Regime (Wetland Type)	Indicator Species
Regularly flooded (salt/brackish marsh)	Smooth Cordgrass (tall form)
Regularly flooded (brackish marsh)	Water Hemp
Regularly flooded (tidal fresh marsh)	Spatterdock, Arrow Arum, Pickerelweed, Soft-stemmed Bulrush
Irregularly flooded (salt marsh)	Salt-hay Grass, Salt Grass, Black Grass, High-tide Bush
Irregularly flooded (brackish marsh)	Big Cordgrass, Narrow-leaved Cattail, Seaside Goldenrod, Olney Three-square
Irregularly flooded (tidal fresh marsh)	Halberd-leaved Tearthumb, Jewelweed
Permanently flooded (nontidal marsh)	Spatterdock, White Water Lily, Pondweeds
Semipermanently flooded (nontidal marsh)	Arrow Arum, Wild Rice, Arrowheads, Bayonet Rush, Water Willow
Semipermanently flooded (shrub swamp)	Buttonbush
Seasonally flooded (nontidal marsh/meadow)	Lizard's Tail, Rice Cut-grass, Sweet Flag, Swamp Milkweed, Arrow-leaved Tearthumb, Boneset, Marsh Fern, Tussock Sedge
Seasonally flooded (shrub swamp)	Highbush Blueberry, Common Winterberry, Pussy Willow, Swamp Rose
Seasonally flooded (forested wetland)	Atlantic White Cedar, Northern White Cedar, Swamp White Oak, Green Ash, Skunk Cabbage, Tussock Sedge
Temporarily flooded (nontidal wet meadow)	Goldenrod, Joe-Pye-weed, Multiflora Rose
Temporarily flooded (forested wetland)	Sycamore, Pin Oak, Box Elder, Silver Maple, Eastern Cottonwood, Black Cherry, May Apple, Trout Lily, Virginia Knotweed
Saturated (shrub bog)	Leatherleaf, Labrador Tea, Pitcher Plant, White Beak-rush, Cranberry

Note: These species are generally reliable indicators of the specified water regime for the wetland type. Shrub swamps and forested wetlands are nontidal wetlands; marshes may be either tidal or nontidal, as designated.

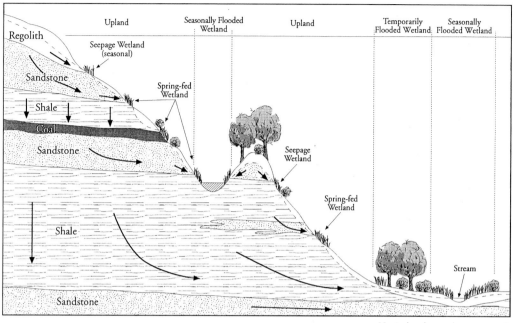

Vertical scale greatly exaggerated.

Figure 2.8. Wetlands form on the landscape wherever there is at least a seasonal excess of water accumulating or being discharged. (Source: Tiner and Burke 1995)

influenced by precipitation and evapotranspiration rates (Figure 2.9). From winter through spring or early summer, the water table is usually at or near the surface in most wetlands. Sometime between May and June, it begins to drop at a noticeable rate. It continues to fall through summer, reaching its low point in September or October. Although this pattern characterizes typical years, abnormally wet years result in higher water tables, and unusually dry years produce lower water tables. This makes attempts to verify the existence of wetlands by the presence and depth of water in the soil at any given moment futile and fruitless if one does not have knowledge of the hydrologic pattern for wetland types and the relationship

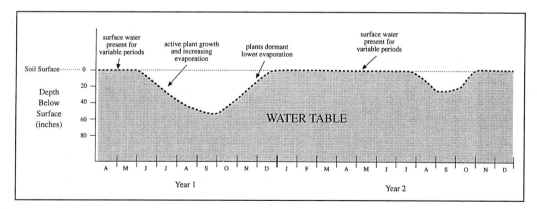

Figure 2.9. Water-table fluctuations in a typical nontidal forested wetland. In general, the water table is at or near the surface through winter and spring, then drops markedly in summer, and rises in the fall. Both seasonal and annual variations are due to changes in precipitation and other factors. (Source: Data from Lyford 1964)

between the current conditions and typical or "normal" conditions.

Determining typical hydrology or normal conditions is not possible for some wetlands. For example, lake levels in the Great Lakes fluctuate considerably over time in response to regional precipitation with no predictable pattern. The effect on Great Lakes coastal wetlands is enormous, causing frequent shifts in vegetation zones (e.g., more marsh and aquatic beds in years of above average precipitation and more wet meadows in dry years). Daily water levels may also change drastically due to wind effects and changes in air pressure. Strong winds blowing across the lakes may push more water to one side of the lake. This "wind setup" can cause short-term, but extremely high water levels (e.g., record increase of 16 feet in Lake Erie water level from Toledo, Ohio, to Buffalo, New York). Wind and changes in air pressure can also create lake seiches where lake water sloshes back and forth for a few days or more.

There are many ways to further categorize nontidal wetlands based on particular hydrologic conditions. Three of the more common approaches are 1) to define them by a combination of their water sources (i.e., surface water v. groundwater) and landform (i.e., depression v. slope), 2) to describe them by the flow of water (inflow, outflow, and throughflow), and 3) to classify them by the degree of flooding or saturation. The first two have traditionally been used by hydrologists and the third by ecologists and soil scientists (see Chapter 4 for soil-drainage classes).

The hydrologist Richard Novitzki developed the first approach while attempting to describe the hydrology of Wisconsin wetlands. Nontidal wetlands are characterized as either 1) surface-water depression wetlands, 2) surface-water slope wetlands, 3) groundwater depression wetlands, or 4) groundwater slope wetlands (Figure 2.10). The first two types obtain their water mainly from precipitation and surface water runoff either from adjacent uplands (overland flow, in the first) or from a combination of overland flow and river overflows or high lake levels (in the second). Floodplains contain examples of both types of surface-water wetlands, with oxbow lakes and sloughs representing the former type and bottomland forests on gently sloping terraces the latter. Groundwater wetlands derive a significant portion of their water from underground sources. Groundwater depression wetlands are directly connected to the underlying water table, which fluctuates in harmony with regional or local water tables. Many wetlands on sandy soils exemplify this behavior (e.g., on Cape Cod in Massachusetts or in the New Jersey Pine Barrens). Groundwater slope wetlands are associated with seeps and springs.

Novitzki's types do not adequately characterize the hydrology of the vast saturated peatlands of the boreal forest (northern U.S. and Canada). These wetlands may be classified as so-called paludified wetlands, where peat moss (*Sphagnum* spp.) plays a prominent role in the formation and expansion of permanently saturated wetlands. Many northern bogs are examples of this type.

Hydrologists have also classified nontidal wetlands on the basis of three water-flow patterns: 1) inflow, 2) outflow, and 3) throughflow. Inflow wetlands are collection basins from which there is no significant outflow. In contrast, outflow wetlands are sources of water, which exits as either groundwater flow or surface water (stream). Throughflow wetlands receive inputs from wetlands topographically above them and discharge water (as groundwater or surface water) to wetlands or other areas downslope. Some groundwater depression wetlands act in different ways seasonally—as outflow wetlands in the spring (recharging local groundwater supplies when basins fill with water) and as inflow wetlands the remainder of the year (receiving groundwater discharge and overland flow).

Ecologists have divided nontidal wetlands into several types based on the frequency and duration of flooding and saturation. This approach is used by the U.S. Fish and Wildlife Service to describe wetland hydrology or "water regime." Flooded types are inundated (covered by surface water) for variable periods, while saturated types are rarely flooded and are best characterized by their waterlogged soils (Table 2.6). Saturated wetlands can be separated into two types: 1) permanently saturated wetlands and 2) seasonally saturated wetlands. The former include floating wetlands (a mat of vegetation floats on the surface of a lake or pond, e.g., floating cattail marshes and floating bogs), paludified bogs and other wetlands with continuously waterlogged soils, and permanent seepage or

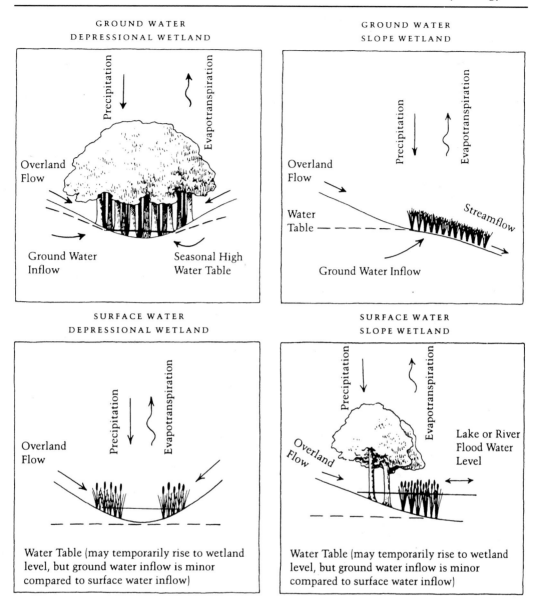

Figure 2.10. Hydrologic inputs and outputs for surface-water and groundwater wetlands. (Source: Tiner 1988; redrawn from Novitski 1982)

spring-fed wetlands. Seasonally saturated wetlands include seepage wetlands (subject to seasonal discharges of groundwater) and perched-water-table wetlands. The latter type occurs in areas with confining subsoils where the local water table is "perched" above this layer, such as groundwater depression wetlands. Seasonally saturated wetlands are among the most difficult to recognize, especially in the summer and early fall, when water is usually absent.

Certain species of plants are reliable indicators of various water regimes, yet many species like red maple and soft rush are wide-ranging and capable of growing under a variety of hydrologic regimes. Table 2.5 gives several examples of wetland plants that are useful for predicting the hydrology of nontidal wetlands. While these plants are generally reliable indicators, exceptions do occur. However, whenever a large stand of these species is observed, the water regime can usually be accurately predicted.

Table 2.6. Definitions of major water regimes used by the U.S. Fish and Wildlife Service for wetland mapping.

Group	Type of Water	Water Regime	Definition
Tidal	Salt/brackish	Subtidal	Permanently flooded by the tides
		Regularly flooded	Flooded daily by the tides
		Irregularly flooded	Flooded less than daily by the tides
	Fresh	Regularly flooded	Flooded daily by the tides
		Seasonally flooded–tidal	Flooded irregularly by the tides and for extended periods by a river
		Temporarily flooded–tidal	Flooded irregularly by the tides and for brief periods by a river
Nontidal	Fresh	Permanently flooded	Flooded constantly
		Semipermanently flooded	Flooded throughout the growing season in most years
		Seasonally flooded	Flooded for extended periods during the growing season
		Saturated	Water is present at or near the surface for most of the year, and surface water is seldom present
		Temporarily flooded	Flooded for brief periods, usually early in the growing season, and water table is usually well below the surface
		Artificially flooded	Hydrology is controlled by pumps or siphons, usually in combination with dikes or dams

Additional Readings

National Research Council, Committee on Characterization of Wetlands. 1995. *Wetlands: Characteristics and Boundaries.* National Academy Press, Washington, DC.

Novitzki, R. P. 1989. Wetland hydrology. In *Wetland Ecology and Conservation: Emphasis in Pennsylvania,* ed. S. K. Majumdar, R. P. Brooks, F. J. Brenner, and R. W. Tiner, Jr., pp. 47–64. Pennsylvania Academy of Sciences, Easton, PA.

Stone, A. W., and A. J. Lindley Stone. 1994. *Wetlands and Ground Water in the United States.* The American Ground Water Trust, Concord, NH.

Winter, T. C., and M.-K. Woo. 1990. Hydrology of lakes and wetlands. In *Surface Water Hydrology of North America. The Geology of North America,* Vol. 0-1, ed. M. G. Wolman and H. C. Riggs, pp. 159–187. Geological Society of America, Boulder, CO.

CHAPTER 3

The Birth and Growth of Swampland
Wetland Formation and Evolution

Wetlands are dynamic natural environments subject to changes through natural processes and human activity. Natural forces have been acting for eons to create, shape, and destroy wetlands as well as to degrade and improve their quality. These events include rising sea level, plant-community dynamics, the hydrologic cycle (long-term changes in rainfall patterns), sedimentation, erosion, beaver-dam construction and damage by animals, hurricanes, and fire. Some of the natural processes are catastrophic events, but most occur more gradually over time or affect only a small portion of the landscape. Deposition of water-borne sediments along rivers and streams usually leads to the formation of new wetlands, while erosion removes wetland acreage. In the past, most fish and wildlife species adapted to these dynamics over time, and natural ecosystems were in balance with these natural forces. More recently, however, the effects of humans on wetlands have become significant. Human actions, in contrast to most natural forces, are usually too rapid and widespread for wildlife to make necessary adjustments, resulting in permanent impacts such as local elimination of species and for some, extinction.

Origins of Swampland

Wetlands form wherever there is water above, at, or near the soil surface for a sustained period at frequently recurring intervals (see Chapter 2). This wetness must be sufficient to create environmental conditions that stress most plants and favor hydrophytes. Climate, topography, soils, and geology play a major role in wetland formation. Humans have also played a role by building wetlands for specific purposes (e.g., rice paddies, farm ponds, wastewater treatment, wildlife management, or mitigation of adverse impacts from development) or by inadvertently causing wetlands to form (e.g., due to irrigation, undersized road culverts, or construction of reservoirs).

Wetter climates favor wetland development, as do cooler climates. The more precipitation, the better the chances for water to collect and form wetlands. Cooler climates produce lower evapotranspiration rates, which can facilitate wetland formation even when annual precipitation is not high. Evaportranspiration rates are higher in warmer climates and can result in a significant lowering of the water table.

Topography is an important factor in wetland formation. Low topography or depressional features promote the collection of surface-water runoff, while sloping terrain usually has better drainage. Wetlands are more abundant in valleys and depressions than on the sides of mountains. The extensive wetlands occurring throughout the flat-to-gently-sloping coastal plain versus a scattering of wetlands in the Appalachian Mountains provides further evidence of this. Depressions can be caused by many processes: glacial activity, landslides (earthquakes), shifting fault blocks (tectonic activity), aeolian processes (e.g., sand dunes), land subsidence (especially sinkholes in limestone regions, caused by dissolution of carbonate rocks underground), meteorite impacts, and animal and human activity (e.g., beaver dams, impoundments, or excavations). Toes of slopes, concave slopes (e.g., drainageways), saddles between mountains, and broad flats are also good places for wetland formation.

The nature of the soils plays an important role in wetland creation. The low permeability of clayey soils often promotes the retention of water (ponding) at the ground surface. In contrast, rapidly permeable sandy soils do not typically support wetlands, except at low elevations subject to flooding or where outlets for surface-water or groundwater discharge are lacking. Poor external drainage causes wetlands to form under these circumstances.

The local geology also influences wetland development. Permeable soils in areas of stratified drift promote rapid infiltration of water, so wetlands may not form in typical locations such as depressions unless the basin intersects the local water table (e.g., kettle holes on Cape Cod). Till and impermeable stratified drift deposits favor wetland development in areas of low relief, even on mountain tops. Areas that were once glacial sea or lake bottoms have marine clays or fine-textured lacustrine sediments that promote wetland formation in low or level landscape positions. Springs and seeps are often common in limestone, giving rise to wetland formation in sinkholes and valleys.

During the millions of years that have passed in Earth's history, there have been enormous changes in land masses and climates. Past climates have been both warmer and colder than today's climate. The land has experienced many cycles of erosion and sedimentation. The geologic record provides evidence of several episodes of the advance of the sea over much of the Northeast. For example, the rich limestone deposits of New Jersey's Kittatinny valley, now deep underground, were formed when the climate was more tropical and shallow seas covered the area more than 300 million years ago. In the Triassic period (about 200 million years ago), much of the Northeast was a lowland with large lakes and extensive swamps. The swamps were overgrown with tree-sized cycads and other conifers, since flowering plants had not yet evolved. Dinosaurs and other reptiles were the dominant creatures; mammals were small, and humans were not present. These swamps are now buried beneath hundreds of feet of sediment. In many parts of the world, such buried wetlands are the source of today's rich oil and gas deposits.

The Big Freeze: Glaciation Effects

While these prehistoric shifts in climate and landscapes are interesting and important, the geologic event most responsible for shaping the current landscape in northern areas and creating today's wetlands is the Wisconsin Ice Age. This event took place 80,000 to 18,000 years ago. The world's climate became increasingly colder, leading to the buildup of arctic ice

and causing the Laurentide Ice Sheet to move southward. Roughly one-third of the world's land area was covered by ice, compared to only 10 percent today. At the glacier's maximum advance, ice covered virtually all of New England, most of New York, the northern quarter of New Jersey, and the northeastern and northwestern corners of Pennsylvania (Figure 3.1). This ice mass was over one-half mile thick.

The glacier destroyed drainage patterns by blocking stream outlets, causing new bodies of water to form and new outlets to open. One look at a map of the major drainage basins of New Jersey clearly reveals the disrupted pattern in the north and a marked difference in the courses of rivers between the glaciated and nonglaciated portions of the state. Throughout the glaciated region, lakes are abundant. South of this region, the few lakes typically represent impounded rivers and river valleys (reservoirs and manmade recreational lakes).

In the glaciated portion of the Northeast, most wetlands, especially the larger ones, originated in former glacial lakes and depressions formed during the postglacial period over 10,000 years ago. Glacial lakes were created in three types of locations: 1) in rock basins scoured by glacial erosion, 2) in depressions (kettles) formed by melting ice blocks left by the retreating glacier, and 3) in basins produced by glacial drift in river valleys. Large glacial lakes developed in major watersheds, such as the Passaic, Hackensack, Hudson, Connecticut, and Merrimack Rivers (Figure 3.1), and in the expanded basins of Lakes Erie and Ontario.

As the glacier receded about 10,000 years ago, most glacial lakes drained, promoting wetland formation in these basins. The exposed clay lake beds restricted the downward movement of water, thereby providing an excellent substrate and environmental conditions for the establishment of wetlands. Large wetland complexes in northern New Jersey like Great Swamp, Great Piece Meadows, Troy Meadows, and Black Meadows formed where glacial Lake Passaic existed.

The Great Lakes began to form as the Laurentide Ice Sheet receded about 14,000 years ago. When glacial ice still covered the area occupied by Lakes Superior, Huron, and Ontario and most of Lake Michigan, the first of the lakes appeared south of the ice front—Glacial Lake Chicago (southernmost end of Lake Michigan) and Glacial Lake Maumee (Lake Erie). As the

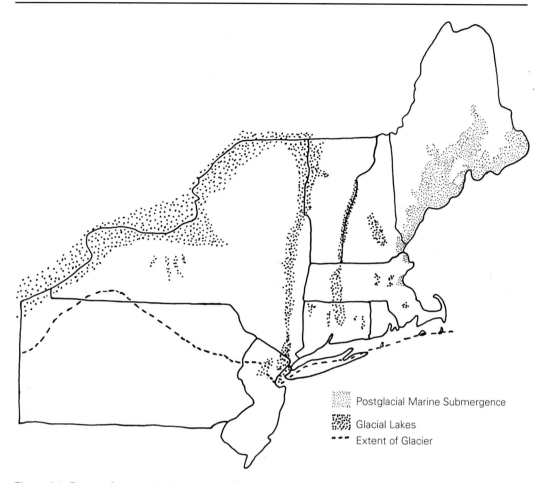

Figure 3.1. Extent of recent glaciation in the Northeast, the general location of major glacial lakes, and areas of postglacial marine submergence. (Note: Vermont's Lake Champlain was first an estuary [Champlain Sea] and then a glacial lake. Also, the Great Lakes formed as the glacier melted and receded.)

ice sheet receded, the glacial lakes grew in size and two lakes emerged—Glacial Lake Iroquois (Lakes Ontario and Erie) and Algonquin (Lakes Michigan and Huron). Further changes occurred as the glacier retreated to the north (see Professor Steve Dutch's webpage for a pictorial history of the Great Lakes at http://www.uwgb .edu/dutchs/glkhist/glkhist0.htm). About 6,000 years ago, the Great Lakes assumed their present-day configuration. Former lakebeds lie inland of today's shorelines and many wetlands formed in these locations (e.g., the Great Black Swamp in western Ohio).

The weight of the glacier had depressed the landmass significantly. In some coastal areas, the land was pushed below sea level, submerging it. Marine clays were deposited in this environment. When the glacier retreated, the landmass rebounded. Low-lying coastal areas were now above sea level, where their dense clay soils promoted the establishment of wetlands. This condition occurred along the seaboard lowland from Boston, Massachusetts, north as well as in the Lake Champlain valley, which was an arm of the Atlantic Ocean (Champlain Sea) about 10,500 years ago. The latter area became Lake Champlain as the glacial rebound of the land eliminated its connection with the ocean and the basin became filled with freshwater.

The glacier also had significant effects on wetland formation south of the great ice sheet. The climate became arctic, and the sea level dropped about 400 feet below its present-day levels. Millions of acres of the continental shelf were dry land. The mouth of the Delaware River was 75 miles east of its current position. Temperatures were below freezing for much of the year during the Wisconsin Ice Age. Permafrost

Figure 3.2. Periodic flooding and the often fine-textured soils deposited on floodplains of rivers and streams make the floodplains ideal places for wetland formation. Differences in elevations on a floodplain create varied flooding and soil-saturation conditions, which lead to the establishment of assorted plant communities, including marshes, wet meadows, seasonally flooded forested wetlands, and temporarily flooded forested wetlands. This aerial photograph shows floodplain wetlands along the Chester River near the Maryland-Delaware border (middle of photo). Also visible are depressional Delmarva bay or pothole wetlands scattered on the broad, flat interstream divides typical of the lower coastal plain.

and tundra developed under these conditions. A boreal forest of spruce, pine, and birch covered the Mid-Atlantic landscape, interspersed with marshes and wet meadows. North of this region was taiga, a swamp forest dominated by spruce and fir. Freeze and thaw (thermokarst) basins formed from alternate freezing and warming. When the climate warmed with the retreat of the glacier, ponds and wetlands formed in basins underlain by impervious clays.

The Floods: Alluvial Processes

Wetlands typically form on floodplains along major rivers and streams (Figure 3.2). A complex interaction among geology, climate, topography, and soils determines where broad, flat plains develop on the landscape. Water reaching the earth's surface either evaporates, percolates into the soil, infiltrates groundwater stores, or runs off the land as surface water. It flows downhill following the path of least resistance, carving out channels in more erodible substrates and creating streams. Small streams join other small streams as water moves down-gradient, forming larger streams and eventually rivers. Flowing water carrying suspended sediments reaches flat areas on the landscape that may have been shaped by the river's erosive action or other processes. These broad plains are subject to periodic flooding, with the frequency and duration determined largely by local elevation and the size of the watershed upstream. Within floodplains, the river channel shifts back and forth through time; a winding, meandering river usually indicates the presence of floodplains and large watersheds. This process affects the landform in ways that create, modify, and destroy wetlands.

Deposition of river-borne sediments leads to a buildup of substrates that initially support wetland vegetation. Sedimentation is most active on the inner banks of river bends, where the water moves more slowly. Here point bars develop, and over time a series of ridges and depressions (sloughs) forms. The latter are prime sites for wetland creation. Continued deposition may ultimately create elevations that are only infrequently flooded, not often enough to support wetlands. A single flood usually deposits a thin veneer of sediment on the broad floodplain. If the flood results from record rainfalls and occurs at the wrong time of the year, when soils have recently been tilled, enormous amounts of sediment can be deposited. One flood in Louisiana's Atchafalya River, a tributary of the Mississippi, deposited a foot and a half of sediment.

Erosion in the river channel is greatest on the outside of bends, where the highest velocities develop. The channel moves laterally over time, eroding one side and building up the other. Bends widen until they are cut off as new channels develop out of the river's natural tendency to take the path of least resistance downstream. This process forms "oxbows": former river bends, now separated from the river's mainstream. As sediments accumulate in these basins, wetland vegetation colonizes shallow water, often inducing further sedimentation.

Depositional and erosional processes continually reshape river channels and floodplains. Poor land-use practices have added much sediment to the nation's rivers. In some places, wetlands and their plant and animal life have benefited from these unintentional additions of sediments. Studies in a few places have shown an increase in floodplain wetlands since colonial times as shallow river basins filled with eroded farm soils. Many freshwater tidal marshes along the Delaware River supposedly developed in this way. Today, improved soil- and water-management practices are better conserving productive topsoils, which ironically may be having a negative effect on wetland formation in some areas.

Early stages of floodplain development are represented by extensive marshes bordering streams. Valleys are gradually filled in by sediments, raising the elevations of associated floodplains. Over time, natural levees form along channels where water-borne coarser materials such as sands and gravels first settle out as floodwater velocity decreases. The increased elevation creates a type of natural berm, limiting flooding. Woody plants tend to occupy these sites. Meanwhile, backwater marshes, usually fed by minor streams, continue to increase in elevation as finer sediments are deposited. They eventually reach a level favoring the growth of woody plants. Shrub swamps become established, which may become forested wetlands under most natural circumstances (assuming no catastrophic events).

Large wetland complexes occur along rivers throughout the country and extensive river swamps typify the coastal region from New Jersey to Texas. Rivers originating in foothills and mountains (e.g., the Piedmont and the Appalachians) meet the relatively flat, easily eroded coastal-plain landscape, and their waters flatten out, creating broad floodplains. Most of the wetlands in the conterminous United States occur in the Southeast because of extensive floodplain development and the region's high rainfall.

The Tides: Effects of Rising Sea Levels

At maximum recent glaciation nearly 18,000 years ago, much of the world's ocean water was stored as ice. Sea levels were about 400 feet lower than their current levels. This means, for example, that the New Jersey shoreline was 80 miles east of its present position and the coastal plain was more than twice its current size. When the glacier began to melt, water was released back to the oceans, thereby raising sea levels dramatically. River valleys were drowned by the rising waters, forming estuaries. The largest estuaries in the Northeast, including Delaware Bay, Port Newark, New York Harbor, and Chesapeake Bay, were all river valleys in glacial times.

About 5,000 years ago, sea levels stabilized, with a much lower annual rise, allowing salt marshes to develop. The Northeast shoreline looked much as it does today. Sea level continues to rise at about one foot per century in much of the region. With this increase in ocean levels, barrier islands continue to gradually migrate landward, while stream valleys are being slowly submerged.

Coastal marshes typically develop in sheltered embayments with heavy sedimentation. These areas are generally located behind barrier islands and along tidal rivers. Sediments are carried downstream by rivers flowing to the sea and along the beaches by longshore ocean currents. When a river meets the sea, its flow slows and sediments begin to settle out to form deltas and bars at the mouth and on intertidal flats between barrier islands and the mainland. Sedimentation also takes place when tidal currents slow during slack-water periods as the tides are changing. The rate and extent of sedimentation depend on the original size and age of the estuary, upstream erosion rates, and deposition by tides, marine currents, and the river.

Salt marsh development has been described by Professor A. C. Redfield of the Woods Hole Oceanographic Institution as follows. Initially, mud and silt are deposited in coastal waters to form tidal flats in the shallow water zone. As elevations increase and exceed mean sea level, salt-marsh vegetation, especially smooth cordgrass, becomes established. The presence of these plants further slows the velocity of flooding waters, promoting more sedimentation. As levels rise above mean high tide, other salt-marsh plants replace smooth cordgrass. Sediments and peat continue to build up until an equilibrium with erosion and oxidation is reached.

In many areas along the Atlantic coast, salt marshes continue to migrate landward, advancing into low-lying uplands or formerly nontidal wetlands (Figure 3.3). This migration is part of a worldwide coastal swamping process that is largely attributable to the rise in sea level. It is most likely accelerating due to the so-called greenhouse effect, the buildup of carbon dioxide in the atmosphere that triggers a rise in the earth's temperature, which in turn increases melting of the polar ice caps. Rising sea levels not only increase the effect of tidal flooding, but also can raise the water tables of surrounding lands, causing changes in plant communities.

Peat deposits exposed on coastal beaches provide stark evidence of the landward movement of the beaches over former salt marshes. Other evidence of rising tides are dead trees and stumps in salt marshes. In many Northeast salt marshes, stumps of Atlantic white cedars can be found. Cedar logs have also been found buried in offshore marine sediments; the remnants of former swamps, such trees were growing on the continental shelf when it was exposed thousands of years ago. Analysis of peat cores in salt marshes shows salt-marsh peats above peaty remains of freshwater plants.

The Eastern Shore of Maryland and Virginia (lower Delmarva Peninsula) provides the most dramatic example of the effects of sea-level rise on coastal plant communities in the Northeast. Here the coastal plain is subsiding as the sea permanently inundates the lowest marshes, changing them to open-water habitats. The added weight on the coastal plain is causing the land to gradually decline. Loblolly-pine forests are being inundated with salt water, changing the plant community from lowland pine forest to salt marsh in a matter of a few decades (see Plate 5). This condition is so widespread that the forests have been mapped as a

Figure 3.3.(opposite) Submergence of upland with rising sea level caused tidal wetlands to form in Connecticut's Pataguanset River estuary during the last 4,000 years (upper left—4,000 years before present; upper right—3,000 years ago; lower left—2,000 years ago; lower right—1,000 years ago). The proposed sequence is based on marsh-accretion (peat-buildup) rates and sea-level-rise rates of about 3.3 feet per 1,000 years, which is much lower than the current rate of sea-level rise. (Source: Orson et al. 1987; reprinted by permission of the Estuarine Research Federation; copyright Estuarine Research Federation.)

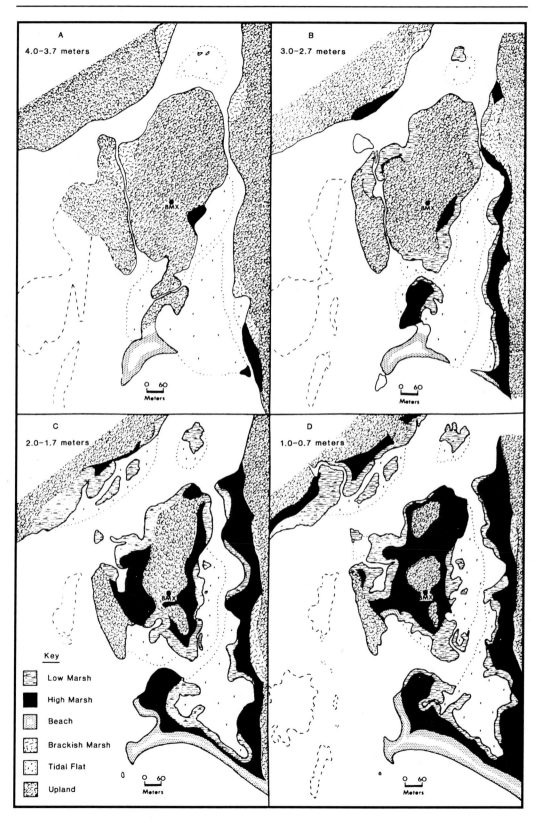

type of estuarine wetland by the U.S. Fish and Wildlife Service's National Wetlands Inventory Project. Maps show former freshwater Carolina bay wetlands that are now salt marshes.

Another example of changing coastal vegetation due to rising tides can be found in the Hackensack Meadowlands in New Jersey. In the 1800s, the Secaucus Bog along the Hackensack River was an Atlantic white-cedar swamp. Black spruce and larch reached their southern limits along the coast here. The last of the cedars died by 1935, and today the former bog is a brackish tidal marsh of cattails and common reed. Rising sea level coupled with dredging of the Hackensack River increased saltwater flooding in the area.

Hydrarch Succession: Plant-Community Dynamics

Filling of lakes, ponds, oxbows, and other bodies of water with the remains of aquatic and wetland plants is a part of a natural process called "hydrarch succession." This process is one of many ways that wetland plant communities change over time. It results in a progressive series of wetland types within the former basin (Figure 3.4), which changes the wildlife habitat and other wetland functions through time.

Hydrarch succession is characterized by an evolution of communities from aquatic beds to forested wetlands. Initially, an open body of water is colonized by submerged aquatic plants. Each year these plants die back, depositing organic material (leaves and stems) on the bottom of the lake or pond. Slowly, over hundreds or thousands of years (depending on the initial water depth), the basin becomes shallow enough to support emergent herbs like cattails, bur-reeds, and bulrushes, or hydrophytic shrubs such as buttonbush and water-willow. These species add more organic matter to the basin, and gradually the basin becomes completely filled with vegetation—it evolves into a marsh. With repeated introductions of organic material, open water disappears and a solid mat of organic soil eventually fills the former basin. A more stable substrate provides support for shrubs and trees. The marsh gradually turns into shrub swamp and eventually into a forested wetland. The open body of water has eventually become a red-maple swamp, for example. It took about

10,000 years for Titcut Swamp in southeastern Massachusetts to change from a lake to a forested wetland.

The change in vegetation due to hydrarch succession is not always a straight line from open water through aquatic beds, marshes, and shrub swamps to forested wetland. Many times the process is interrupted by fire during droughts, by animal and human activity, and by floods during a series of extremely wet years. This dynamism makes wetlands particularly challenging and interesting places for study and observation (Figure 3.5).

In the first half of this century, some ecologists mistakenly believed that a swamp would eventually become an upland forest as elevations increased due to a buildup of organic matter and as drainage improved. This false notion has been corrected in current ecology textbooks. Although peat does accumulate in certain wetlands, the areas typically remain wetlands unless drained, filled, or drastically altered by catastrophic natural processes, such as a major climatic change. Poorly drained soils simply do not become better drained unless the water source is altered. Accounts of upland forest being the end result or climax community in hydrarch succession may have been due, in large part, to an incomplete understanding of the concept of wetland.

Paludification: Climbing Bogs

In northern climes and glacially formed landscapes, bog vegetation—especially peat moss (genus *Sphagnum*)—has played a major role in hydrarch succession of lakes and kettle ponds. Peat moss has an amazing ability to absorb water: it can retain about twenty-five times its dry weight. This allows it to wick up water and even grow up slopes, converting upland spruce forests to boggy swamps, a process called "paludification." This swamping process is a major force shaping landscapes in the boreal forest and subarctic regions of the Northern Hemisphere. Noticing this phenomenon in the late 1800s, Nathaniel Shaler, the author of *General Account of the Fresh-water Morasses of the United States* referred to peat mosses as "the most important swamp-breeding plants."

Peat moss initially colonizes the shores of a kettle pond and gradually advances as a mat to-

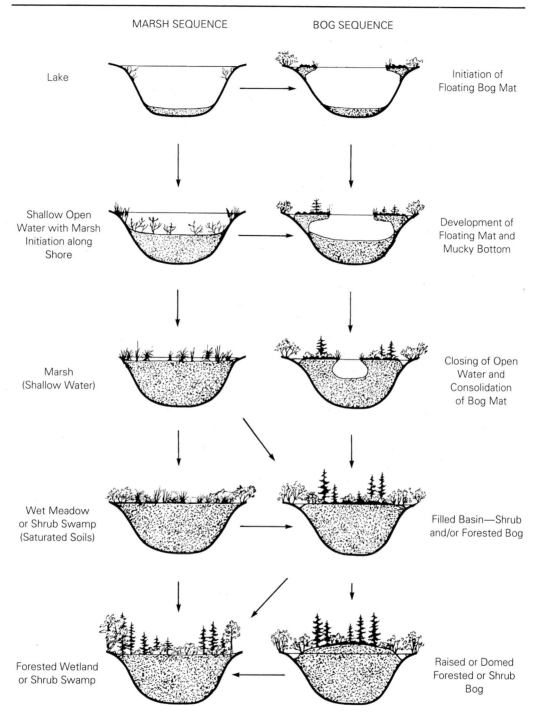

MARSH SEQUENCE　　　　BOG SEQUENCE

Lake — Initiation of Floating Bog Mat

Shallow Open Water with Marsh Initiation along Shore — Development of Floating Mat and Mucky Bottom

Marsh (Shallow Water) — Closing of Open Water and Consolidation of Bog Mat

Wet Meadow or Shrub Swamp (Saturated Soils) — Filled Basin—Shrub and/or Forested Bog

Forested Wetland or Shrub Swamp — Raised or Domed Forested or Shrub Bog

Figure 3.4. Hydrarch succession, a sequence of wetlands developing from the gradual filling-in of lakes. This is one way that nontidal wetlands form. (Source: Tiner 1985; adapted from Dansereau and Segadas-Vianna 1952.)

ward the center of the pond. As the mat thickens, it provides support for other plants, first sedges or water-willow and later bog shrubs, especially leatherleaf. Following the pattern of hydrarch succession, the open body of water be-

comes smaller and smaller, until only a small pond called an "eyelet pond" is left near the center of the basin. Eventually the basin is completely covered by bog vegetation and over time becomes a black-spruce, balsam-fir, or larch

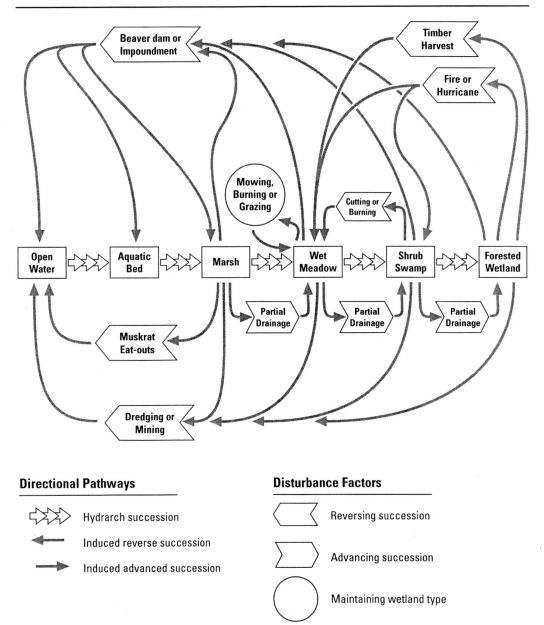

Figure 3.5. Rather than being an orderly succession of wetland types from lakes to forested wetlands, plant-community change is dynamic, due to a host of factors including beaver activity, grazing by various animals, fire, and human actions.

swamp. This type of process is still occurring in glaciated regions.

Paludification is the predominant factor creating wetlands in regions where extensive peatlands exist (e.g., cool, coastal regions and inland boreal regions). Restricted drainage and cool temperatures (low evapotranspiration) promote this process. Once a depression has been filled with peat (initial paludification phase

—a type of hydrarch succession; see Figure 3.6), the peat moss begins to advance into adjacent uplands in relatively flat terrain (often old glacial lake beds) forming a blanket over the land surface (advanced phase). This blanket can expand across watersheds forming peatland landscapes as evidenced by the Agassiz Lowlands of northwestern Minnesota (Glacial Lake Agassiz). This process has also been called

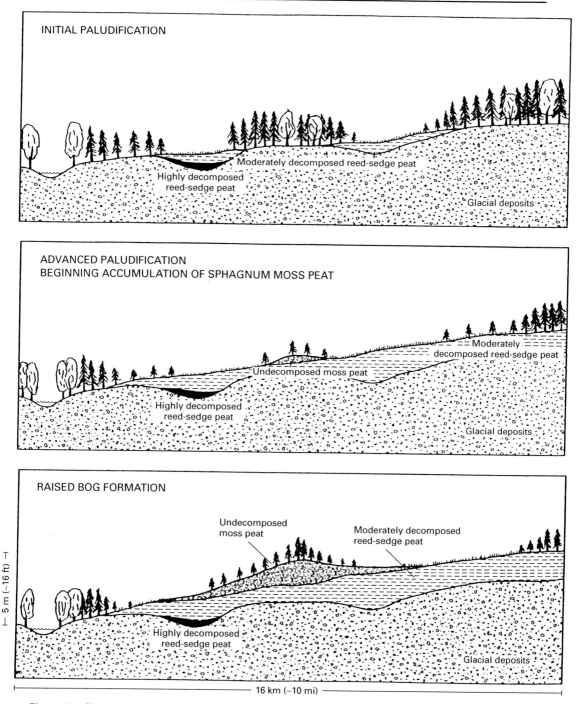

INITIAL PALUDIFICATION

Moderately decomposed reed-sedge peat

Highly decomposed
reed-sedge peat

Glacial deposits

ADVANCED PALUDIFICATION
BEGINNING ACCUMULATION OF SPHAGNUM MOSS PEAT

Moderately
decomposed reed-sedge peat

Undecomposed moss peat

Highly decomposed
reed-sedge peat

Glacial deposits

RAISED BOG FORMATION

Undecomposed
moss peat

Moderately decomposed
reed-sedge peat

Highly decomposed
reed-sedge peat

Glacial deposits

5 m (~16 ft)

16 km (~10 mi)

Figure 3.6. Stages in paludification of a northern landscape. (Source: Minnesota DNR 1987)

"swamping" as the entire landscape is engulfed by sphagnum and associated bog vegetation. The peatland landscape may be represented by a complex mosaic of fens, bogs, raised bogs, swamps, water tracks, and variously shaped is-

lands that are quite impressive on aerial photographs as well as on the ground. In the Agassiz Lowlands, peat deposits may be up to 30 feet thick; these wetlands are about 5,000 years old. Paludification can also be initiated on wet slopes (blanket bogs) and even, under certain conditions, when northern forests are harvested or tree cover is removed by fire. In the absence of trees, water loss by transpiration decreases and soil water content may increase to the point where tree growth is inhibited. The area gets increasingly wetter promoting peat moss colonization and bog formation.

The Hydrologic Cycle: Changing Water Levels

Precipitation amounts and patterns have obvious effects on wetlands. Regions of higher precipitation tend to have more wetlands than drier regions, except where paludification is a dominant process. Annual and long-term fluctuations in precipitation can have a significant impact on vegetation in some areas. The Midwest prairies (e.g., Dakotas) have a subhumid climate typified by a ten- to twenty-year cycle of wet and dry years, with severe droughts occurring periodically. In wet years, marshes and aquatic beds dominate the pothole wetlands, while in dry years, wet meadow and low prairie vegetation abound. Pothole soils store the seeds of all the species and the seeds wait for the proper conditions to germinate. (These natural seed banks have facilitated restoration of formerly drained wetlands.) During extreme droughts, upland prairie species may become abundant in pothole wetlands or they may be tilled for crop production.

The Great Lakes have experienced significant fluctuations in lake levels since their formation around 6,000 years ago. In wet years, lake levels rise, whereas in years of low precipitation, levels fall affecting coastal vegetation patterns, beach development, fish habitat, and other features. Differences in lake levels between the extremely wet and extremely dry years may range from 3.5 to 6.5 feet. Extremely wet years occurred in the 1870s, early 1950s, early 1970s, mid-1980s, and mid-1990s, whereas extremely low water years were recorded in the late 1920s, mid-1930s ("dust bowl years"), mid-1960s, and late 1990s. Today, locks and dams affect water levels in two lakes (Lake Superior and Lake Ontario), but their levels still change with precipitation rates, although at lower ranges. Fluctuating lake levels cause a shifting of coastal wetland vegetation zones. In wet years, there are more extensive shallow water aquatic beds (submerged and floating-leaved aquatics) and marsh acreage than in dry years, when there is more wet meadow, shrub swamp, forested wetland, and exposed shoreline. Inundation for extended periods kills most shrub and tree species and drowns out meadow species, whereas exposed substrates favor their colonization (Figure 3.7). As much as 40 percent of the wetland vegetation of some individual Great Lakes coastal wetlands may be lost to open water during extremely wet years. Similar responses to wet and dry years may also be observed on a much smaller scale around the margins of other lakes and ponds (e.g., coastal plain ponds). Years of low precipitation (natural drawdowns) create a wider belt of emergent wetland along the shores, while in extremely wet years high water may increase aquatic beds at the expense of emergents. Consequently, these wetlands do not tend to follow the typical hydrarch successional path exhibited by many wetlands in the Northeast and Great Lakes region. Instead, plant communities move back and forth or up and down the gradient depending on the hydrologic conditions.

Nature's Hydraulic Engineer: The Beaver

The beaver has a long history of shaping North America's landscape. Among all wild animals, it probably has had the greatest influence on this continent's wetlands. Its activities have changed the pattern of the landscape by adding marshes and ponds to a vast expanse of forest. This added diversity has benefited many species that utilize such habitats for food, water, or breeding.

Prior to the colonial period and heavy trapping, beavers were abundant across the continent, with a population estimated at 11 million. Many of today's wetlands owe their existence to these animals, which are well known for building dams of sticks and mud across streams, creating ponds. Beavers cut down trees and use branches and trimmed sticks to make dams that may exceed 1,500 feet in length (most are less than 1,000 feet) and 6 feet in height (Figure 3.8). The dams are packed with mud, stones, and aquatic-plant remains. Beavers use their feet and noses to tamp down the mud; contrary to popular belief, they do not apply the mud with their

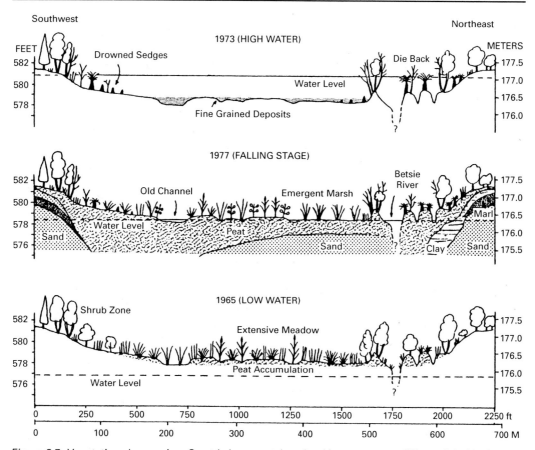

Figure 3.7. Vegetation changes in a Great Lakes coastal wetland in response to different lake levels. (Source: Herdendorf et al. 1986)

tails. They also construct stick lodges that range from 15 to 40 feet wide at the base and may extend 5 or 6 feet above a pond's surface. Some beavers live in dens along river banks or lake shores.

Besides creating new wetlands, beavers may also alter the hydrology of existing wetlands, setting back hydrarch succession. Forested wetlands may be converted to marshes and open water, providing preferred habitat and food for beavers. Selective feeding by beavers on quaking aspens can affect the plant community of adjacent forests by favoring the growth of non-food species such as paper birch and spruce.

From the 1600s to the 1800s, beavers were heavily trapped to the point of extinction in most of the Northeast region. Today, due largely to reintroductions, these wetland builders are making a comeback in many areas. Beavers play a prominent role in wetland creation in Maine and other parts of New England and upstate New York. The return of the beaver benefits many wildlife, especially wood ducks, hooded mergansers, and tree swallows, which build nests in cavities in standing dead trees found in beaver ponds. Flooded trees also serve as prime nesting spots for colonies of great blue herons.

The Human Factor

For the past two hundred years, human influences have been the most significant factor affecting the condition of wetlands (see Chapter 8). Although far more responsible for the destruction of wetlands than for their creation, people have made new wetlands by accident or design. In most cases, wetlands have formed where people have created depressions on the landscape such as ponds or where there is water running off the land due to seepage or irrigation.

Figure 3.8. Beaver play a major role in wetland creation and modification through the construction of dams that block streams and create ponds and marshes.

Many mill ponds, farm ponds, artificial lakes, and reservoirs have wetlands developing in shallow water and along their shorelines. The shallow-water zone of these bodies of water may gradually fill in with sediments, become completely overgrown with marsh plants, and proceed down the path of hydrach succession. Marshes now occupy the entire basins of some abandoned mill ponds in New England. Wetlands have become established along highways where surface-water drainage is impeded by undersized culverts, in roadside ditches, and in stormwater detention basins. Wherever there is more water than can drain off the land, from natural or human-induced causes, wetlands will form.

In some places, wetlands are being built to provide secondary treatment for wastewater. This practice, which is widespread in Europe, may also have significant value in rural areas that lack the financial resources for modern treatment plants. Such constructed wetlands provide important water-quality benefits and help reduce water pollution.

Recently, wetlands have been purposely cre-ated in conjunction with government-financed construction projects such as highways, port expansion, and flood-control impoundments, or by private developers to mitigate unavoidable losses of natural wetlands. The Army Corps of Engineers has successfully built tidal wetlands in conjunction with dredging needed to maintain navigable waterways. Overall, however, attempts to create wetlands are fraught with problems. It is not a simple feat to make a wetland where one never existed. Except for building ponds and establishing wetlands next to existing natural wetlands or excavating down to the water table, establishing wetland hydrology at a nonwetland site is often a difficult task. Moreover, creating a wetland is one thing, but replacing the functions of an existing wetland is quite another.

Additional Readings

Heinselmann, M. L. 1970. Landscape evolution, peatland types, and the environment in the Lake Agassiz peatlands natural area, Minnesota. *Ecological Monographs* 40: 235–261.

Herdendorf, C. E., S. M. Hartley, and M. D. Barnes (editors). 1981. *Fish and Wildlife Resources of the Great Lakes Coastal Wetlands within the United States.* Volume 1: Overview. U.S. Fish and Wildlife Service, Biological Services Program, Washington, DC. FWS/OBS-81/02-v1.

Jackson, S., and T. Decker. 1993. *Beavers in Massachusetts.* University of Massachusetts Cooperative Extension System and Massachusetts Division of Fisheries and Wildlife, Amherst, MA. Publication C-213.

Minc, L., and D. A. Albert. 2004. *Great Lakes Coastal Wetlands: Abiotic and Floristic Characterization.* U.S. Environmental Protection Agency, Great Lakes National Program Office, Chicago, IL.

Mitsch, W. J., and J. G. Gosselink. 1986. *Wetlands.* Van Nostrand Reinhold, New York.

Niering, W. A. 1989. Wetland vegetation development. In *Wetlands Ecology and Conservation: Emphasis in Pennsylvania,* ed. S. K. Majum-

dar, R. P. Brooks, F. J. Brenner, and R. W. Tiner, Jr., pp. 103–113. Pennsylvania Academy of Sciences, Lafayette College, Easton, PA.

Niering, W. A., and R. S. Warren. 1980. Vegetation patterns and processes in New England salt marshes. *BioScience* 30: 301–307.

Niering, W. A., R. S. Warren, and C. G. Weymouth. 1977. *Our Dynamic Tidal Marshes: Vegetation Changes as Revealed by Peat Analysis.* Connecticut College, New London, CT. Connecticut Arboretum Bulletin 22.

Pielou, E. C. 1991. *After the Ice Age: The Return of Life to Glaciated North America.* University of Chicago Press, Chicago.

Redfield, A. C. 1965. The ontogeny of a salt marsh estuary. *Science* 147: 50–55.

Swan, J. M. A., and A. M. Gill. 1970. The origins, spread, and consolidation of a floating bog in Harvard Pond, Petersham, Massachusetts. *Ecology* 51: 829–840.

Van der Valk, A. 1989. *Northern Prairie Wetlands.* Iowa State University Press, Ames, IA.

CHAPTER 4

Swamp Earth
Hydric Soils

The combination of landscape position, underlying geology, vegetation, and climate influences the flow of water across and through the land surface. Out of all the factors affecting soil formation, or pedogenesis, water exerts the greatest effect. Differences in the frequency, duration, and seasonality of water flows typically cause a variety of different features to develop in soils of the same parent material. Certain soil properties develop in response to recurrent prolonged wetness, serving as reliable predictors of the long-term hydrology. This makes interpretation of soil morphology vital for wetland identification and delineation, especially in the more problematic situations.

What Is Soil?

In the United States, "soil" is defined by scientists as a collection of natural bodies, made up of mineral and organic materials, that supports or is capable of supporting the growth of land plants "out of doors." The upper limit of soil is air or shallow water. Land plants are self-supporting, free standing plants, such as trees, shrubs, and robust nonwoody species like grasses, sedges, and flowering herbs. Forests, meadows, lawns, and gardens all possess soils. Other unconsolidated materials supporting different plants or lacking vegetation are not soil. Areas vegetated by water lilies, submerged aquatic plants, free-floating species, and algae do not have soils—their supporting medium is a nonsoil called substrate. Other nonsoil areas include glaciers, rock outcrops, salt flats, and barren lands.

Soil is usually made up of a combination of sand, silt, clay, and organic material. Some soils have various amounts of gravel, stones, and rocks. Soils are separated into two general types based on the amount of organic matter in the upper layer: 1) organic soils and 2) mineral soils. Organic soils are dominated by the remains of plants—leaves, stems, twigs, and roots—that accumulate in significant amounts at the soil surface. These soils are commonly called mucks and peats. Mineral soils are mainly composed of mixtures of sand, silt, and clay, often with some enrichment of the surface layer with organic matter. They are further classified by texture based on the relative proportions of sand, silt, and clay (Figure 4.1).

Effects of Waterlogging on Soil Development

Flooding and prolonged waterlogging significantly affect soil characteristics. With few exceptions, soils associated with wetlands possess properties unlike those of dry-land soils. Extended wetness inhibits the natural breakdown of leaves and other plant parts by aerobic bacteria, causing a buildup of organic matter on the soil surface. Shorter periods of wetness are not sufficient to stop oxidation of these materials, but alternate wetting and drying lead to the development of other unique soil properties.

Flooding and soil saturation for long periods (a couple of months or more) during the growing season create long-term anaerobic soil conditions sufficient to prevent aerobic decomposition or oxidation of leaves, stems, roots, and other dead plant parts. Aerobic bacteria that are responsible for the oxidation of these materials cannot survive under such wet conditions, so organic materials accumulate at the surface. This buildup causes the formation of peats and mucks. Over a period of 10,000 years, peat layers 50 feet or more thick develop in some waterlogged depressions, changing lakes to forested wetlands in the process called hydrarch succession (see Chapter 3).

Where soils are not sufficiently wet to promote this organic accumulation, mineral soils typically develop. They exhibit a wide range of

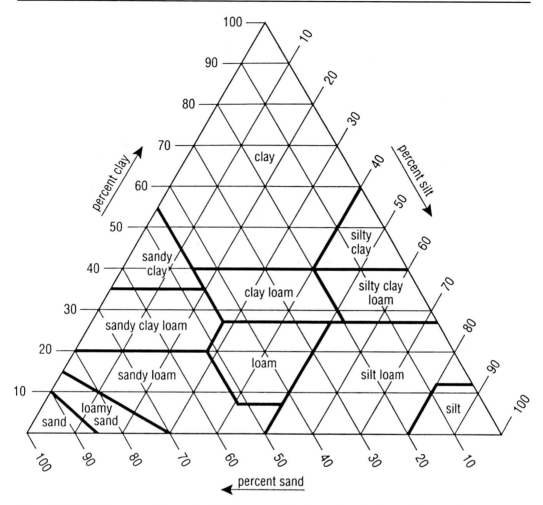

Figure 4.1. Soil textures are the result of different combinations of sand, silt, and clay particles, as illustrated by the textural triangle. (Source: Soil Survey Division Staff 1993)

properties related to differences in parent material, climate, topography, age, and other factors. Changes in the frequency, duration, and depth of saturation produce soil characteristics that help separate wetlands from uplands. Wetter mineral soils subject to significant flooding often develop thick organic surface layers (up to 16 inches thick). Most mineral soils are not wet enough for these layers to develop. Instead, organic matter enriches the surface layer but does not accumulate as a peat or muck layer. This often produces a very dark brown to blackish surface layer in wetland soils. Below the topsoil, the subsoil is typically grayish (see Chapter 11).

Inundation of a soil effectively eliminates gas exchange between the atmosphere and the soil. Existing oxygen in soil pores is quickly consumed by soil microbes, creating anaerobic conditions. Some gas exchange still occurs at the soil-water surface, but the little gas exchange between water in the pores and the soil is extremely slow—ten thousand times slower than gas exchange from air-filled pores to the soil. Anaerobic conditions can develop within one day in some soils and within a few days in most soils. Temperature affects the rate at which such conditions become established: the colder the temperature, the slower the rate. Soil scientists have long considered 41° F measured at a depth of 20 inches to be "biological zero," the temperature at which biological activity in the soil ceases. Recent studies in Alaska have found that permafrost soils never get above this temperature at that depth, yet hydric soils still develop above permafrost when saturated for prolonged periods. Microbial activity has also been detected in these soils and at lower temperatures. Thus the concept of biological zero needs revision since it is likely that soil is biologically active to some degree whenever it is not frozen.

Anaerobic bacteria are important agents in the formation of soil properties associated with wetlands. These microbes are well adapted to the low oxygen conditions resulting from sustained wetness. They derive their energy from the oxidation of organic matter. A series of biochemical events called the oxidation-reduction (redox) process takes place in flooded or saturated soils. In a flooded soil, the oxygen that was present prior to flooding is consumed by microbes in a couple of days provided there is organic matter present for them to digest. When dissolved oxygen is removed, the soil becomes chemically reduced. This means that various elements will become more soluble in sequence, beginning with nitrate. After converting nitrate to free nitrogen in a process called "denitrification" (an important part of the water-quality renovation provided by wetlands), selected microbes reduce manganese from the manganic (oxidized) to the manganous form. Iron is the next element to be reduced from ferric iron (oxidized) to ferrous iron. The process continues with microbes reducing the sulfates and carbon and producing respectively hydrogen sulfide (giving the smell of rotten eggs) and methane (odorless) as by-products. Reduced compounds are often soluble in water and available for plant uptake. These and other mobilized elements (e.g., aluminum) are toxic to most plants in large quantities, so they limit the types of plants that can live in wetlands. The reduction process, therefore, has a profound effect on plant composition as well as on soil chemistry and morphology.

Morphological properties called "redoximorphic features" indicate varying amounts of soil wetness. The biochemical processes causing iron and manganese reduction have a great effect on soil color and morphology. Iron is typically the most abundant element in the soil. In its oxidized (ferric) state, iron gives well-drained soils their characteristic yellowish, reddish, orangish, or brownish colors. When iron is reduced (ferrous), it is soluble and usually moves within or out of the soil, leaving sandy soils grayish and finer-grained soils bluish, grayish, or greenish (gleyed). These colors are considered "redox depletions" since they are caused by iron depletion; the process is called "gleization." Gray reflects the natural color of soil particles (sand, silt, and clay), while blue usually indicates the presence of ferrous iron. If the soil conditions are such that free oxygen is present, organic matter is absent, or temperatures are below freezing and thus too low to sustain microbial activity, gleization will not begin and redox depletions (gleyed colors) will not appear, even though the soil may be saturated for long periods. Soils saturated only during the coldest part of winter do not develop gleyed colors.

Redox depletions may also occur in clay through similar processes. Clay depletions appear as gray coatings on either channels through or the outer surface of natural soil macroparticles (peds). The adjacent soil matrix and the underlying soil layers will have a higher clay content, and so clay coatings can be found on outer layers of peds.

Mineral soils that are wet for most of the year usually have reduced colors dominating the subsoil layer immediately below the surface layer and often within one foot of the surface. Soils exposed to shorter periods of wetness are variously colored. The wetter ones are grayish, with spots or blotches of yellow, orange, or reddish brown (high-chroma mottles). These brighter colors represent concentrations of iron oxides—"redox concentrations"—and a fluctuating water table. Many times, they develop along channels in wet soils where plant roots are leaking oxygen and soluble iron has combined with the oxygen to form ferric iron. These orangish coated channels, called "oxidized pore linings" or "oxidized rhizospheres," are evidence of a plant living under anaerobic conditions. Drier soils subjected to brief periods of wetness are brighter colored overall (red, yellow, brown, or orange) with grayish mottles (low chroma mottles). Soils with short-term wetness may have only high-chroma mottles, while soils lacking significant wetness typically are not mottled. Chapter 11 tells how to recognize hydric soils.

What Is Hydric Soil?

Over time, several terms have been used to describe wetland soils. In characterizing soils for agricultural uses, scientists have divided them into seven drainage classes: (1) excessively drained, (2) somewhat excessively drained, (3) well drained, (4) moderately well drained, (5) somewhat poorly drained, (6) poorly

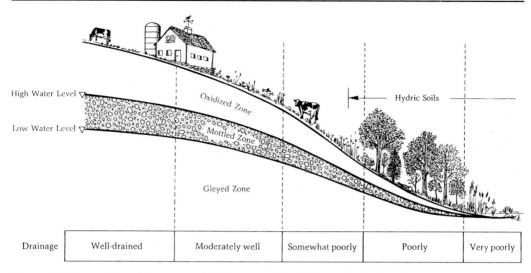

Figure 4.2. The position of the seasonal high water table relative to the land surface is strongly influenced by slopes. Hydric soils typically develop in depressions or at the bottom of slopes. These soils also form on steeper slopes in areas of heavy groundwater discharge, along drainageways, or in depressions on the slopes. Very poorly drained soils typically occur in depressions that are inundated for much of the growing season. Poorly drained soils are flooded for shorter periods or saturated for extended periods. (Source: Tiner and Veneman 1995)

drained, and (7) very poorly drained. The latter two types are characteristic of wetlands. Poorly drained soils remain waterlogged at or near the surface long enough during the growing season that most crops cannot be grown unless the land is artificially drained. Very poorly drained soils are saturated for extended periods, are frequently inundated, and have the same limitation for crop production. Under natural conditions, both of these soil types typically support the growth of hydrophytes (wetland vegetation). The other drainage classes are too dry to produce wetlands, with two exceptions: 1) better-drained soils that are frequently flooded for long periods and 2) somewhat poorly drained soils in lower landscape positions that occasionally are wet enough to become wetlands (Figure 4.2). The latter soils are typically wet only for short periods, usually early in the growing season. It is important to recognize that these drainage classes were not developed for wetland determinations; rather, they were created to separate different soils into groups for agriculture. Moreover, the terms have been interpreted locally, so what one state or county considers to be somewhat poorly drained may be poorly drained from another state's or county's view. This has caused some confusion in their use for wetland identification.

Wetland soils have also been called "hydromorphic," "halomorphic," and "alluvial." The first term includes waterlogged soils associated with marshes, swamps, and bogs: peats, mucks, and gleyed soils. The third term includes frequently inundated floodplain soils, whereas the second term is applied to the wet soils of coastal salt marshes and inland alkaline wetlands (arid to subhumid regions in the West).

Wetland soils are now called "hydric." The term was coined as part of the national wetland classification system developed by the U.S. Fish and Wildlife Service (FWS) in the late 1970s (see Chapter 1). The "predominance of undrained hydric soils" is one of the indicators for identifying wetlands. During the past twenty years, the Department of Agriculture's National Technical Committee for Hydric Soils has worked to refine the term and maintain county, state, and national lists of hydric soils (Chapter 11). The latest technical criteria for defining hydric soils are listed below.

2000 technical criteria for hydric soils (http://soils.usda.gov). Aquic soils are saturated and reduced for a significant period during the growing season, which is defined by soil temperature ("biologic zero" $\geq41°$ F measured at 20 inches below the surface); the water-table depths in the criteria identify those soils with water tables at or near the surface. Ponding is surface water derived from high water tables and runoff from adjacent uplands, while flooding involves overbank inundation from high river flows. Long and very long duration are greater than one week and greater than one month, respectively. "Frequently," in terms of inundation, means occurring more than fifty times in 100 years.

1. All histels except folistels, and histosols except folists, or
2. Soils in aquic suborders, great groups, or subgroups, albolls suborder, historthels great group, histoturbels great group, pachic subgroups, or cumulic subgroups that are:
   a. somewhat poorly drained, with a water table equal to 0.0 foot from the surface during the growing season, or
   b. poorly drained or very poorly drained, and have either:
      (1) water table equal to 0.0 foot during the growing season if textures are coarse sand, sand, or fine sand in all layers within 20 inches, or for other soils:
      (2) water table less than or equal to 0.5 foot from the surface during the growing season, if permeability is equal to or greater than 6.0 inches/hour in all layers within 20 inches, or
      (3) water table less than or equal to 1.0 foot from the surface during the growing season, if permeability is less than 6.0 inches/hour in any layer within 20 inches, or
3. Soils that are frequently ponded for long or very long duration during the growing season, or
4. Soils that are frequently flooded for long or very long duration during the growing season.

Hydric soil is now defined as *"soil that is saturated, flooded, or ponded long enough during the growing season to develop anaerobic conditions in the upper part."* This means simply that wetness is of sufficient duration for the soil to lose oxygen long enough for the growth of most plants to be curtailed in favor of hydrophytes (Chapter 5). Some hydric soils are submerged for the greater part of the year, while those intergrading toward better-drained upland soils are wet for shorter periods and dry at the surface for much of the growing season. Hydric soils typically consist of very poorly drained and poorly drained soils having a water table within a foot of the surface for two weeks or more during the growing season. Very poorly drained soils are flooded for much of the year or permanently saturated. Poorly drained soils are usually saturated to the surface at some time

during the year, especially winter in the Northeast. Hydric soils also include soils ponded or frequently flooded for seven or more consecutive days during the growing season.

Many hydric soils are drained by open ditches, tiles, or combinations of dikes, ditches, and pumps. Water is also diverted from wetlands through other means, such as by government-financed channelization projects, upstream dams, or groundwater extraction (e.g., municipal or industrial well fields). Soils that were formerly wet but are now effectively drained, no longer having wetland hydrology and therefore not providing wetland functions, are among the most difficult in which to make wetland determinations, usually requiring measurement of the current hydrology. Where drainage fails, such soils usually revert to hydric conditions and return to wetland. This simple fact makes wetland restoration of these sites feasible and cost effective.

Key points to remember are that 1) soil classified as hydric must be saturated or flooded long enough during the growing season to produce the low oxygen conditions that damage most plants but favor hydrophytes; 2) soil that is well drained but frequently flooded or saturated for short periods of time is not hydric; and 3) soil that was formerly wet but is now effectively drained is considered drained hydric soil, and the area is not considered wetland.

Distribution of Hydric Soils

Across the country, hydric soils are most abundant in glaciated regions, along the coastal plain, and in major river valleys. As of September 2004, roughly 182 million acres of hydric soils have been mapped in about 90 percent of the coterminous U.S. (excluding some areas of significant wetland acreage such as northern Maine and Minnesota) by the U.S. Department of Agriculture. From these data, we find that the Northeast and Great Lakes region contain over 74 million acres of drained and nondrained hydric soils, representing about 41 percent of the hydric soil acreage mapped in the lower forty-eight states (Table 4.1). The nineteen states in these two regions also possess about 64 percent of the organic soils mapped in the coterminous United States. Minnesota alone has over 19 million acres of hydric soils, while Michigan and Illinois each has over 10 million acres.

Table 4.1. Acreage of potential hydric soils for the Northeast and Great Lakes region based on soil mapping.

State	Organic Soil Acreage (% of total)	Hydric Mineral Soil Acreage (% of total)	Total Acreage
Connecticut	162,765 (17)	819,608 (83)	982,373
Delaware	13,211 (10)	119,032 (90)	132,243
Illinois	97,104 (1)	10,658,446 (99)	10,755,550
Indiana	333,608 (4)	7,510,882 (96)	7,844,490
Massachusetts	231,189 (34)	447,997 (66)	679,186
Maryland	120,223 (10)	1,067,197 (90)	1,187,420
Maine	575,974 (44)	719,196 (56)	1,295,170
Michigan	3,559,942 (28)	9,200,640 (72)	12,760,582
Minnesota	3,852,774 (20)	15,294,783 (80)	19,147,557
New Hampshire	152,119 (26)	424,267 (74)	576,386
New Jersey	257,801 (16)	1,315,941 (84)	1,573,742
New York	687,522 (24)	2,123,040 (76)	2,810,562
Ohio	82,243 (1)	5,577,899 (99)	5,660,142
Pennsylvania	39,396 (2)	2,147,717 (98)	2,187,113
Rhode Island	32,832 (24)	106,414 (76)	139,246
Virginia	70,677 (5)	1,320,145 (95)	1,390,822
Vermont	69,615 (21)	266,781 (79)	336,396
Wisconsin	2,183,069 (46)	2,605,869 (54)	4,788,938
West Virginia	—	153,781 (100)	153,781
Area Total*	12,522,064 (16)	61,880,616 (84)*	74,402,680*
Lower 48 States Total	19,451,824 (11)	162,647,687 (89)	182,099,511

*includes 981 acres of hydric mineral soils in the District of Columbia.
Note: Figures include drained phases that no longer support wetlands. (Source: USDA Natural Resources Conservation Service, National Soil Information System)

Additional Readings

Mausbach, M. J. 1994. Classification of wetland soils for wetland identification. *Soil Survey Horizons,* Spring 1994:17–25.

Mausbach, M. J., and J. L. Richardson. 1994. Biogeochemical processes in hydric soil formation. *Current Topics in Wetland Biogeochemistry* 1: 68–127.

Ponnamperuma, F. N. 1972. The chemistry of submerged soils. *Advances in Agronomy* 24: 29–96.

Richardson, J. L., and M. J. Vepraskas (editors). 2001. *Wetland Soils: Genesis, Hydrology, Landscapes, and Classification.* Lewis Publishers, CRC Press, Boca Raton, FL.

Tiner, R. W., and P. L. M. Veneman. 1995. *Hydric Soils of New England.* University of Massachusetts Cooperative Extension Service, Amherst, MA.

Vepraskas, M. J. 1996. *Redoximorphic Features for Identifying Aquic Conditions.* North Carolina State University, North Carolina Agricultural Research Service, Raleigh, NC. Technical Bulletin 301.

CHAPTER 5

Swamp Plants
Hydrophytes and Wetland Plant Communities

What Is a Hydrophyte?

Wetland plants, technically called *"hydrophytes," are plants growing in water or on a substrate that is at least periodically deficient in oxygen as a result of excessive water content.* These conditions typically create anaerobic environments that require plants to develop specialized adaptations for growth and reproduction (see following two sections). Hydrophytes are not restricted to true aquatic plants growing in bodies of water such as ponds, lakes, rivers, and estuaries, but also include plants morphologically or physiologically adapted to periodic flooding or saturated soil conditions typical of marshes, swamps, and other wetlands.

The concept of a hydrophyte applies to individual plants and not simply to species of plants, although some genera and many species are strictly hydrophytic. Among the hydrophytic genera are *Typha* (cattails), *Sparganium* (bur-reeds), *Sagittaria* (arrowheads), *Nuphar* and *Nymphaea* (water lilies), and *Potamogeton* (pondweeds). Species that are entirely hydrophytic include smooth cordgrass, skunk cabbage, buttonbush, and bald cypress. Yet most hydrophytes are adaptable species with broad ecological ranges, occurring in both wetlands and nonwetlands to varying degrees. For example, certain individuals of American holly and pitch pine are considered hydrophytes when growing in undrained hydric soils, while those growing on uplands are not. To survive in wet habitats, they possess specialized adaptations that other populations of their species do not. These "wetland ecotypes" have evolved to live a semiaquatic existence in periodically flooded or saturated, anaerobic soils. They are usually not morphologically distinguishable from their nonhydrophytic relatives. Exceptions are designated as subspecies or varieties: for example, trident-leaved red maple (*Acer rubrum trilobum*) and swamp black gum (*Nyssa sylvatica biflora*). Most wetland ecotypes are recognized simply by their occurrence in wetlands. The existence of these wetland ecotypes is the reason that some principally upland species, such as American holly, white pine, pitch pine, white oak, and black cherry, are cited as potentially hydrophytic species on the national list of wetland plants. (Chapter 10 contains illustrations and descriptions of most of the Northeast's common hydrophytic species.)

Water Stress on Plants

Plants react to a host of environmental factors: light, temperature, nutrients, soil chemistry, and substrate type. Soil conditions are crucial to growth, survival, and reproduction. All plants require oxygen in their root zone for water and nutrient uptake. The lack of oxygen is what limits survival of most plants in waterlogged soils. It is common knowledge that through hydroponics, many nonwetland plants like tomatoes can be grown in water, but this water solution must be highly aerated (oxygen-rich). If not, the plants would die from oxygen starvation. Studies have shown that the roots of tomato plants cannot survive when flooded for three days in an oxygen-deficient environment. Corn roots can withstand a similar amount of flooding, but rye and pumpkin can tolerate less than a day, and peas, less than ten hours of inundation. In contrast, yellow flag, a marsh iris, can withstand a month of flooding. Red maple can withstand a few weeks of flooding—especially in the early part of the growing season—and months of saturation. It takes a couple of years of continuous flooding to kill red maple.

Conditions in the root zone determine survival. Prolonged exposure of roots to anaerobic conditions exerts a severe stress on plants, one that most species cannot tolerate and find fatal. Adaptive plants find ways of getting oxygen to their roots or getting their roots to oxygen. Some plants are more opportunistic, colonizing wetlands only during dry periods, but once flooding and normal soil saturation occur, they are eliminated. This happens in impoundments and reservoirs during summer drawdown. If plants are to survive and reproduce in wetlands, they must be adapted for life in water or saturated soils and to the accompanying anaerobic conditions.

The lack of oxygen is not the only problem facing plants growing under conditions of prolonged saturation or flooding. As we saw in Chapter 4, certain elements in the soil change from an insoluble (oxidized) form to a soluble (reduced) form that is ready for plant uptake. Under oxygen-deficient conditions, oxides of nitrogen, iron, manganese, and sulfur become reduced and thus toxic to most plants. Excessive wetness also lowers soil temperatures retarding germination, growth of seedlings, and nutrient uptake. Nonsandy soils swell upon flooding, and their changing structure has adverse effects on root growth. This is especially true for clay soils.

Plant Adaptations for Life in Water or Wetlands

Prolonged flooding or soil saturation causes the leaves of most plants to turn yellow and drop off, and the roots to die from oxygen starvation. Adaptable plants find ways either to relieve oxygen deficiency or to avoid it. Of the nearly 21,000 plant species occurring in the United

States and its territories, only about 7,000 species (or one-third of the nation's flora) are likely to be found growing under wetland conditions. The remaining 14,000 species rarely or never occur in wetlands. Thus about two-thirds of the plant species in the United States are not hydrophytic—they simply avoid wet areas.

Most plants growing in water and wetlands adapt physiologically to oxygen deficiency and associated toxic by-products. These internal mechanisms involve rather complex chemical processes requiring technical explanations that are beyond the scope of this book. But plants have also developed specialized structures or morphological adaptations (Plate 1; Table 5.1) and, in some cases, reproductive mechanisms to colonize wetlands and aquatic habitats. The former are of particular value for wetland identification since they may be readily observed in the field.

The most common morphological adaptation for avoiding water stress in wetlands is a shallow root system. Prolonged wetness causes a change in root direction. Rather than growing deep to obtain water as they do in well-drained, oxygen-rich soils, most roots of wetland plants grow laterally or even upward to the surface. This happens because roots near the surface can readily obtain oxygen during most of the year. Also, since there is usually plenty of water, at least seasonally, in wetlands, deep roots aren't necessary to find water. Red maple, for example, has an adaptable root system. When it grows in saturated soils, it develops a shallow

Table 5.1. Examples of morphological and reproductive plant adaptations for living in water or in wetlands.

Life Form	Adaptations
Floating-leaved aquatic plants	Leathery leaves with waterproof upper surfaces, leaves filled with large air spaces, stems growing in as much as 6.6 feet of water, large air spaces in roots and rhizomes, surface flowers, long-lived seeds, buoyant fruits and seeds, underwater germination
Emergent herbaceous plants	Air-filled tissue (aerenchyma) in stems, hollow stems, adventitious water roots, soil water roots, broader emergent leaves and more linear submerged leaves (heterophylly), large air spaces in roots and rhizomes (up to 70% of root volume), spongy roots, shallow roots, ability to oxidize the environment surrounding roots (oxidized rhizospheres), swollen stems (stem hypertrophy), long-lived seeds, buoyant fruits and seeds, underwater germination
Woody plants	Hypertrophied lenticels, swollen stems or trunks (stem hypertrophy), adventitious water roots, shallow roots, roots that regenerate in soil after flooding (soil water roots), aerial growths of roots (knees or pneumatophores), internal tissue with large air spaces, increased air space in roots, relatively pervious cambium, seedlings developed on parent (vivipary), underwater germination

Note: Wetland plants often possess one or more of these characteristics in addition to physiological adaptations.

lateral root system. In dry upland soils, a deep-penetrating tap root is formed. In contrast, eastern hemlock has shallow roots wherever it grows, allowing colonization of wetlands, shallow soils, and rocky areas where little soil exists.

Prolonged anaerobic conditions kill the roots of most plants. In many cases, there is an annual process of root dieback and regrowth. Hydrophytes can develop other roots, including water roots and adventitious roots, to compensate for the dieback of primary roots. Studies in the Netherlands have examined the effect of prolonged waterlogging on three species of dock (genus *Rumex*) that occupy different elevations on floodplains. All species slowed their growth during the first twelve days of saturation, but two of the species recovered and developed flood-resistant roots. Two types of water roots were observed: 1) thin, many-branched roots growing at the ground surface, and 2) thick, white, weakly branched roots that penetrated the saturated soil. Most of the new root growth was in the upper four inches of the soil. The roots grew horizontally, enabling the plant to get air near the ground surface. The degree of tolerance to extended waterlogging in the three species was consistent with their distribution on the floodplain. The most flood-resistant species occurred at the lowest elevations, where it was subjected to longer flooding and waterlogging. In other plants, including green ash, willows, and water primroses, adventitious roots may emerge above the ground surface in response to similar conditions (Plate 1c). These roots have the same survival effect as water roots.

True aquatics and emergents living in nearly permanent water cannot use this strategy, since their substrates are always flooded or saturated and anaerobic. Other mechanisms and structures have evolved in plants to obtain oxygen and deal with toxic by-products of anaerobiosis. Some plants facilitate the movement of air from aboveground parts (leaves and stems) to their roots through a network of enlarged internal cells called "aerenchyma," which serve as open conduits to move oxygen downward. Up to 70 percent of the wetland plant's volume may consist of these air channels. When oxygen arrives in the roots, some of it leaks into the surrounding anaerobic soil, creating an oxygenated microenvironment immediately surrounding the roots (oxidized rhizospheres, root channels, or pore linings). This permits the roots to take up nutrients and water from the soil.

What happens when an emergent plant is flooded above its leaves or stem? Many marsh plants rapidly increase their stem and leaf growth. A complex set of biochemical reactions stimulates shoot elongation, allowing the plants to move their leaves and stems out of the water to photosynthesize and to move oxygen down to their roots.

Woody plants have special organs called "lenticels" on their bark to facilitate gas exchange between the inner tissues and the environment. Lenticels permit a plant to take in air and carbon dioxide and to release gases like oxygen, a by-product of photosynthesis (the process by which plants convert sunlight, water, and carbon dioxide into sugar and oxygen). When flooded, the lenticels of many trees and shrubs become fleshy and enlarged, or "hypertrophied," to increase gas exchange. These morphological features can be observed as scars when water is not present, thereby providing indirect evidence of prolonged flooding.

Another important adaptation in some species relates to reproduction in an aquatic environment. Most wetland plants reproduce through flowers, an adaptation required for life on land. It is particularly interesting to note that wetland plants are thus representatives of an evolutionary process: they are land plants that have reverted to life in the water. The most evolved of the group are the aquatic flowering plants that live in ponds, lakes, and rivers.

Since flowers must be pollinated by insects or by wind, how do plants living underwater become pollinated? Many aquatic plants, like water lilies, spatterdock, and pondweeds, produce flowers at the water's surface, which are pollinated by these agents. A most curious and elaborate reproductive mechanism has evolved in wild celery or tape-grass, a submergent plant of tidal and nontidal fresh waters. This plant produces separate male and female flowers. The female flower is attached to a stalk that elongates until the flower reaches the surface. The stalk has been observed growing at a rate of about ¾ inch per hour. At the surface, the flower opens and is ready to be pollinated. Meanwhile, the male flower breaks off from the plant below and floats up. Reaching the surface, it opens and develops into a sort of "love boat" that is carried by the currents to a waiting female flower. Eventually the male flower unites with the female flower, and pollination takes place.

Factors Affecting Wetland Plant Communities

Plant communities are affected by many factors, both natural and human-induced. Waterlogging is the principal factor promoting hydrophytes over other species. Plants respond positively or negatively to this condition, whether it results from natural processes or human influence. The wide range in the frequency and duration of wetness yields a multitude of environments and a correponding diversity of plant communities. Some of the more important physical factors related to hydrology are water depth, fluctuation of the water table, long-term soil moisture, and salinity. Soil type is another important natural factor influencing plant communities. Certain plants grow only on peaty soils, while some are restricted to sandy or other soils. Regional and local climates, properties of physiographic regions with their underlying geologic deposits, plant competition, animal actions (including grazing and insect infestation), fire, and human disturbance also influence plant communities.

Soil chemistry is an important property affecting plant growth. Salt stress occurs along the coast, where tides bring in salt water that stops the growth of most plants and favors the colonization of salt-tolerant species (halophytes) with specialized mechanisms to deal with this stress. Hydrogen-ion concentration (pH) determines the acidity or alkalinity of a given wetland and thus plays a major role in determining plant species composition. Acidity-loving species, including ericaceous shrubs like blueberries, cranberries, and leatherleaf, are characteristic of nutrient-poor bogs. Calcareous plants such as shrubby cinquefoil and sedges like yellow sedge typify alkaline fens that receive significant nutrient input from groundwater inflow in limestone regions. The underlying geologic formations and the groundwater flow patterns create this condition.

Other plant communities require periodic burning to become established. In the New Jersey Pine Barrens, the presence of plants like sand myrtle (which resembles big cranberry when flowers and fruits are absent) and turkey beard (a white-flowering member of the lily family) suggests recent fires. Pitch pine, the characteristic plant of this area, is a fire-dependent species, requiring a hot burn to open its cones and release seeds. Elsewhere, the establishment of cinnamon fern (a typical wetland species) on up-lands is sometimes aided by fire. Fire suppression is leading to significant changes in plant-community structure in many areas.

Herbivory (the grazing of animals on living plants) affects wetland plant communities to a lesser degree. Yet in places, animals can wipe out marsh vegetation. For example, heavy feeding by migratory snow geese on rhizomes of smooth cordgrass has created barren tidal flats in some coastal marshes, especially along the Delaware Bay in southern New Jersey. Muskrats and nutria are known to eat virtually all vegetation in some marshes. Muskrat eat-outs are particularly widespread in the Prairie Pothole region of the upper Midwest (the Dakotas and western Minnesota). In the Northeast, the effect of muskrats is often more localized. Heavy grazing by livestock can significantly alter the vegetation of wet meadows. Overgrazing can lead to low diversity in a community dominated by soft rush, an unpalatable grasslike plant.

Over the past 150 years, human activities have become the dominant force shaping wetlands in most of the country (Chapter 8). Many construction projects have altered the hydrology of wetlands directly through channelization and drainage or indirectly by changing surface-water runoff and groundwater flow patterns. These actions have exerted a profound effect on the plant composition of the remaining wetlands. For example, mosquito ditching of salt marshes in the 1930s increased the abundance of high-tide bush or marsh elder, especially on spoil mounds next to the ditches. Salt and brackish marshes have migrated upstream with increased saltwater intrusion and reduced freshwater flows due to reservoir construction and the diversion of river flows for industrial and other uses. Channel deepening of coastal rivers for navigation (e.g., the Delaware River) has had similar effects. Land development throughout the New Jersey Pine Barrens and the accompanying enrichment of surface and groundwater have changed the water chemistry sufficiently to alter wetland plant communities in affected areas: native plants adapted to the nutrient-poor, acidic conditions of the Pine Barrens are being replaced by nonnative species.

Wetland Types

Wetlands can be categorized in many ways, with the use of differences in hydrology (including

Table 5.2. Wetland types described in this book, with their corresponding technical classification according to the U.S. Fish and Wildlife Service.

Wetland Type Used in This Book	U.S. Fish and Wildlife Service Classification			
	System	*Subsystem*	*Class*	*Water Regime(s)*
Salt/brackish marsh	Estuarine	Intertidal	Emergent wetland	Regularly flooded
				Irregularly flooded
Tidal fresh marsh	Riverine	Tidal	Emergent wetland (nonpersistent)	Regularly flooded
	Palustrine		Emergent wetland	Seasonally flooded-tidal
Tidal swamp	Estuarine	Intertidal	Forested wetland	Irregularly flooded
	Palustrine		Forested wetland	Seasonally flooded-tidal
				Temporarily flooded-tidal
Marsh (nontidal)	Palustrine		Emergent wetland	Semipermanently flooded
				Seasonally flooded
	Lacustrine	Littoral	Emergent wetland (nonpersistent)	Semipermanently flooded
	Riverine	Lower perennial	Emergent wetland (nonpersistent)	Semipermanently flooded
Savanna/Carolina bay	Palustrine		Emergent wetland	Seasonally flooded
Wet meadow	Palustrine		Emergent wetland	Seasonally flooded
				Temporarily flooded
				Saturated (seasonally)
Fen	Palustrine		Emergent wetland	Seasonally flooded
Shrub bog	Palustrine		Scrub-shrub wetland	Saturated
Forested bog	Palustrine		Forested wetland	Saturated
Shrub swamp	Palustrine		Scrub-shrub wetland	Semipermanently flooded
				Seasonally flooded
				Temporarily flooded

water chemistry) and vegetation being the most common practice. Northeast wetlands can be divided into two general types based on hydrologic differences: 1) tidal wetlands and 2) nontidal wetlands (Chapter 2). Only nontidal wetlands occur in the Great Lakes region. Most wetlands are vegetated by herbaceous (nonwoody) plants or woody plants (trees and shrubs). Exposure to waves or large quantities of salt precludes the establishment of rooted plants, and nonvegetated wetlands form in these locations. Vegetated wetlands can be distinguished by the predominant form of the vegetation: marshes and wet meadows (emergent wetlands) by herbaceous flora, shrub swamps (scrub-shrub wetlands) by low-to medium-height woody plants, forested wetlands by trees, and aquatic beds by floating or floating-leaved plants. Additional subdivisions are possible based on water chemistry, soil types, dominant plant species, and other characteristics. Table 5.2 correlates the U.S. Fish and Wildlife Service classification system with wetland types used in this book.

Tidal Wetlands

Tidal wetlands include both vegetated and nonvegetated wetlands subject to periodic tidal flooding. Most tidal wetlands are vegetated, and four major types are found in the Northeast: 1) salt marshes, 2) brackish marshes, 3) tidal fresh marshes, and 4) tidal swamps. Submerged aquatic beds are also common in marsh pools,

Table 5.2. (continued)

Wetland Type Used in This Book	U.S. Fish and Wildlife Service Classification			
	System	Subsystem	Class	Water Regime(s)
Hardwood swamp	Palustrine		Forested wetland	Seasonally flooded
Larch swamp	Palustrine		Forested wetland	Seasonally flooded
				Saturated
Floodplain forested wetland	Palustrine		Forested wetland	Temporarily flooded
Evergreen forested wetland	Palustrine		Forested wetland	Seasonally flooded
				Temporarily flooded
				Saturated (seasonally)
Vernal pool	Palustrine		Forested wetland	Seasonally flooded
			Shrub-shrub wetland	Semipermanently flooded
			Emergent wetland	Seasonally flooded
				Semipermanently flooded
			Unconsolidated bottom or shore	Seasonally flooded
				Semipermanently flooded
Aquatic bed	Lacustrine	Littoral	Aquatic bed	Permanently flooded
				Semipermanently flooded
	Palustrine		Aquatic bed	Permanently flooded
				Semipermanently flooded
	Riverine	Lower perennial	Aquatic bed	Permanently flooded
				Semipermanently flooded

Note: This system was developed primarily for mapping the nation's wetlands. Subclasses are not shown here with the exception of nonpersistent for emergent wetland in the Riverine and Lacustrine systems, which is used to separate those types from Palustrine emergent wetland. (See Chapter 14 for more information.)

but are more common in shallow coastal waters. Nonvegetated tidal wetlands, including coastal beaches, rocky shores, mud flats, sand flats, and gravelly shores, are most common in more exposed situations and at lower elevations (subjected to longer flooding). They are more abundant than vegetated wetlands in areas with high tidal ranges such as the Maine coast.

Salt Marshes

Salt marshes are grasslands periodically flooded by salt water (Plate 2). They develop along tidal rivers and in sheltered areas behind barrier islands and beaches and along coastal bays. They are quite extensive along the Atlantic coast from Long Island south, where barrier islands are the dominant shoreline feature.

The presence of salt water creates saline soil conditions that are lethal to most plants. Salt marshes are therefore colonized by a group of plants called "halophytes," a name meaning "salt-loving plants." Of course, these plants do not really love salt. Their existence in salty environments is born of necessity rather than love. They simply have adapted different ways to handle salt stress, including 1) mechanisms to reduce salt uptake by roots (salt exclusion), 2) development of salt-secreting glands to get rid of excess salt, 3) salt-concentrating organs (e.g., fleshy leaves) that are periodically shed, 4) succulence (fleshiness) to keep internal salt concentrations at acceptable levels (Figure 5.1), 5) waterproof leaf surfaces with a waxy coating, 6) reductions in leaf area for evapotranspiration and saltwater contact, and 7) isolation of salt within certain internal organs. Halophytes use one or more of these strategies to solve the salt-stress problem. Some plants simply avoid salt stress by colonizing the upper reaches of salt marshes, where infrequent tidal flooding and

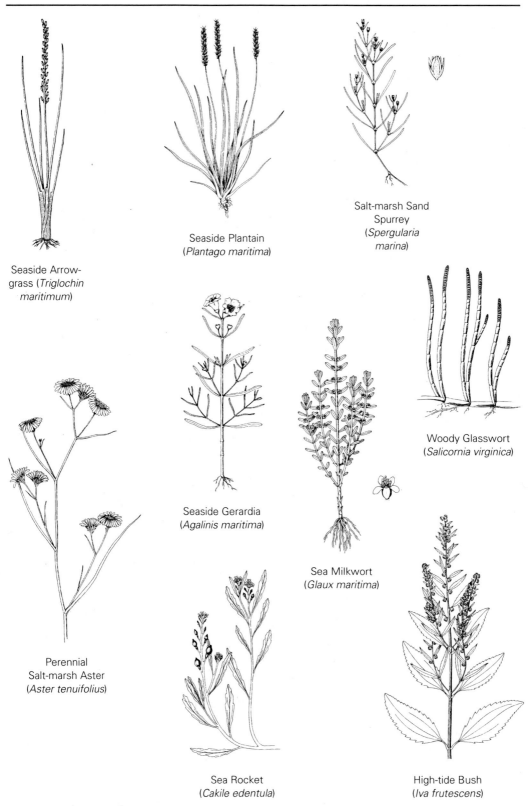

Seaside Arrow-
grass (*Triglochin
maritimum*)

Seaside Plantain
(*Plantago maritima*)

Salt-marsh Sand
Spurrey
(*Spergularia
marina*)

Perennial
Salt-marsh Aster
(*Aster tenuifolius*)

Seaside Gerardia
(*Agalinis maritima*)

Woody Glasswort
(*Salicornia virginica*)

Sea Milkwort
(*Glaux maritima*)

Sea Rocket
(*Cakile edentula*)

High-tide Bush
(*Iva frutescens*)

Figure 5.1. Some salt marsh plants exhibiting succulent leaves and/or stems. Sea Rocket is more common on sandy beaches near the primary dunes. See also Marsh Orach, Common Glasswort, and Seaside Goldenrod (Species 1–3).

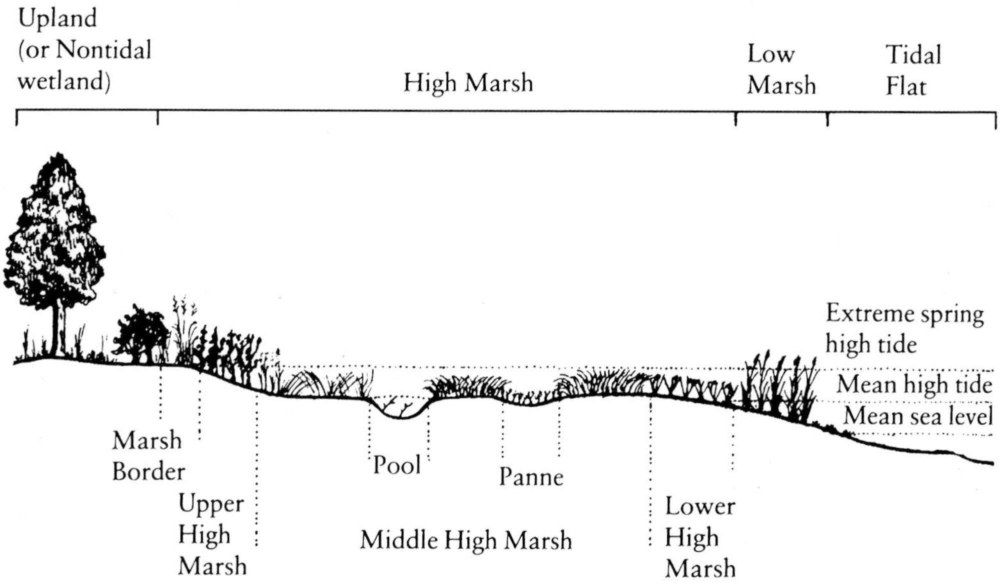

Figure 5.2. Generalized zones of a Northeast salt marsh. Note that the high marsh is separated into several subzones. Within individual wetlands, these zones are often intermixed, forming a complex mosaic pattern. (Source: Tiner 1987; copyright 1987 Ralph W. Tiner, Jr. Reprinted with permission.)

significant freshwater runoff from the adjacent upland create a more hospitable environment.

Most salt-marsh plants can grow even better in freshwater areas, but they cannot successfully compete against other species. Consequently, they do not grow naturally in freshwater habitats, but instead find homes in the saline intertidal soils along the coast. Interestingly, of all the plants that occur in salt marshes (more than fifty species), only glassworts (three species of *Salicornia*) can actually be considered obligate halophytes—that is, plants requiring salt water for growth and survival. Californians call these plants "pickleweeds," and the name is apt, for the fleshy shoots can be pickled or eaten raw. Euell Gibbons, the author of many books on edible wild plants, used to dress up his wild-herb salads with glassworts.

Low Salt Marshes. Two distinct vegetation zones exist according to the frequency of tidal flooding: 1) the low marsh, and 2) the high marsh (Figure 5.2). The low marsh is flooded at least once daily by the tides. Here, smooth cordgrass, one of the dominant salt grasses and one of four species of cordgrasses (genus *Spartina*) that occur in eastern salt marshes, grows to a height of 6 feet or more. The frequent tidal flushing improves drainage, providing exchange of nutrients and removal of salts that might otherwise accumulate in the soil. The lush vegetation growing along the banks of tidal streams

throughout these marshes is evident. The tall form of smooth cordgrass is generally considered a good indicator of the mean high-tide line (the average height of all tides combined). The significance of this mark is that, in most states, it represents the dividing line between private wetlands and state-owned wetlands. Of course, there are exceptions, such as places where 250 years ago the king of England, or more recently state legislatures, gave away or sold public wetlands to certain individuals to promote maritime commerce. The mean high-tide line or the low marsh is an important habitat for the clapper rail, which lives and nests in the streamside cordgrass marsh. Its clucking call is a familiar sound in salt marshes in summer.

High Salt Marshes. The zone above the low marsh is flooded less often than once a day. The high marsh may actually be further divided into subzones, such as the lower high marsh, upper high marsh, panne, pool, and marsh-upland border (Figure 5.2). These subzones are variously flooded and salt stressed, due to natural drainage patterns, groundwater inflows, and surface runoff, conditions that have profound effects on the plant communities.

Pannes are the most salt stressed habitats in the Northeast. These high-marsh depressions temporarily fill with salt water brought in by higher-than-daily tides. As the water gradually evaporates, sea salt accumulates. Summer salinities in pannes far exceed those of ocean water (over 100 parts per thousand compared to 35 in sea water). This salt-rich habitat is inhospitable to most plants, and only the most salt-tolerant hydrophytes grow here. Glassworts love these places, and many pannes are covered with them. Common associates include marsh orach, sea blites, salt-marsh fleabane, and a short form of smooth cordgrass that attains a height of only about one and a half feet in these severely salt-stressed environments. In New England, seaside plantain and arrow-grass may also be common. Some pannes have a dense mat of blue-green algae (periphyton) covering the ground surface. When the pannes are flooded, these mats may be suspended in the water for some time. Pannes are most striking in the fall, when common glasswort turns a brilliant red, showing as red-pockmarked depressions on the otherwise golden yellow salt marshes. In southern marshes, salt flats dominated by Bigelow's glasswort take on a yellowish orange color in the fall, with similar visual effect.

The lower high marsh may have more pannes than the upper high marsh. It occurs just above the mean high-tide mark and is typically dominated by the short form of smooth cordgrass, which provides a marked contrast to the adjacent tall form growing to 6 feet or more along creek banks and tidal shores. Green algae are common in the lower high marshes, as are fly larvae (dipteran maggots) in the soil. This zone is flooded more often than the other high-marsh zones and doesn't drain well, so it's quite soggy, even at low tide.

Two low-growing grasses characterize the middle high marsh: salt hay grass and salt grass. They are often intermixed, forming a turflike grassland (less than 2 feet high) that is easy to walk on. Salt hay is probably the more common of the two. Wetter areas tend to have more salt grass, and drier areas more salt hay grass. As the name implies, salt hay grass has been harvested (along with salt grass and black grass) for winter livestock fodder since colonial times. Many marshes have been diked and fitted with tide gates for this purpose. Diked marshes are especially common along Delaware Bay and the southern New England coast around Newburyport and Rowley, Massachusetts, where salt farming remains active. After years of decline in value, salt hay is now regaining popularity among urban and suburban gardeners as a weed-free winter mulch. Other common high-marsh plants are salt-marsh aster and marsh orach.

Toward the upland border, the marsh surface rises slightly in elevation. Flooded only by the highest tides (storm tides and spring tides), this area receives freshwater runoff from the upland. Salt stress is lowest here, and as a result, plant diversity is the highest. Black grass (not a grass at all, but a rush) becomes dominant as the upper high marsh grades into the border slope. Vegetation here typically includes halophytes like seaside goldenrod, Olney three-square, rose mallow, salt-marsh bulrush, and narrow-leaved cattail, and species with freshwater affinities, such as switchgrass, groundsel bush, bayberry, marsh fern, and grass-leaved goldenrod. Switchgrass forms the upper border in many areas and may extend into adjacent fields. Eastern red cedar and poison ivy can also be found along the marsh border.

Many marshes were ditched in the 1930s for mosquito control as part of a government-sponsored program to employ people laid off during the Depression. The common practice at that time was to deposit excavated material on the marsh surface right next to the ditch. This often left a pile of material that raised the elevation of the marsh, and when elevations in any wetland, especially a salt marsh, are changed, there will often be a noticeable change in vegetation. In this case, a fleshy-leaved shrub aptly named high-tide bush colonized the mounds, along with black grass. The dominance of this shrub often indicates altered hydrology.

Brackish Marshes

Upstream in tidal rivers, salinity is reduced by mixing with fresh water, leading to brackish conditions. Here salinity may range from slightly more than half sea strength to almost fresh, varying markedly with the seasons. Many brackish marshes are fresh in the spring due to high river discharge, whereas they are saltiest in summer due to maximum upstream saltwater penetration during low river flows. This periodic salt intrusion may be responsible for keeping trees from invading the uppermost brackish marshes. Over the range of brackish waters, there is a substantial change in salt stress, going from permanent moderate stress downstream to periodic minor salt stress upstream. Salt stress is much less than in a salt marsh, so plant diversity tends to be much greater in most brackish marshes.

Strongly Brackish Marshes. The more seaward of the brackish marshes are characterized by salt-marsh species. Smooth cordgrass, salt hay grass, and spike grass are joined as dominants by several other species, including Olney three-square, switchgrass, seaside goldenrod, high-tide bush, and common reed. From Maryland south, black needlerush forms broad swards that are almost impenetrable due to the sharp-tipped stems growing to almost 6 feet (Plate 3). It is particularly abundant on the lower Eastern Shore of Maryland and Virginia and dominates similar marshes from there to Florida. Other interesting plants often make their appearance in the more seaward brackish marshes, such as salt-marsh loosestrife, the pink-flowering seashore mallow, mock bishopweed, and water pimpernel.

Moderately Brackish Marshes. At least five plants reach their maximum abundance in brackish marshes: narrow-leaved cattail, big cordgrass, common reed, Olney three-square, and black needlerush (from Maryland south). All of these species are either grasses or grasslike plants (graminoids). The first three species usually attain heights well above 6 feet, while the latter two usually are about 4 to 6 feet tall. This gives moderately brackish marshes a very different appearance from the salt marshes dominated by short grasses (Plate 4). Other characteristic plants include water hemp, salt-marsh bulrush, rose mallow, seashore mallow, spikerushes, seaside goldenrod, and salt-marsh fleabane.

Common reed deserves special mention because it has recently invaded many former salt marshes where tidal flow has been restricted by tide gates or undersized culverts, or cut off entirely by dikes or road and railroad embankments. Here common reed grows to a height of 12 feet or more, replacing more valuable salt-marsh species and blocking channels, further limiting access of estuarine fish and invertebrates to these marshes. The Hackensack Meadowlands, which is the largest estuarine wetland complex in northern New Jersey, and many marshes along Delaware Bay are dominated by common reed (Figure 8.2a). This plant is also frequently observed in ditches along highways, along the upper edges of salt marshes, and in areas of degraded water quality or soil disturbance.

Slightly Brackish Marshes. The farthest upstream of the brackish marshes are referred to by several names: "slightly brackish marshes," "oligohaline marshes" (due to their low salt content), or "transition or intermediate marshes" (since they occur between salt and fresh marshes). These marshes have an interesting flora, a mixture of halophytes characteristic of brackish marshes with typical freshwater-marsh species. As a result, they are the sites of some of the most diverse plant communities in the region, and their preservation is especially important for maintaining biodiversity.

Wild rice, arrow arum, spatterdock, arrowheads, common three-square, pickerelweed, soft-stemmed bulrush, sweet flag, big cordgrass, sedges, smartweeds, spikerushes, and beggar-ticks are common in these marshes. Other plants of interest are rice cut-grass (named for its rough leaves and stems), greater bur-reed (a plant that superficially resembles cattail early in the year), swamp dock, and water parsnip (a tall herb growing to more than 6 feet, bearing many small white flowers in branched terminal clusters and resembling a giant Queen Anne's lace, the familiar upland field weed).

Tidal Fresh Marshes

Coastal rivers with large watersheds and correspondingly heavy seasonal discharges of fresh water provide opportunities for the development of tidal fresh marshes. The Delaware River from Trenton to the Memorial Bridge is fresh water under tidal influence: the water levels rise and fall with the tides. Here the mean tide range (the difference between mean high and mean low tide) is 5.6 feet (Oldsman Point) to 6.8 feet (Trenton). At spring tides, the range is even greater, 5.9 feet and 7.1 feet, respectively. These tidal fluctuations expose vast areas at low tide. Since marsh plants colonize areas above mean sea level, there is much room for marsh formation. The Delaware River and its tributaries support some of the most significant stretches of tidal fresh marsh in the Northeast. Some other Northeast rivers where such conditions exist are the Merrimack, Connecticut, Hudson, and the many tributaries to Chesapeake Bay, including the Potomac, Patuxent, Nanticoke, Pocomoke, Choptank, and Chester Rivers.

Low Tidal Fresh Marshes. Tidal fresh marshes exhibit a zonation pattern similar to that of the salt marshes, but the contrast is sometimes more striking due to the plant species differences (Figure 5.3). The low marsh (flooded twice daily) is colonized by a broad-leaved plant

Figure 5.3. Tide-induced plant zonation is evident in this tidal fresh marsh along Crosswicks Creek, a tributary of the Delaware River. Spatterdock dominates the low marsh, while a diverse community occupies the high marsh.

called spatterdock, or yellow pond lily. At high tides, the leaves float at the water's surface, while at low tide, they stand erect supported by the mainstem. Spring and storm tides may completely inundate these plants, depositing a thin coating of fine silt on the leaves. A good indicator of recent high water, these silt-covered leaves are used by government regulators to verify wetland hydrology.

Joining spatterdock in the low marsh are wild rice, water hemp, water smartweed, pickerelweed, bur marigold, and broad-leaved arrowhead. Pickerelweed and spatterdock are restricted to the lowest areas and to tidal ponds. Wild rice occurs in all zones except ponds. Its seeds can germinate under a broad range of flooding, from less than daily to 12 hours per day. In Minnesota, Native Americans still harvest the ripe seeds; this delicious grain can be bought at major supermarkets or gourmet food stores. In the Northeast, wild-rice seeds are left as food for red-winged blackbirds, common grackles, bobolinks, and many other birds.

High Tidal Fresh Marshes. Immediately above the low marsh is the stream-bank levee. Arrow arum and giant ragweed make their ap-

pearance here, joining most of the low-marsh species. Beyond the levee is the high marsh, commonly inhabited by arrow arum, sweet flag, broad-leaved arrowhead, broad-leaved cattail, bur marigold, halberd-leaved tearthumb, jewelweed, and wild rice. River bulrush is abundant in some marshes, as is sensitive fern.

An interesting feature of tidal fresh marshes is the seasonal turnover in plant composition. Their appearance changes more dramatically than that of any other type of wetland in the Northeast. Throughout the winter, many tidal fresh marshes look like mud flats. The plants are overwintering either as underground rhizomes (e.g., spatterdock, arrow arum, and pickerelweed) or as annual seeds (e.g., wild rice, tearthumbs, and bur marigold). In spring, warming of the tidal muds germinates the annuals, and their seedlings cover the high marsh. By late spring and early summer, the broad-leaved emergents—especially spatterdock, arrow arum, and arrowhead—dominate. As the season progresses, tall-growing plants like water hemp, wild rice, and bur marigold, along with jewelweed and smartweed, become visibly dominant. A close look at the ground will often reveal an understory of spatterdock. In late summer, the wild rice is impressive with its masses of yellowish seeds that attract thousands of birds. In late September and early October, bur marigold, a yellow-flowered member of the aster family, blooms, drawing attention to these marshes.

Tidal Swamps

Tidal wetlands dominated by trees and shrubs also occur in the Northeast. Most are freshwater swamps occurring on the floodplains along the freshwater reaches of tidal rivers, but some are saltwater swamps. The former have plant life like that of nontidal swamps (see the discussion later in this chapter), whereas the latter possess a curious mixture of freshwater and saltwater species. While it is clearly too far north and too cold for mangroves, saltwater swamps are developing in places where rising sea level and subsidence of the coastal plain have combined to bring salt water to former freshwater swamps or low-lying upland forests. This process is perhaps best seen on the Eastern Shore of Chesapeake Bay, where salt marshes are steadily moving landward into low-lying loblolly-pine forests. Marsh burning for wildlife management and extensive ditching may have accelerated this transgression. Such wetlands represent habitats in transition, since they truly embody the dynamics of wetland plant succession, demonstrating how plant communities change in response to environmental change, either natural or human induced. These former forests of towering pines now have salt-marsh plants growing beneath salt-stressed and dying pines (Plate 5). Beneath the dominating loblolly pines is an understory of salt hay grass, salt grass, switchgrass, common reed, high-tide bush, groundsel bush, poison ivy, and others. In some places, dead trunks or stumps in the salt marsh provide the last remaining evidence of former forests. Examples can be found throughout the Northeast, with the dead trees in many northern marshes being Atlantic white cedar. Cedar wood is highly resistant to weathering and decay, so stumps remain in place for centuries.

Submerged Aquatic Beds

Underwater beds of sea grasses and other aquatic plants are common in Northeast tidal waters. Although not considered wetlands by most definitions, they are productive aquatic habitats, providing some of the same functions—substrate stabilization, nursery grounds for fish and shellfish, and food for wildlife—that wetlands do. Some of the species are also abundant in marsh pools and tidal ditches, and on exposed tidal flats (Figure 5.4).

Eel-grass is perhaps the most common submerged plant along the coast, preferring waters at or near sea strength. While it is typically submerged in most of the region, in Maine it also occurs in the intertidal zone, being frequently seen at low tide on exposed mud flats. Eel-grass beds are critical habitat for the bay scallop, whose larvae settle upon the leaves and develop into the familiar seashell and tasty seafood that sea birds and many people enjoy.

Another abundant grass of saline waters is widgeon-grass, a favored food of waterfowl. It grows in high marsh pools and in tidal creeks and ditches, giving it the nickname "ditch grass." Sea lettuce, a bright green algae, is locally abundant in sounds and shallow rivers. Sea lettuce is a preferred food for brant, sea geese resembling Canada geese, which can been observed in winter feeding by the hundreds in coastal bays.

With decreasing salinities in the upper portions of tidal rivers, other aquatic plants replace eel-grass. These species include sago pondweed, clasping-leaved pondweed (redhead-grass), and horned pondweed in moderately brackish waters. In slightly brackish waters, numerous pondweeds (genus *Potamogeton*) form aquatic beds, along with wild celery (also called tape-grass due to its flattened leaves), naiads (bushy pondweeds), and submergent forms of pipeworts and arrowheads. These aquatics and bur-reeds form submerged beds in tidal fresh waters.

Nontidal Wetlands

Beyond the reach of the tides, other hydrologic forces—surface-water runoff, river floods, and groundwater—prevail. Nontidal wetlands are far more abundant than their coastal counterparts. They develop along floodplains and the shores of lakes and ponds, on isolated depressions surrounded by upland, in areas with seasonal high water tables, and in slopes and lower areas where groundwater is discharging at or near the land surface.

Given the extent of these wetlands and the varied landscapes in which they occur, nontidal

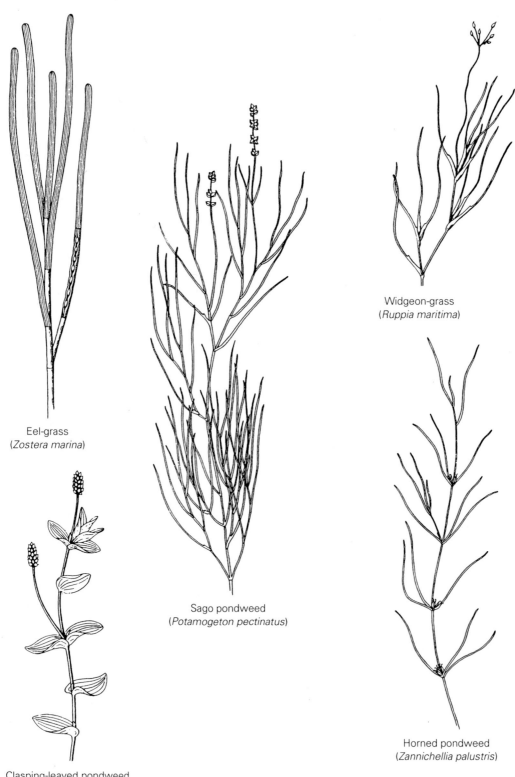

Eel-grass
(*Zostera marina*)

Widgeon-grass
(*Ruppia maritima*)

Sago pondweed
(*Potamogeton pectinatus*)

Horned pondweed
(*Zannichellia palustris*)

Clasping-leaved pondweed
(*Potamogeton perfoliatus*)

Figure 5.4. Examples of salt and brackish aquatic vegetation. (Source: Tiner 1987; copyright 1987 Ralph W. Tiner, Jr. Reprinted with permission.)

wetlands represent a diverse lot of marshes, swamps, and bogs.

In the Northeast and Great Lakes region, nine major types are present: 1) marshes, 2) wet meadows, 3) fens, 4) bogs, 5) shrub swamps, 6) hardwood swamps, 7) floodplain forested wetlands, 8) evergreen swamps, and 9) aquatic beds. Vernal pools are a minor type of lesser abundance, but they are essential breeding grounds for many salamanders and woodland frogs. The major types intergrade between one another, especially forested bogs and evergreen forested swamps, so in some cases the types given may not be distinctive on the ground. Despite this, they are useful for discussion purposes. Major types can be further divided into specific wetland types often named after their characteristic plants, such as Atlantic white-cedar swamps, red-maple swamps, pitch-pine lowlands, loblolly-pine flatwoods, alder swamps, leatherleaf bogs, buttonbush swamps, and cattail marshes.

Marshes

Marshes are one of the more easily recognized and familiar wetland types (Plate 6). They are open grasslands adjacent to lakes, ponds, and rivers, flooded by a foot or more of water through most of the year. Medium-to-tall grasses and grasslike species (graminoids) such as cattails, common reed, wool grass, bulrushes, sweet flag, wild rice, rice cut-grass, bluejoint, reed canary grass, bayonet rush, and bur-reeds characterize marshes.

Herbs with showy flowers accent some marshes. Pickerelweed, a fleshy-leaved herb with many small purplish flowers borne on a terminal spike, grows in shallow water along the margins of lakes, ponds, and streams. Arrowheads may also occur; look for their single flowering stalk bearing many three-petaled white flowers on whorled branches. Blue flag, the purple iris of freshwater wetlands, and its introduced relative, yellow flag, brighten the shores of many waters. Swamp milkweed, with its many small pinkish flowers borne in terminal umbrella-like clusters, is a magnet for monarch butterflies. Purple loosestrife, an invasive and undesirable exotic with spectacular pinkish purple blooms in late summer is codominant in many cattail marshes.

Other common marsh plants include arrow-leaved tearthumb, halberd-leaved tearthumb, arrow arum, smartweeds, jewelweed, tussock sedge, other sedges, and spikerushes. Several shrubs can also be found in marshes. Buttonbush, perhaps the most flood-tolerant shrub in the North, frequently occurs in the wetter marshes, especially along lake and river shores. Water-willow (also called swamp loosestrife), a semishrub with persistent woody bases and herbaceous stems, also grows in shallow-water marshes. Its unique habit of stem-tip rooting gives it the appearance in winter of arches leap-frogging across the shallows. Swamp rose is another common marsh shrub. Although not common, poison sumac prefers organic soils and can be encountered in some marshes.

Great Lakes Coastal Marshes. The Great Lakes (Ontario, Erie, St. Clair, Huron, Michigan, and Superior) represent the world's largest freshwater lake ecosystem. The shoreline of this lake system resembles that of an ocean with large embayments, river deltas, barrier beaches and islands, and rocky shores. In fact, many lakeshore ecosystems have been called "freshwater estuaries" because of these similarities. Due to the enormous size of these lakes, their waters are even subject to minor tidal influence and sometimes devastating wind-driven waves (lake seiches). Wetlands form in sheltered coves and lagoons, and along rivers emptying into the lakes. Water levels in the Great Lakes fluctuate causing marshes to change either to open water and aquatic beds in wet years and to wet meadows and shrub swamps in dry years (see Chapter 3). These coastal marshes are perhaps the most dynamic wetlands in the eastern half of the country and clearly represent the greatest expanse of such wetlands in this region.

The marshes are dominated by emergent species such as soft-stemmed bulrush, hard-stemmed bulrush, sweet flag, cattails, big arrowhead, common three-square, bur-reeds, spikerushes, arrow arum, pickerelweed, water-willow, and river bulrush. Aquatics beds may be interspersed with the marshes. Typical species include pondweeds, water milfoil, spatterdock (yellow pond lily), white water lily, duckweeds, coontail, waterweed, wild celery (tape-grass), water smartweed, water shield, muskgrass, and water bulrush. An aquatic shrub, buttonbush, may also be abundant in places.

Wet meadow species such as tussock sedge, other sedges, bluejoint, reed canary grass, fowl meadowgrass, common reed, rice cutgrass, rattlesnake or manna grass, panic grass, boneset, jewelweed, smartweeds, hedge bindweed, nodding bur marigold or beggar-ticks, marsh fern, and royal fern occur in the zone above the marshes. Disturbed meadows may be dominated by cattails, purple loosestrife, common reed, or reed canary grass.

Shrubs invade the exposed marshes during dry periods. Willows (especially silky willow), dogwoods (silky, red osier, and gray), shadbush or serviceberry, wild raisin, speckled alder, meadowsweets, swamp rose, sweet gale, winterberry, and saplings of trees such as green ash, red maple, quaking aspen, and eastern cottonwood are among these opportunistic woody invaders. Their existence, however, is short-lived as they are killed by high water during wet periods when the marsh vegetation reestablishes itself.

Savannas and Carolina Bays

Along the coastal plain from New Jersey south, marshes called "savannas," "glades," and "Carolina bays" can be found. These marshes often are surrounded by dense forests, and their openness provides a welcome break in the landscape for people and animals alike. In the New Jersey Pine Barrens, such marshes are dominated by species preferring acidic waters, including bull sedge, pipeworts, twig rush, cottongrass, coast sedge, white beakrush, redroot, and twisted yellow-eyed grass. These types of marshy areas are not flooded as long as cattail marshes, being more seasonally wet and saturated. The name "savanna" comes from their low grassland appearance complemented by a scattering of shrubs like highbush blueberry and dense St. John's-wort (Plate 7). Associated herbs include lowland broomsedge, Virginia meadow beauty, three-way sedge, Canada rush, bog rush, bladderworts, and marsh and dwarf St. John's-worts. Some of the more unusual savanna species are purple pitcher plant, bog asphodel, slender blue flag, gold crest, bartonia, twisted yellow-eyed grass, ten-angled pipewort, and several orchids. Drier savannas typically have more woody plants present, including pitch pine, inkberry, bayberry, sheep laurel, fetterbush, chokeberries, northern wild raisin, and leatherleaf. Savannas were more abundant in the Pine Barrens at the turn of the century. Today, fire suppression has allowed them to become overgrown with leatherleaf, highbush blueberry, red maple, black gum, and Atlantic white cedar. Periodic fires appear necessary to maintain this wetland type.

Similar savanna-like marshes occur on the Delmarva Peninsula in somewhat egg-shaped depressions called "Delmarva bays," "Carolina bays," "glades," or simply "potholes." Walter's sedge is a frequent dominant, along with plants of southern affinity such as giant beard grass, maiden-cane, and warty panic grass. Buttonbush and scattered saplings of red maple, sweet gum, and a wetland ecotype of persimmon may be present in these glades. Fetterbush typically forms a shrubby border between the marsh and the adjacent woods. These types of marshes develop on broad interfluvial flats occurring between major drainages from the Delmarva Peninsula to Georgia and are most common in South Carolina and North Carolina; hence the nickname "Carolina bays."

Wet Meadows

Wet fields called "wet meadows" are seasonally flooded for varying periods or saturated near the surface due to local high water tables. Some wet meadows occur along streams and lakes, while others develop in depressions or river valleys maintained as open land by human intervention. In agricultural areas, wet meadows often exist on lands once occupied by forested wetlands. In colonial times, these forests were cleared for pastures; they were too wet to grow crops, but they produced good hay for livestock. Such wetlands have been maintained by grazing or periodic mowing.

Wet meadows can be dominated by a single species or characterized by a diverse assemblage of plants. Sedges, sweet flag, and soft rush are chief dominants along with various grasses. Sedge meadows are dominated by species of the genus *Carex*. Some sedge meadows are called fens (see the following subsection). Tussock sedge is perhaps the most abundant wet-meadow sedge, forming nearly pure stands from New England to northern New Jersey (Plate 8). These are the wettest of the meadows, with standing water present for much of the year. Even when exposed in late summer, they remain saturated to the surface. The tussocks of tussock sedge serve as microsites for colonization by other species including false nettle, meadowsweets, and saplings of red maple. Sweet flag with irislike leaves is particularly common along seepage areas, narrow drainageways, and depressions near dairy farms. It may have been valuable at some time to farmers; perhaps the sweet smell of its crushed or trampled leaves improved the air in the cow barn. In England, dried leaves were sold to churches to be put on their floors and provide a more heavenly scent. Sweet flag also has been used medicinally since ancient times: chewing the raw root helps stop an upset stomach. In agricultural areas, soft rush, a leafless grasslike species with needlelike stems up to 3½ feet tall, dominates many wet pastures because livestock find it unpalatable. Although it seems to be the only plant remaining in many grazed meadows, a closer examination often reveals the presence of other species, many of which have been heavily browsed and are short in stature.

Many wet meadows are diverse plant communities with an abundance of grasses, sedges, and flowering herbs (Plate 9). Common wet-meadow grasses are reed canary grass, bluejoint, and bent-grass. Lurid sedge is probably the most frequently encountered sedge in these mixed meadows. Flowering herbs provide seasonal colors. In spring, yellow flowers of winter cress or yellow rocket, marsh marigold, and golden ragwort are most conspicuous. Purple-flowered and white-flowered violets and the pinkish wild geranium also bloom at this time. In summer, tall flowering plants (over 4 feet tall) visually dominate. Blue vervain (purplish flowers), New York ironweed (purplish to pinkish flowers), Joe-Pye-weeds (pinkish), boneset (white), flat-topped white aster, blue flag (bluish to purplish), jewelweed (orange), and swamp milkweed (pinkish) are among the more prominent species. Jewelweed is of particular interest since its juices have been touted as a folk remedy for poison ivy; if you have inadvertently touched poison ivy, crush some jewelweed stems and rub the juices on the exposed area.

Three members of the composite family—asters, beggar-ticks, and goldenrods—bloom in the fall. The blue-to-purplish flowers of swamp, New York, and New England asters liven up many wet meadows. Small white-flowered aster also is common, along with several yellow goldenrods, especially grass-leaved goldenrod. Beggar-ticks produce yellow or orangish flowers, and their barbed seeds, called "stick-tights," are likely to be familiar: they are the seeds (about ¼ inch long) with two to five barbs that stick to your jeans and socks in late summer and fall.

Other typical wet-meadow plants include skunk cabbage, angelica (a relative of Queen Anne's lace), marsh fern, and sensitive fern. Many shrubs occur in wet meadows, including silky dogwood, arrowwoods, and the pink-flowering steeplebush and its white-flowering cousins, broad-leaved meadowsweet and narrow-leaved meadowsweet. Two thorny shrubs, blackberry and multiflora rose, also may be present, especially in drier meadows.

Fens

Fens are variably mineral-rich peatlands that develop in areas of groundwater discharge and along rivers, streams, and lakes in the glaciated regions (Plate 10). In the Northeast, they are most abundant in northern New England and upstate New York, predominantly in limestone regions, but they also occur as rare habitats in the Berkshires of western Massachusetts and Connecticut. Minnesota possesses 20 percent of the world's calcerous fens. They are most common in the western half of the state. In forested regions, fens are usually surrounded by an acidic landscape, and they may lie adjacent to a bog.

Depending on the relative wealth of minerals, three types of fens are recognized: 1) rich fens, 2) medium fens, and 3) poor fens. The major differences among them can be determined by soil chemistry—namely pH (hydrogen-ion concentration)—and by vegetation. Rich fens are more alkaline, with pH values above 6.0, while poor fens are more acidic, with pH values between 3.5 and 5.0. Medium fens have intermediate pH values. Soil-chemistry differences produce a noticeable change in plant communities. Rich fens have more calcium-loving species and little or no peat moss, while poor fens have significant numbers of peat mosses and other bog species mixed with a variety of sedges, including few-seeded sedge, poor sedge, and coast sedge.

The availability of minerals such as calcium creates conditions for the establishment of an unusual assemblage of plants, especially in rich fens. Woolly-fruited sedge and other sedges are the dominant species forming a quaking mat that advances into open water. In many cases, the former body of water is completely filled in with vegetation, giving the fen a grassland appearance much like that of a wet meadow. Other species found in rich and medium fens include buckbean, three-way sedge, marsh cinquefoil, shrubby cinquefoil, cotton-grasses, white beak-rush, alders, and cattails. Plants considered calciphiles (calcium-loving plants) found in rich fens are the sedge *Carex sterilis*, yellow sedge, the rush *Juncus stygius*, false asphodel, grass-of-Parnassus, brook or Kalm's lobelia, hoary willow, arrow-grass, and twig-rush (Figure 5.5). It is interesting to note that the latter two species also occur in salt and brackish marshes, respectively, which are rich in mineral salts. Peat moss is conspicuous by its absence or low abundance in fens.

Some fens represent pioneer communities that invade bodies of water and create marshy habitats, paving the way for bog communities. Often, fens and bogs are closely intertwined, with fens occupying the mineral-rich portion of a basin, presumably at the point of groundwater discharge. Given their pioneer nature, many fens are destined to become something else, either bogs or forested wetlands. In some places, herbaceous fens are maintained by land-use practices (mowing or burning), natural fires, and possibly by beaver activity. Shrub fens are characterized by several shrubs, including shrubby cinquefoil, hoary willow, red osier dogwood, speckled alder, swamp birch, swamp fly honeysuckle, and alder-leaved buckthorn (Figure 5.5). Tree fens are dominated by two trees, northern white cedar and larch. Red maple, black spruce, black chokeberry, sweet gale, and rarely swamp birch are woody associates. Herbs forming the groundcover may include blue flag, skunk cabbage, starflower, goldthread, blue bead-lily, and fowl manna grass. The aromatic, tiny-leaved trailing vine creeping snowberry (a cousin of wintergreen) may also be present. Its crushed leaves have a smell reminiscent of the muscle salve Ben-gue (or Ben-gay).

Bogs

Bogs are one of the more familiar wetland types in North America, where they form an important part of the boreal landscape. They are mostly permanently saturated, nutrient-poor peatlands dominated by shrubs of the heath family (shrub bogs) or by evergreen trees (forested bogs) growing in peat moss–based organic soils. Precipitation may be the only source of water for some bogs (ombrotrophic bogs), while other bogs are fed by precipitation, nutrient-poor surface waters, and seepage. The latter are sometimes called "poor fens" due to limited groundwater influence. Various types of bogs have been described based on either their topographic position or their vegetation.

Bogs are most characteristic of the glaciated region, where they are frequently found in isolated depressions. During the past 10,000 years, these "lake-fill bogs" have wholly or partially filled former glacial kettle ponds and lakes with plant remains through the process of hydrarch succession (Plate 11; see Chapter 3). Bogs also occur on seepage slopes and along nutrient-poor streams and lakes behind narrow streamside or lakeshore fens.

The maritime influence and cool, wet climate along the coast from eastern Maine north create a frequently fog-shrouded shoreline favoring the rapid buildup of peat and the formation of two other types of bogs: raised bogs and sloping bogs. Raised bogs may form a broad plateau above the surrounding landscape. At the edge of these bogs, there is usually a water-filled moat (lagg). When the buildup of peat is markedly greater in the center, the bog takes on a domed or convex appearance (in cross section), giving it the name "domed bog."

Sloping bogs, as their name implies, develop on slopes where an excess of water promotes the growth and spread of peat moss. Many sloping bogs have their origins in a lake-fill bog at the bottom of a slope. With a great abundance of water in the basin, the peat moss begins to climb the slope, drastically altering both the vegetation and character of the land. Peat moss can ascend 25° slopes. This process of paludification is responsible for building peatlands in the northern latitudes. In places like Newfoundland, Scotland, Ireland, Wales, Minnesota, and Canada, the entire landscape has been reshaped by this process. Low-rolling hills are now covered by peat moss and bogs. The presence of impermeable subsoils (hardpans) in these regions has facilitated the formation of these "blanket bogs." In the Arctic and subarctic, permafrost serves as a similar catalyst.

Less common in the Northeast are "patterned peatlands," with their characteristic pattern of ridges (strings) and depressions (flarks) extending perpendicular to the flow of water. These wetlands develop in relatively flat landscapes where seepage water is abundant. Their appearance from the air is spectacular (Figure 5.6). Freezing and thawing combined with oxidation of organic matter seem to heighten the elevational differences between the strings and flarks. Different species of peat moss colonize these varied habitats. Patterned peatlands occur

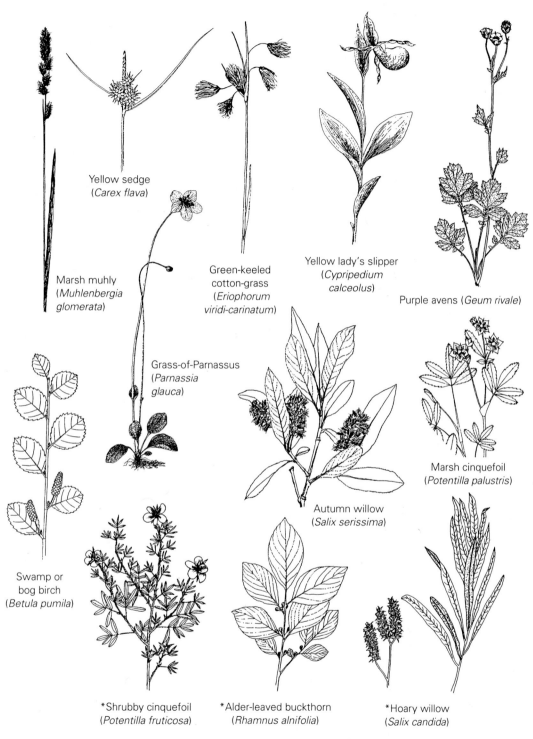

Figure 5.5. Examples of calcium-loving plants (calciphiles) in the Northeast. These plants are good indicators of groundwater-influenced wetlands in limestone regions. An asterisk (*) indicates a low-growing shrub (usually less than 4 feet tall). (Sources: Crum 1988 [copyright University of Michigan Press], Magee 1981, Tiner 1994. Reprinted with permission.)

Figure 5.6. Aerial view of a patterned bog in northern Maine. (Hank Tyler)

from Maine to Minnesota where extensive peatlands exist. Minnesota has more peatland than any state except Alaska.

Northern Shrub Bogs. Bogs are frequently named after their characteristic vegetation, such as leatherleaf bog, cranberry bog, and black-spruce bog, or more generally by the dominant life-form (shrub bog v. forested bog). In shrub bogs, the dominant plants are usually members of the heath family (Ericaceae), with leatherleaf, sheep laurel, bog laurel, bog rosemary, Labrador tea, rhodora, huckleberries, blueberries, and cranberries being typical species. Heath species are noted for their ability to thrive in nutrient-poor conditions, which makes bogs an ideal habitat. Many ericads are evergreen, a characteristic that helps them conserve energy but at the same time makes them vulnerable to drying winter winds. Thick and leathery leaves aid their survival. In addition, the leaves often curl inward to protect them further against damaging winds, as commonly observed in the garden varieties of rhododendron in winter. Heath leaves are also toxic to animals, thereby reducing the effects of browsing. All ericads have root fungi (mycorrhizae) that assist in mineral absorption. Larch and black spruce also have these fungi.

Another interesting observation about heath plants is that some begin flowering when their roots may still be frozen; leatherleaf is the earliest blooming ericad, starting in late April.

Nonheath plants are also widespread in bogs. Sweet gale, a relative of bayberry, is a nitrogen-fixing plant, with specialized bacteria in its root nodules that convert nitrates and ammonia to usable nitrogen. This symbiotic relationship with bacteria (benefiting both organisms) helps sweet gale colonize low-nutrient sites. Alders can also do this by similar means. A thornless member of the rose family (Rosaceae), black chokeberry is another characteristic bog shrub. From the Down East coast of Maine into Canada's Maritime Provinces, two berry-producing, somewhat creeping ground shrubs are locally abundant: black crowberry (with black berries) and baked appleberry (with its tasty orangish raspberry-like fruits). Tufted bulrush forms conspicuous grassy clumps with the distinctive, somewhat cottony tips of its stems. Few-seeded sedge may be the dominant herb in open bogs. Shrub bogs often have stunted black spruce or larch scattered on the bog mat.

Bogs possess some of the most interesting plants found in North America (Figure 5.7). Es-

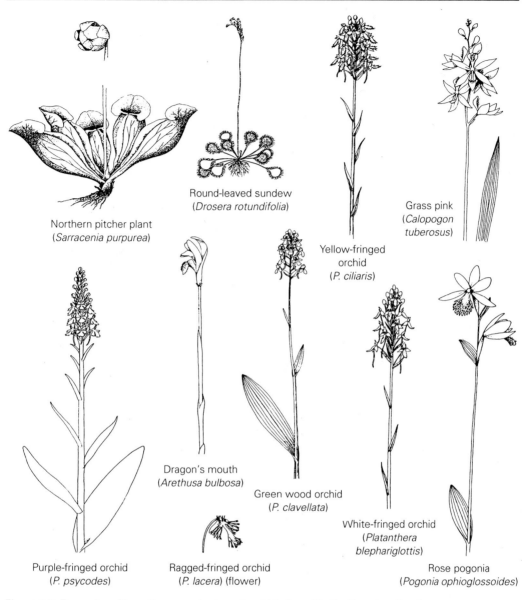

Northern pitcher plant
(*Sarracenia purpurea*)

Round-leaved sundew
(*Drosera rotundifolia*)

Grass pink
(*Calopogon tuberosus*)

Yellow-fringed
orchid
(*P. ciliaris*)

Dragon's mouth
(*Arethusa bulbosa*)

Green wood orchid
(*P. clavellata*)

White-fringed orchid
(*Platanthera blephariglottis*)

Purple-fringed orchid
(*P. psycodes*)

Ragged-fringed orchid
(*P. lacera*) (flower)

Rose pogonia
(*Pogonia ophioglossoides*)

Figure 5.7. Examples of insectivorous plants and orchids found in Northeast wetlands.
(Source: Magee 1981 and Tiner 1988)

pecially noteworthy are the carnivorous plants that obtain some of their nutrients from captured insects and spiders. The purple or northern pitcher plant catches unsuspecting invertebrates that easily enter its leafy tube but once there are prevented from climbing out by stiff hairs. Sundews use a different strategy: insects encountering their sticky hairs are destined to become the next meal. Some of nature's most beautiful flowers, members of the orchid family (Orchidaceae), grow in bogs. It's surprising to see these wonderful flowers in such a nutrient-

poor environment, but it probably takes that much beauty to attract a pollinator. Other abundant bog plants are cotton-grasses (actually sedges with cottony white flower-seed masses), white beak-rush, the purple-flowering bog aster, and the yellow-flowering bog goldenrod.

The lagg surrounding some bogs receives more nutrients from adjacent runoff, and other species are found here. Water lilies grow in permanently flooded moats, while other moats are more shrubby, marshy, or sedgy. Speckled alder is abundant in many laggs, along with other shrubs like northern wild raisin, winterberry, mountain holly, and alder-leaved buckthorn. Characteristic herbs include wild calla, cinnamon fern, royal fern, sedges, and grasses. In some places, northern white-cedar trees may grow in the lagg.

Northern Forested Bogs. Forested bogs are dominated by trees such as black spruce, balsam fir, and larch (Plate 12). These bogs occur mostly from northern New England and the Adirondacks to Minnesota, but also are found farther south to northwestern New Jersey and west to northern Illinois. Red maple and yellow birch are frequent associates, while hemlock, red spruce, white pine, and pitch pine may also occur. Black gum has been observed in some Pocono bogs where red spruce predominates. Balsam-fir stands can be found in boggy-to-marshy areas in West Virginia's Canaan valley.

When the canopy is more open, the peaty soils of forested bogs support plants characteristic of shrub bogs such as leatherleaf, sheep laurel, Labrador tea, rhodora, mountain holly, winterberry, highbush blueberry, sweet gale, small cranberry, and creeping snowberry. Northern wild raisin may also occur. The principal herb is often three-seeded sedge. Otherwise, the herb layer is sparse, especially in densely shaded sites. A scattering of other herbs, including goldthread, starflower, three-leaved false Solomon's seal, cinnamon fern, pitcher plant, and bunchberry (a dwarf species of dogwood, usually less than 6 inches high) may colonize the mossy substrate. Bunchberry is easily recognized in late summer and fall by its terminal cluster of red berries.

Southern Bogs. The term "bog" has also been applied to certain shrub swamps in the nonglaciated region of the Northeast. They too have a dominant groundcover of peat mosses, but their vegetation is considerably different due in large part to the climatic dissimilarity. A comparison of two New Jersey bogs found that the coastal-plain bog had much higher plant diversity: 149 species compared to 43 for the northern bog. Only nine plants were common to both bogs, including sweet pepperbush, swamp azalea, red maple, highbush blueberry, and big cranberry. Bogs are rare on Maryland's Western Shore, but they contain an interesting combination of species with northern and southern affinities. Some of the typical northern-bog plants present were leatherleaf, big cranberry, white beak-rush, cotton-grass, and sun-

dews, while southern species included Virginia meadow beauty, Virginia chain fern, and even giant cane, a dominant plant in the Great Dismal Swamp.

Bogs are also uncommon in western Maryland and West Virginia. Besides peat moss, haircap moss is abundant. This moss is interesting in that it is also common on uplands throughout the Northeast, often on sandy soils. Haircap moss forms the dominant groundcover for much of the Canaan valley wetlands, the largest wetland complex in West Virginia. Dense St. John's-wort is a dominant bog shrub that adds yellow color to these bogs from midsummer to fall. Red spruce, a relative of the black spruce, is found in these southern bogs, in contrast to its preference for uplands in New England. Larch reportedly occurs at its southernmost location in the Craneville Pine Swamp on the Maryland–West Virginia border. Many northern-bog species can be found as far south as Georgia. These populations are apparently holdovers from earlier times, when the climate was colder. The high elevations and cool temperatures maintain an environment suitable for their continued survival. Species include mountain holly, rose pogonia, bog goldenrod, and velvet-leaf blueberry.

Shrub Swamps

Wetlands dominated by woody plants less than 20 feet tall are considered shrub swamps. Shrub bogs growing in nutrient-poor conditions on peat soils are one type of shrub wetland, while other shrub swamps develop in richer sites on either mucky soils or hydric mineral soils. Shrub swamps occur over a wide range of water regimes, from permanently flooded to temporarily flooded or seasonally saturated. Some dominant shrubs in the Northeast and Great Lakes region are buttonbush, water-willow, alders, willows, arrowwoods, and dogwoods.

Buttonbush is among the shrubs that are most tolerant of wetness, rivaling sweet gale, water-willow, and leatherleaf. Its ability to germinate underwater allows it to take root in the shallow bottoms of lakes and ponds, along slow-flowing rivers, especially in oxbows, and in marshes. If you see a shrub thicket growing in water during the summer, it is most likely a buttonbush swamp. On the Delmarva Peninsula, buttonbush grows in the depressions called "Delmarva bays" or "potholes." Associated species include shrubs like swamp rose, willows, and dogwoods and marsh herbs such as arrow arum, spatterdock, manna grass, three-way sedge, and cattail.

Water-willow or swamp loosestrife is a shrublike species that grows in similar locations to buttonbush. This plant is not a true willow, but a member of the loosestrife family (Lythraceae)—a cousin of purple loosestrife and swamp candles. It has been called a "semi-shrub" because its upper parts are herbaceous while its bases are woody. It colonizes shallow water through the ability of its stems to root at their tips. A precursor to bog formation, it commonly forms the leading edge of bogs, facilitating the advance of the bog mat into open water. Water-willow is also frequently found in tidal fresh marshes.

Low thickets of sweet gale line the shores of many New England and upstate New York lakes. From a distance, these thickets along the water's edge may be mistaken for leatherleaf, but the leaves of leatherleaf are not aromatic when crushed like those of sweet gale, and in winter the evergreen leaves of leatherleaf persist while those of sweet gale do not. Sweet gale may grow by itself or in the company of other species. Among its associates are tussock sedge, blue flag, cattail, soft-stemmed and hard-stemmed bulrushes, bur-reeds, twig-rush, three-way sedge, pickerelweed, speckled alder, and leatherleaf .

Many other shrubs form wet thickets or mixed meadow-thickets (Plate 13). Among the more widespread are alders, silky dogwood, red osier dogwood (easily recognized in winter by its bright red stems, it seems to like mineral-rich sites), winterberry, highbush blueberry, steeplebush (named for its prominent flowering spike), meadowsweets, arrowwoods, willows, swamp rose, swamp azalea, sweet pepperbush, maleberry, and chokeberries. These shrubs grow under many hydrologic conditions, ranging from seasonally flooded to seasonally saturated (water table at or near the surface for extended periods, usually from winter to early spring).

Alders typically form dense thickets throughout the region, frequently on hillside seeps. Alder swamps are great places to look for the woodcock, a stubby woodland bird that loves to feed on the earthworms that grow in abundance in these wetlands. Meadowsweets are often intermixed with bluejoint, a common wet-meadow grass, forming open shrubby meadows. Many shrub swamps are represented by more than one shrub species and support a variety of herbs.

Willows are common wetland shrubs. Some willow thickets are successional communities following abandonment of farming. Sandbar and other willows often dominate what appear to have been cultivated or pastured wet meadows. Hoary willow is indicative of calcium-rich conditions. Many people associate willows with wet areas. In fact, dowsers often use willow branches to find water for private wells. (I'm told that my grandfather did this; maybe that's where I get my urge to locate and map wetlands.) The familiar pussy willow is noted for its "pussies," the emerging whitish, hairy, bud-shaped catkins that seasonally adorn many a vase in the Northeast. To some, the occurrence of pussies heralds the beginning of spring.

Other shrub swamps are dominated by saplings of trees (less than 20 feet tall). Red-maple-sapling swamps are the most common and widespread. These swamps will eventually become forested wetlands. Some are successional communities that have arisen from sedge meadows, for maple saplings can be found growing on the tops of tussock-sedge hummocks.

Hardwood Swamps

Hardwood swamps dominated by deciduous trees are the most common wetland type in the Northeast, especially from central New England south. Deciduous trees lose their leaves in the fall. These swamps occur mainly in depressions of variable size that are either isolated or drained by a small stream. They are usually flooded for extended periods (more than two weeks) during the year, and the soils are variously saturated when not exposed. While red maple is the predominant species, a variety of other trees coexist or dominate in many forested wetlands. Factors such as climate, soils, geology, hydrology, and human activities influence the composition of individual swamps. No two swamps are the same. In order to provide a meaningful summary of the various types, the following discussion has been intentionally simplified. To adequately characterize these communities would fill the pages of another book, and there are many useful texts already written on specific types (see Additional Readings at the end of the chapter).

Northern Hardwood Swamps. From central New Jersey north, common wetland trees are red maple, yellow birch, black ash, green ash, and American elm. They combine in different ways to form a variety of swamp forests, with red maple usually dominant or codominant (Plate 14). Yellow birch and black ash are northern associates. Red maple–black ash swamps are common in limestone regions. Swamp white oak, with its shallow lobed leaves and acorns borne on long stalks, is locally abundant in some places. White pine and hemlock are evergreen trees that frequently codominate red-maple swamps in parts of New England and New York. Pin oak may be common in old glacial lake-bottom deposits, as in northern New Jersey. Black gum is codominant in some red-maple swamps.

Shrubs are quite abundant in these swamps, many producing berries. The tiny yellow flowers of spicebush bloom in early spring, giving a yellow tint to the swamp understory. This shrub also has a distinctive lemony scent when its twigs or leaves are crushed. Its red berries can be dried and used as a substitute for allspice. Highbush blueberry is one of the more common shrubs. Its creamy white urn-shaped flowers bloom before its leaves unfold. Its tasty blueberries can be harvested in summer. Female shrubs of winterberry are conspicuous in fall and winter, as their orangish red berries can be seen at great distance. Other common shrubs include northern arrowwood, northern wild raisin, silky dogwood, red osier dogwood, maleberry, and alders. Two fragrant white-flowering shrubs are prominent in red-maple swamps, especially near the coast: swamp azalea (with its spectacular showy midsummer blooms) and sweet pepperbush (a great nectar bush for bees in late summer).

The herb flora of red-maple swamps often varies seasonally. In spring, it's hard to miss the large leaves of skunk cabbage, a member of the arum family (Araceae). The cabbage-like leaves emits a foul odor when crushed, making the plant easy to recognize. This odor has a specific purpose: not so much to reduce its appeal to would-be grazers, but to attract insect pollinators to minute flowers borne on a fleshy appendage (spadix) within a surrounding hood (spathe). The flowers are known to generate an enormous amount of heat, which increases the odor and their attractiveness. On a cold blustery February day on Cape Cod, the flowers can produce enough heat to keep the inside of the hood as comfortable as your living room in winter (around 68°). Skunk cabbage is probably the best indicator of a swamp, since it grows only in wetlands and is often abundant. In spring, it virtually covers the swamp floor (Plate 15). Other prominent spring flowers include marsh marigold and golden ragwort (yellow), violets (white or purple), and spring beauty (pinkish white).

Ferns (e.g., cinnamon, royal, sensitive, and marsh) are often abundant throughout the growing season. Grasses and sedges also occur, with manna grasses, wood reed, tussock sedge, and three-way sedge being common throughout the Northeast. Jack-in-the-pulpit, wild calla, and sweet flag, all cousins of skunk cabbage grow in maple swamps. Jack is the flowering fleshy spadix within the hood-like spathe.

In the summer, cardinal flower, aptly named for its brilliant crimson blossoms, jewelweed with its uniquely shaped tubular, saclike orangish flowers, and less commonly the white-flowered turtlehead are distinctive. Jewelweed also goes by the name "spotted touch-me-not" because its ripe fruit capsules burst open when disturbed, launching seeds. In fall, asters and goldenrods can be seen. Peat moss and other mosses occur in variable quantities and can be observed year-round, as can crested fern, an evergreen species.

Coastal-Plain Swamps. Many swamps occur along the large rivers that drain the coastal plain. Here southern trees such as sweet gum, sweet bay, willow oak, and basket oak (closely resembling swamp white oak) flourish alongside red maple (Plate 16). Red maple–sweet gum communities are particularly widespread. Black gum and green ash are also abundant. Stands of red maple mixed with either of these species are fairly common. Three pines may codominate or be common associates with red maple: pitch pine (in the New Jersey Pine Barrens) and loblolly pine and pond pine (farther south). Atlantic white cedar may be present in some swamps. Curiously, American holly, a typical upland plant in most of the Northeast, grows in many coastal-plain swamps, usually on more sandy soils. In some places, it is codominant with red maple or other deciduous trees, or may even form small pure stands.

Many herbs and shrubs of coastal-plain swamps are the same as in northern hardwood swamps. Southern arrowwood, smooth alder, and southern wild raisin replace their northern counterparts. Fetterbush and Virginia sweetspires are other southern shrubs that join sweet pepperbush, highbush blueberry, silky dogwood, winterberry, and swamp azalea in the shrub understory. Red chokeberry and dangleberry are also common. Sweet bay is usually a shrub or small tree, but sometimes rather large trees can be found. Common elderberry, which produces small blackish purple berries used for wine and jelly, occurs but is not particularly abundant. Lizard's tail, a broad-leaved herb with tiny fragrant flowers borne on a long erect stalk, may dominate wetter swamps and seasonally flooded depressions in drier swamps. While various ferns are present, net-veined chain fern is often most abundant. Its sterile fronds resemble those of sensitive fern, but its leaf margins are finely toothed, in contrast to the smooth (entire) margins of sensitive fern. The fertile fronds of these two species are also different: compare the beadlike clusters of sensitive fern to the open featherlike leaflets of chain fern. Peat mosses are abundant in the wetter swamps or in wet depressions in drier swamps.

Some coastal-plain swamps are almost impenetrable due to the presence of common greenbrier along their edges. This member of the lily family (Liliaceae), a climbing woody vine with numerous ¼-inch thorns, is so abundant in some places that it is virtually impossible to get through the tangled mass. Other woody vines include Japanese honeysuckle, poison ivy, Virginia creeper, and trumpet creeper. All of these vines seem to be more plentiful in drier swamps.

Drier hardwood swamps on the Delmarva Peninsula are more appropriately called "wet hardwood flatwoods" rather than swamps due to their flat terrain and hydrology. They form on broad interstream divides (flats) where drainage is poor. The groundwater table is near the surface during the winter and early spring, but these areas are dry at the surface for most of the growing season. They have been locally referred to as "winter wet woods" because of their hydrology. These wetlands are often characterized by a mixture of trees, including red maple, sweet gum, basket oak, willow oak, white oak, southern red oak, black gum, and loblolly pine.

From Maryland south, bald cypress enters diverse river-swamp communities of red maple, green ash, black gum, swamp black gum, and other species. This hardwood community marks the beginning of a rich southern swamp flora that extends into Florida and around the Gulf of Mexico into Texas and north in the Mississippi drainage into southern Illinois and Indiana. The Pocomoke River swamp on the Del-

marva Peninsula probably represents the northernmost southern river swamp on the East Coast. Two members of this community (bald cypress and swamp black gum) probably possess the easiest-to-recognize morphological adaptations of woody plants to prolonged inundation—swollen bases of trunks (buttressing)—while bald cypress also has conspicuous knees (pneumatophores, extensions of the root system). Although it is a matter of debate among scientists, the latter structures seem to aid in respiration, as evidenced by their occurrence on trees in ponded or flooded depressions and their absence from trees in less frequently flooded soils beyond the depressions.

Larch Swamps

From northern New England and upstate New York to northern Minnesota, boreal swamp forests are sometimes dominated by a species of larch, the tamarack. With its needle-leaves and cones, this species resembles an evergreen in summer, but it is deciduous, being leafless in winter. Larch swamps are most conspicuous in fall when the larch needles turn yellow. This community resembles a forested bog in many respects, but it also contains swamp plants. The microtopography is hummocky due to the presence of peat mosses. Black spruce, balsam fir, and red maple are associated trees, with black spruce or red maple codominating. Where red maple is abundant, poison sumac, the taller cousin of poison ivy, may be found. In more acidic sites, the shrub layer may include both bog and swamp species, such as rhodora, speckled alder, northern wild raisin, mountain holly, Labrador tea, leatherleaf, sheep laurel, and steeplebush. Three-seeded sedge, bladder sedge, and cinnamon fern are among the few herbs found in this community. At richer sites, the herb layer is more diverse and may include marsh marigold, tall meadow rue, skunk cabbage, royal fern, and several sedges. Shrubs at these sites are fen species such as red osier dogwood, alders, swamp birch, and alder-leaved buckthorn.

Floodplain Forested Wetlands

Alluvial processes create an environment characterized by periodic disturbances. Variable loads of sediments are deposited on floodplains annually: heavy in some years, but light in most. Some places are scoured by the strong erosive forces accompanying record floods. Species living here must be able to tolerate these disturbances. The hydrology of such wetlands is better defined as temporarily flooded, since they are usually inundated for brief periods (less than a week) in average-rainfall years. Although brief, flooding may occur several times during the year. Forests dominate the more mature, well-developed floodplains (Plate 17), except where the land is cultivated or in pasture.

Floodplain trees must have the ability to withstand burial by a foot or more of sediment from record floods. The ubiquitous red maple is still here, especially along narrow floodplains and floodplains along slow-flowing coastal-plain streams, but other species become important components of the plant community. Two other maples replace red maple on broad floodplains associated with major drainage basins, like the Connecticut, Susquehanna, Patuxent, and Potomac Rivers. Silver maple, the floodplain cousin of red maple, often predominates, while box elder (also known as ash-leaved maple due to its compound ashlike foliage) may be locally abundant. Cottonwood, green ash, American elm, sweet gum, ironwood, and sycamore are other common floodplain trees. Black willow thrives in wet sloughs. Tulip poplar is a significant component of some temporarily flooded alluvial woods in the Potomac River drainage. Pin oak dominates floodplain forests in northern New Jersey. River birch is more typical of southern floodplains, but occurs as far north as southern New England and southern Wisconsin. It seems to prefer disturbed areas on riverbanks, where its reddish brown peeling bark is easy to spot.

The shrub layer in many floodplain wetlands is sparse, since most shrubs seem intolerant of periodic sedimentation or scouring. Spicebush, elderberry, and arrowwoods are among the shrubs present. Pawpaw may occur on riverbank levees in Maryland and Virginia. Multiflora rose, Japanese barberry, and Japanese honeysuckle are introduced woody plants that have become well established on floodplains.

Herbs make up for the paucity of shrubs. Trout lily and spring beauty are conspicuous in the spring. Their yellow and pinkish white flowers, respectively, add color to the floodplain floor. Spring beauty is usually the more abundant species, forming a floral carpet on many Northeast floodplains. False nettle, sensitive fern, and ostrich fern are common along the Connecticut River. False nettle seems to prefer the wetter sites, while ostrich fern occupies drier, but still moist, sites. At higher levels subject to less-frequent flooding, the stinging wood nettle can be found, growing to 6 feet high on fertile-soiled levees. Grasses like white grass (a relative of rice cut-grass), reed canary grass, wood reed, rye grass, and deer tongue (a broad-leaved grass related to switchgrass) are also frequently encountered along with other floodplain herbs, including jewelweed, jack-in-the-pulpit, clearweed, white avens, smartweeds, and Virginia knotweed (or jumpseed, named after the ripe seeds that seem to jump off the flower spike when touched). Clumps of green needle-like stalks of field garlic, reminiscent of scallions in form and in odor when broken, can be seen in some forests.

A few introduced herbs can be found in floodplain swamps. Japanese stilt grass, an invasive species from Asia, covers the forest floor of many New Jersey floodplains. Garlic mustard, a European herb, is easily recognized in spring and early summer by its terminal cluster of small four-petaled white flowers; its crushed leaves smell like garlic. Moneywort, another European transplant, is a trailing herbaceous vine with round, coin-shaped leaves and bright yellow five-petaled flowers.

Grape vines are the characteristic lianas (vines) of many floodplain forested wetlands. Their thick, dark brown woody stems resemble ropes and often tempt one to try a Tarzan swing. Vines typical of swamps also grow on floodplains, including Virginia creeper, trumpet creeper, and poison ivy.

Evergreen Forested Swamps

Six conifers characterize evergreen forested swamps in the Northeast and Great Lakes region—1) northern white cedar, 2) Atlantic white cedar, 3) eastern hemlock, 4) eastern white pine, 5) pitch pine, and 6) loblolly pine—as opposed to the three other species that typify forested bogs (black spruce, red spruce, and balsam fir; see bog subsection above). Northern white-cedar swamps differ from the rest in that they often characterize mineral-rich sites in limestone regions, while the others occur at acidic, nutrient-poor sites.

Northern White-Cedar Swamps. These swamps are most common in limestone regions, but some develop along the edges of bogs, where nutrients are more abundant (Plate 18). The occurrence of northern white-cedar swamps on limestone formations gives them a distinctive flora, including herbs such as foamflower, mitrewort, tufted loosestrife, star-flowered false Solomon's seal, and the sedge *Carex leptalea*. Red osier dogwood and swamp fly honeysuckle may be present. Black ash is another possible associate that seems to like limestone regions. The remaining species have mostly deciduous swamp affinities, including red maple, yellow birch, elm, highbush blueberry, arrowwood, ironwood, sensitive fern, cinnamon fern, royal fern, marsh fern, and golden ragwort. Surprisingly, paper birch may be found in some cedar swamps. Mountain holly and sheep laurel can be found in some swamps. The rest of the associates are either other evergreen trees, such as hemlock, balsam fir, white pine, and black spruce, or the deciduous conifer larch.

Atlantic White-Cedar Swamps. These swamps are most abundant near the coast, a situation that has placed some cedar swamps at risk of elimination from frequent saltwater flooding due to rising sea level. Many former cedar swamps are now salt marshes, as evidenced by the presence of cedar logs and stumps in salt marshes. Cedar swamps usually develop on organic soils, and peat moss and other mosses are characteristic of the groundcover (Plate 19). Despite an apparent affinity for the coast, some cedar swamps can be found inland, well beyond any maritime influence.

The big differences in plant-community composition between coastal and inland swamps seem to be 1) the presence of hemlock and occasionally larch, yellow birch, black spruce, and balsam fir at inland sites and 2) the presence of coastal-plain species in the coastal swamps, notably Collin's sedge, Walter's sedge, and fetterbush. Otherwise, they are home to many of the same species, including red maple, black gum, pitch pine, white pine, sweet pepperbush, dangleberry, swamp azalea, highbush blueberry, mountain laurel, and skunk cabbage. Rosebay rhododendron is another common understory species. Surprisingly, black birch, usually characteristic of well-drained uplands, can be found in cedar swamps. Due to dense shading, the herb layer is sparse. Starflower, goldthread, Canada mayflower, and cinnamon fern are typical herbs, along with insectivorous sundews and pitcher plants (see Figure 5.6).

Like their nutrient-poor bog cousins, Atlantic white-cedar swamps harbor many rare species. Orchids—including dragon's mouth, swamp pink, white-fringed orchid, rose pogonia, and green-rein orchid—may be present. The endangered swamp pink, a member of the lily family, makes its home in cedar swamps; if you see a dense egg-shaped cluster of minute pink flowers at the top of a long naked flowering stalk, call your state natural heritage program. Another rare plant is curly grass fern, a small fern (less than 5 inches tall) with a comb-like appearance, found on hummocks near the base of cedar trunks. It has a disjunct distribution, probably the result of glaciation, occurring in New Jersey's Pine Barrens, then jumping to Nova Scotia and Newfoundland. Northern populations apparently survived in ice-free coastal refugia, whereas populations in between were eliminated by the glacier.

Hemlock Swamps. These swamps are among the least studied of the evergreen forested wetlands (Plate 20). Their distribution is not well known, but they have been observed in much of the Northeast. They develop on both organic and mineral soils, which they profoundly influence with their acidic leaf litter. Hemlocks are widely recognized for their ability to promote a soil-forming process known as "podzolization." The high acidity of their decomposing needles creates an organic-acid front that moves down through the soil, stripping the soil particles of whatever organic matter, iron, aluminum, and other minerals they might have. As a result, the stripped (eluvial) layer takes on a more uniform gray color, while a layer of accumulation of an organic-mineral complex (spodic horizon) forms at some depth. This layer, which may be cemented in whole or part, is the source of "bog iron." Podzolization occurs in both uplands and wetlands, wherever hemlocks and other acidic species dominate.

Hemlock swamps often have a pit-and-mound microtopography. The mounds are colonized by various plants, and the pits are usually barren due to prolonged inundation. The canopy of these swamps is often so dense as to prevent much light from reaching the forest floor, which also hinders understory growth. While hemlock forms nearly all of the canopy, other trees may be intermixed. Red maple is frequently a codominant. Yellow birch is also common and abundant enough in places to form mixed hemlock-birch stands. White pine, black gum, and gray birch can be found in some swamps. Red spruce is an associate in western

Maryland. Rosebay rhododendron is an important shrub there and in Pennsylvania, where it occurs along many mountain streams in hemlock ravines. The shrub and herb layers are usually absent or poorly represented, except where the canopy is open.

Mixed hemlock-hardwood stands usually have a richer flora. Spicebush, alders, mountain laurel, mountain holly, highbush blueberry, and winterberry may be present. The herb layer may include cinnamon fern, sensitive fern, skunk cabbage, goldthread, starflower, false hellebore, wild calla, Canada mayflower, marsh marigold, jewelweed, and blue flag. Pale touch-me-not, the yellow-flowering cousin of jewelweed, may also occur in these mountain swamps.

Pine Swamps, Lowlands, and Flatwoods. Pine swamps are quite interesting because they are dominated by species that also characterize the adjacent upland. White pine, pitch pine, and loblolly pine not only grow both in wetlands and uplands, but are dominant species in contiguous wet and dry areas, a situation that often makes pine swamps more difficult to recognize than other swamps with distinctly different vegetation from their surroundings. This is especially true when there is not an abrupt change in surface elevation. The relatively flat coastal plain poses some of the more challenging situations where "flatwoods" are the rule. There are wet flatwoods and dry flatwoods, but they are both flat, are often intermixed, and usually require an examination of the soil to distinguish. Flatwoods are also very important silviculture sites for pine plantations, which brings forestry interests to bear when it is proposed to include these forests as regulated areas.

White-pine swamps appear to be most abundant from eastern Massachusetts and Rhode Island to southern Maine but are scattered elsewhere in New England and in upstate New York. They grow on a range of soils from organic to sandy. The occurrence of white pine in wetlands surprises many people who recognize it as a species of dry upland sites. Yet studies have shown that white pine actually grows better on imperfectly drained sites in southeastern New England. White pine is an excellent example of a species that may have genetically different populations with varying adaptability to site wetness. Wetland ecotypes of white pine are well adapted for life in wetlands, as evidenced by the existence of quite impressive mature stands of white pines in a variety of hydric soils

and of young pines growing alongside cattails in mucky soils. Another possible explanation for the widespread distribution of white pine is ecological plasticity. Some species are simply highly adaptable by nature and can successfully adapt to a wide range of environmental conditions.

The ever-present red maple is the common tree associate in white-pine swamps. Many swamps can be considered white pine–red maple swamps or vice versa, depending on which species is more abundant. Other trees in lesser numbers include Atlantic white cedar, ash, red oak, white oak, and yellow birch. The shrub layer has the typical complement of swamp species, such as sweet pepperbush, highbush blueberry, swamp azalea, spicebush, winterberry, rosebay rhododendron, and others. Skunk cabbage, goldthread, cinnamon fern, royal fern, starflower, and club mosses are some of the herbs encountered, and peat mosses may be present in variable amounts.

Pitch-pine lowlands probably include some of the driest wetlands in southern New Jersey, where they predominate. Pitch pine, like its cousin white pine, grows under a wide range of conditions. It grows in seasonal ponds in the Pine Barrens and in New England bogs, in the driest sites in the region, and virtually everywhere in between. In wetlands, pitch pine is most abundant in seasonally saturated areas that have high seasonal water tables from late fall into spring. These "wet pitch-pine lowlands" represent a transition between other forested wetlands (Atlantic white-cedar swamps and hardwood swamps) and dry pitch-pine lowland forests (Figure 5.8).

Wet pitch-pine lowlands represent about 10 percent of the Pinelands National Reserve, which encompasses much of the Pine Barrens. Red maple and black gum are important associates, sometimes codominating to form pine-hardwood lowlands. Sweet gum becomes more significant in southernmost New Jersey. An occasional Atlantic white cedar, sassafras, American holly, gray birch, or willow oak may also colonize these lowland wetlands. The shrub layer may be quite thick in places. Leatherleaf may dominate low spots. Sheep laurel, dangleberry and other huckleberries, swamp azalea, sweet pepperbush, fetterbush, highbush blueberry, and other shrubs are often present. Mountain laurel and red chokeberry may occur in lesser numbers. Common greenbrier is also common. The presence of sand myrtle

Figure 5.8. A pitch-pine lowland, a type of wet flatwood, common in the Pine Barrens of southern New Jersey.

may indicate drainage. Herbs occur in variable amounts, but generally are not abundant. Ferns may be the most common herbs, including cinnamon, royal, sensitive, and chain ferns. Peat moss occurs in patches, especially at wetter sites. Recently burned or cleared areas often have bracken fern and turkey beard (a tall, handsome, small-white-flowered lily with elongate needlelike leaves in clumps that resemble a pine seedling).

Loblolly pine dominates many flatwoods along the coastal plain from the Delmarva Peninsula to Louisiana. Some wet loblolly flatwoods are seasonally flooded with pools of water in the depressions (Plate 21), while the majority are seasonally saturated with little or no evidence of surface ponding. The name "loblolly" is derived from a Native American word meaning "mud puddle," a most appropriate name for a pine found among mud puddles in winter. The wetter loblolly stands are usually well mixed with red maple and sweet gum. Shrubs include most of the species found in their northern counterparts, the pitch-pine lowlands. Royal fern, cinnamon fern, net-veined chain fern, slender spike grass, skunk cabbage, and some sedges are among the few herbs found

in these swamps. Peat mosses are common in depressions, where in winter and spring they are the green masses floating near the water surface in temporary pools.

Seasonally saturated loblolly flatwoods are wet during the winter and early spring, giving them the nickname "winter wet woods." Their summer dryness has put them at the center of controversy over the limits of wetlands, particularly regulated wetlands. Studies in Louisiana have shown that winter wetness enhances loblolly growth. Most of these pine flatwoods are periodically harvested for timber, so they are more or less managed systems. By early botanical accounts, these woods originally had a mix of associated hardwoods, chiefly three oaks (willow, basket, and white), American holly, and sweet bay. Today, red maple, black gum, sweet gum, and southern red oak have joined these species in mixed loblolly-hardwood stands or as associated species in wet loblolly flatwoods. Flatwood shrubs are typical coastal-plain swamp shrubs. Wax myrtle, a relative of sweet gale and northern bayberry, may form part of the shrub layer. Slender spike grass is a notable herb. Poison ivy and common greenbrier are also frequently present.

Figure 5.9. This Massachusetts vernal pool surrounded by upland woods is the breeding ground for hundreds of salamanders and woodland frogs.

Vernal Pools

Vernal pools are a rather select group of wetlands, depressions that are usually filled with water in winter and spring and often are dry by midsummer. They may be found within wetlands or as isolated basins surrounded by upland (Figure 5.9). They vary in size: many are smaller than a swimming pool, and some are as small as a child's wading pool. While many are shallow ponds, most pools are seasonally flooded just long enough for salamander eggs and larvae to mature. Year after year, generation after generation of salamanders and wood frogs come back to these pools to breed. The best time to see these creatures is when they're breeding in early spring (March and April), since they congregate by the hundreds in woodland vernal pools. To locate a breeding pool, listen for the quacklike calls of wood frogs in the evening or on cloudy days in late March or early April, or the high-pitched calls of spring peepers thereafter.

Buttonbush is one of the more common plants in vernal pools surrounded by deciduous woods or within red-maple swamps. Various marsh plants may also be found in low numbers in these fluctuating pools. Tussock sedge can occur in places. Willows and highbush blueberry may also grow along the pool margins. In pine stands, three-way sedge, wool grass, cinnamon fern, and various ericaceous shrubs, such as leatherleaf, may be found. Peat mosses may carpet the bottom of these pools.

Besides their significance as amphibian breeding sites, some vernal pools support rare plants. This is especially true of those on the coastal plain, which are characterized by widely fluctuating water levels and sandy soils. Many have enough water in most years to be called ponds. Dry years lead to a profusion of plants on the exposed shores. Plants like pipeworts, Walter's sedge, tall beak-rush, bog rush, spikerushes, sundews, yellow-eyed grass, meadow beauty, golden pert, false pimpernel, and St. John's-worts may be present. In wet years, only floating-leaved aquatics like water lilies and a few emergent species may exist.

Aquatic Beds

Many ponds, lakes, and slow-flowing rivers have aquatic beds in shallow water. Common

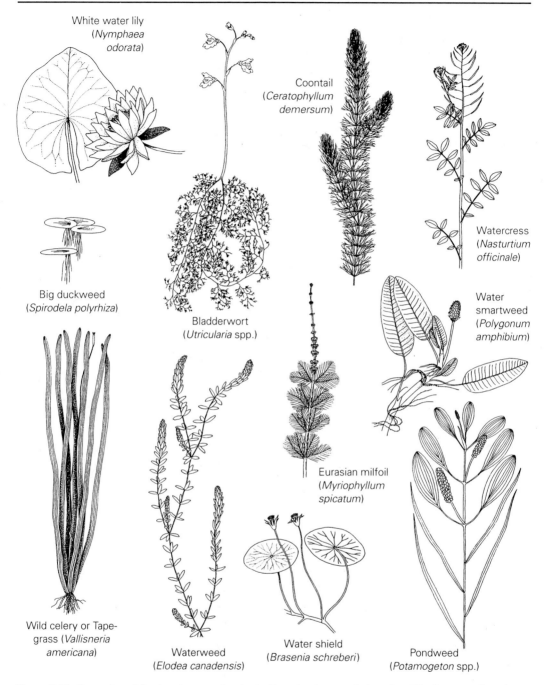

Figure 5.10. Examples of freshwater aquatic plants. See also bur-reeds (species 49), pipeworts (species 51), and spatterdock or yellow pond lily (species 58) in Chapter 9 for other common aquatics. (Source: Magee 1981 and Tiner 1988)

floating-leaved and free-floating plants include white water lily, spatterdock (yellow pond lily), water shield, pondweeds, and duckweeds (Figure 5.10). Submerged aquatic beds are not as readily observed, but most species send flowers to the surface for pollination. These species include wild celery (tape-grass) and naiads, and submergent forms of bur-reeds, pondweeds, pipeworts (in acidic streams), and arrowheads.

Additional Readings

Albert, D. 2003. *Between Land and Lake: Michigan's Great Lakes Coastal Wetlands.* Michigan Natural Features Inventory, Michigan State University Extension, East Lansing. Extension Bulletin E-2902.

Crum, H. 1988. *A Focus on Peatlands and Peat Mosses.* University of Michigan Press, Ann Arbor.

Dennison, M. S., and J. F. Berry (editors). *Wetlands: Guide to Science, Law, and Technology.* Noyes Publications, Park Ridge, NJ.

Golet, F. C., A.J.K. Callhoun, W. R. DeRagon, D. J. Lowry, and A. J. Gold. 1993. *Ecology of Red Maple Swamps in the Glaciated Northeast: A Community Profile.* U.S. Fish and Wildlife Service, Washington, DC. Biological Report 12.

Herdendorf, C. E., S. M. Hartley, and M. D. Barnes (editors). 1981. *Fish and Wildlife Resources of the Great Lakes Coastal Wetlands within the United States.* Volume I: Overview. U.S Fish and Wildlife Service, Biological Services Program, Washington, DC. FWS/OBS-81/02-v1.

Herdendorf, C. E., C. N. Raphael, and E. Jaworski. 1986. *The Ecology of Lake St. Clair Wetlands: A Community Profile.* U.S. Fish and Wildlife Service, National Wetlands Research Center, Washington, DC. Biol. Rep. 85(7.7).

Johnson, C. W. 1985. *Bogs of the Northeast.* University Press of New England, Hanover, NH.

Kozlowski, T. T. (editor). 1984. *Flooding and Plant Growth.* Academic Press, Orlando, FL.

Laderman, A. D. (editor). 1987. *Atlantic White Cedar Wetlands.* Westview Press, Inc., Boulder, CO.

Majumdar, S. K., R. P. Brooks, F. J. Brenner, and R. W. Tiner, Jr. (editors). 1989. *Wetlands Ecology and Conservation: Emphasis in Pennsylvania.* The Pennsylvania Academy of Sciences, Easton, PA. (See Chapters 10 through 12)

Metzler, K., and R. W. Tiner. 1992. *Wetlands of Connecticut.* State Geological and Natural History Survey of Connecticut, Department of Environmental Protection, Hartford, CT. In cooperation with the U.S. Fish and Wildlife Service, National Wetlands Inventory. Report of Investigations No. 13.

Mitsch, W. J., and J. G. Gosselink. 1993. *Wetlands.* Van Nostrand Reinhold Company, Inc., New York.

Sculthorpe, C. D. 1985. *The Biology of Aquatic Vascular Plants.* Koeltz Scientific Books, Konigstein, West Germany.

Teal, J., and M. Teal. 1969. *Life and Death of the Salt Marsh.* National Audubon Society and Ballantine Books, Inc., New York.

Tiner, R. W. 1991. The concept of a hydrophyte for wetland identification. *BioScience* 41: 236–247.

Tiner, R. W. 1985. *Wetlands of New Jersey.* U.S. Fish and Wildlife Service, Hadley, MA.

Tiner, R. W. 1985. *Wetlands of Delaware.* U.S. Fish and Wildlife Service, Hadley, MA, and Delaware Department of Natural Resources and Environmental Control, Dover, DE.

Tiner, R. W. 1989. *Wetlands of Rhode Island.* U.S. Fish and Wildlife Service, Hadley, MA.

Tiner, R. W., and D. G. Burke. 1995. *Wetlands of Maryland.* U.S. Fish and Wildlife Service, Hadley, MA, and Maryland Department of Natural Resources, Annapolis, MD.

Weller, M. W. 1981. *Freshwater Marshes.* Ecology and Wildlife Management. University of Minnesota Press, Minneapolis.

Wright, H. E., Jr., B. A. Coffin, and N. E. Aaseng (editors). 1992. *The Patterned Peatlands of Minnesota.* University of Minnesota Press, Minneapolis.

CHAPTER 6

Swamp Things
Wetland Wildlife

Wetlands are valuable habitats for many animals that have successfully adapted to their alternate wetting and drying. This function of wetlands has been recognized for some time as one of their more important contributions to human society. Millions of acres of wetlands have been purchased by government because they are important fish and wildlife habitats. For example, the U.S. Fish and Wildlife Service's National Wildlife Refuge system was originally established to protect critical wetland habitat for migratory bird species. Likewise, many state fish and wildlife management areas are wetlands. States have passed wetland laws in large part to protect these valuable habitats from destruction. The dependence of many commercially and recreationally important marine fishes on estuarine wetlands was a major reason for enactment of tidal-wetland protection laws by coastal states.

Thousands of animals have either adapted to the rigors of life in wetlands or modified their behavior to frequent wetlands at favorable times. Wildlife uses of wetlands include obtaining food, securing shelter, seeking protection from predators, reproducing, and rearing young. Each animal species has developed its own adaptations—behavioral, morphological, or physiological—to utilize wetlands.

Although insects and other invertebrates (animals without backbones) are probably the most abundant wetland animals, vertebrates are emphasized in this chapter because they are more familiar to people. This review is not comprehensive, but it should provide a good idea of how these creatures have adapted to wetlands and how they depend on wetlands.

Wetland Dependency

Animals requiring wetlands for survival are called "wetland specialists" or "obligate wetland animals." These animals either spend their entire lives in wetlands or have life stages that depend on wetlands (e.g., breeding). The latter species can be found in uplands during a good portion of their lives, while the former occur only in or in close proximity to wetlands. Wetland specialists include salt-marsh snail, aquatic insects, crayfish, fiddler crabs, dragonflies, damselflies, killifishes, spring peeper, bull frog, green frog, eastern tiger salamander, painted turtle, snapping turtle, bog turtle, water snake, bitterns, egrets, herons, ducks, geese, rails, seaside sparrow, long-billed marsh wren, water shrew, star-nosed mole, beaver, and muskrat. Most salamanders and wood frogs spend much of their lives away from wetlands, but require vernal pools for breeding. Wetland specialists are sensitive to environmental changes, and monitoring their populations may provide an indication of how wetland resources are faring in the face of development pressures and other disturbances.

Other wetland-dependent species derive substantial benefits from wetlands and are usually found in or near them. They are called "facultative wetland animals," for they spend most of their lives in and around wetlands. Many of these animals come to wetlands to feed, while others may nest more in wetlands than in similarly vegetated uplands. Facultative wetland species include diamond-backed terrapin, American toad, yellow warbler, yellowthroat, red-winged blackbird, raccoon, snowshoe hare, moose, and black bear (in the Poconos and New England). In some parts of New England and New York, white-tailed deer may be considered wetland-dependent because their survival depends on the availability of winter deeryards in evergreen forested wetlands, namely Atlantic white-cedar and northern white-cedar swamps. Without these wetlands, populations of deer would be eliminated or seriously reduced. Many northeastern bats are considered wetland-dependent because they feed heavily on insects flying over marshes at night. This is also

Figure 6.1. Wild ponies on the Delmarva Peninsula depend on salt marshes for food. These ponies were made famous by the book *Misty of Chincoteague.*

true for many swallows that use wetlands as important feeding grounds during daylight hours.

Many other animals frequent wetlands for cover, to obtain food, or for nesting (Figure 6.1). The last include birds that nest in wetlands but show no preference for them over uplands. All of these animals are considered facultative species since they use but do not require or prefer wetlands. Some birds in this category visit wetlands only during migration. Facultative species include box turtles, garter snakes, eastern moles, song sparrows, catbirds, robins, opposums, foxes, cottontail rabbits, black bear, and white-tailed deer. The full list is far too lengthly to enumerate. Many of these animals are familiar fauna of upland fields and forests. In urban areas, wetlands often provide the last remnant of wild habitat and serve as vital retreats for many animals that do not regularly frequent wetlands elsewhere.

Animal Adaptations for the Wet Life

Water on the surface or in the soil has a profound effect on animals. Many simply avoid these habitats, while others are attracted to wetlands for drinking water and the abundance of edible plants, insects, and other food. Still others have developed the necessary adaptations to cope with permanent or periodic inundation or saturated soil conditions and have successfully colonized wetlands. The obligate wetland animals have taken the challenge and have specifically adapted to aquatic or semiaquatic life.

Wildlife adaptations may be separated into three main types: 1) morphological (changes in body structure), 2) physiological (changes in internal biological processes), and 3) behavioral (changes in animal movement or activity, including reproduction). Most animals display more than one of these adaptations, especially those animals that are most dependent on wetlands. A few examples of morphological and behavioral adaptations are given below.

Morphological Adaptations

Morphological adaptations for life in and around water and swampland are perhaps the most obvious of the three kinds of adaptations. Many wetland animals have developed specialized structures to aid in swimming, diving, and feeding. Fish obviously have streamlined body shapes for efficient swimming. To promote faster and more efficient swimming, waterfowl, muskrat, otter, turtles, and frogs, among others, have developed webbed feet, whereas coot and pied-billed grebe have inflated pads on their toes. The wide-spreading toes of herons, egrets, and moorhens aid these birds in walking quickly and easily across unstable oozy mud flats, mucky wetland soils, and large floating leaves. The tails of muskrat and river otter serve as rudders to guide them through water. The otter's undercoat of thick, soft fur traps air bubbles that increase buoyancy, helping keep the otter afloat. The narrow body of the clapper and other rails facilitates their movement between cattail stems and other reeds. The bittern's camouflage coloration (brownish stripes), coupled with its behavior of "freezing" (thrusting its bill to the sky and standing still), helps this bird blend naturally into the cattails and other reeds.

Aquatic insects have specialized features for swimming. Diving beetles, whirligig beetles, and backswimmers have oarlike hind legs. Water striders have legs with modified claws (at the ends) and a covering of short hairs that help keep them on the water's surface. Moreover, the unique body shape of the water strider distributes its weight to minimize surface pressure. Aquatic nymphs of dragonflies swim by jet propulsion (taking water into their abdomen and releasing it quickly).

The beaver is a remarkably adapted wetland creature (Figure 6.2). Its webbed feet are clearly designed to aid in swimming. Its flattened

Figure 6.2. The beaver, nature's engineer, is one of the best-adapted wetland mammals. (Susie von Oettingen)

tail is a curious structure that not only serves as a rudder but helps regulate body temperature. The beaver also uses its tail to give warning to fellow beavers by slapping the water's surface. As noted before, the beaver does not use its tail to tap down mud when building dams and lodges, but instead uses its front paws and nose. Beavers have large livers and lung capacities, allowing them to hold enough air and aerated blood to stay underwater for more than 15 minutes! Their ears and nostrils can be closed when diving, and special transparent lids cover the eyes, protecting them while allowing the beaver to see underwater. To enable them to feed when submerged, beavers have lips that they can close behind their teeth. Musk glands produce oils that make their fur waterproof.

Several types of bills evolved in birds to maximize feeding success in water and mud. The broad flat bills of ducks are designed for feeding on plants and aquatic invertebrates, while the extremely long pointed bills of herons and egrets are for catching fish, crabs, frogs, and other species. The long curved bills of ibises and curlews are ideal for probing the soil for invertebrates, as are the long thin bills of woodcock and snipe. Many wetland birds have developed long necks for reaching food underwater. Swans, geese, storks, herons, and egrets are striking examples. The long legs of wading birds allow them to hunt in shallow water.

The whirligig beetle has specialized features for feeding at the water's surface. Its eyes are divided into two parts, one above the other. One set is for looking at the water's surface, whereas the other is for viewing underwater. Its upper body is also water-repellent, while the lower half is not; this permits the whirligig to live and feed

on the water's surface without expending much energy.

To survive in cold water for long periods, animals must have some means of keeping warm and resisting water penetration. Some animals, like many aquatic invertebrates and turtles, have hard shells. Water birds have a dense set of fine feathers next to their bodies that keeps water out. Other animals have developed waterproof skin. Beaver, muskrat, river otter, mink, and their relatives have fur that is waterproofed with oils from their musk glands.

Most animals that spend their entire lives in water have gills: fish, clams, oysters, crabs, crayfish, aquatic snails, and larval stages of aquatic insects (e.g., dragonflies) and amphibians. Surprisingly, many aquatic animals lack gills but are still able to spend considerable time underwater. Numerous aquatic invertebrates have water-repellent hairs covering their bodies that enable them to carry air bubbles below. For example, the fisher spider surrounds its hairy body with a diving bell–like air bubble and can stay underwater for more than half an hour. Backswimmers can stay submerged for up to 6 hours with just one bubble. The eggs of the bog copper (a butterfly common in cranberry bogs) have an internal air space that allows them to withstand short periods of flooding. Some insects have developed specialized structures for breathing underwater: the air tubes of mosquito larvae, water scorpions, and rat-tailed maggots are examples. Other aquatic insect larvae can pierce submerged parts of wetland plants and obtain air from aerenchyma (air-filled) tissue, as the cattail mosquito larva does.

Most amphibians have lungs and moist skin to breathe air and exchange gases with the atmosphere. Some salamanders lack lungs, breathing instead through skin and membranes in the mouth and throat. Because their moist skin may dry when exposed to air for extended periods, most amphibians live in wetlands, near bodies of water, underground, or beneath rocks and woody debris, or are active primarily at night, especially in hot, dry weather.

Diving birds, including grebes, loons, and mergansers, have developed specialized lungs that are inflated for buoyancy when the bird is swimming on the surface but deflated when the bird is diving. This coupled with physiological adaptations such as hemoglobin-rich blood (increases oxygen-holding capacity) and lowered heart rate and metabolism permits longer dives than birds can typically do.

Behavioral Adaptations

Behavioral adaptations are perhaps the most widespread animal adaptations to wetlands. Only a few examples are given here, although a book could be written on this topic. Many animals frequenting wetlands are in search of food, just like people who venture into swamps to pick blueberries. Wild turkeys may feed in winter on spore capsules of sensitive fern. Besides occasional feeding, some species visit wetlands for temporary shelter or for specialized activities. Certain fish spawn in floodplain forested wetlands during spring floods. Terrestrial wildlife simply avoid the aquatic or semiaquatic conditions by living above ground in the vegetation or by using wetlands only during dry or low-water periods. Some obligate wetland animals must escape high water, as demonstrated by the air-breathing salt-marsh snail, which eludes high tide by climbing up the stems and leaves of smooth cordgrass. This may be called "stress avoidance." Pheasant may seek winter cover in frozen wetlands, and use wetlands at other times when they are not flooded. White-tailed deer frequently use wetlands for food and cover but avoid them during high water. Most songbirds are wetland generalists, using wetland vegetation for nesting but also nesting in upland communities of similar vegetative structure and showing no distinct preference for wetlands. There are many more examples of terrestrial animals using wetlands, yet these creatures have not really adapted to life in wetlands so much as they have adapted their behavior to reap the bounty of wetlands while avoiding the seasonal stresses. Interestingly enough, as suggested above, people exhibit the same types of behavior (see Chapter 7).

The most specialized wetland animals are stressed more by dry conditions. While clearly suited for an aquatic life, they must adapt to the seasonal or occasional drawdown typical of many wetlands. Frogs, turtles, and killifish, for example, burrow in mud (aestivation, summer hibernation) to survive droughts.

Northern winters are harsh for all animals, wetland and upland creatures alike. Most animals either hibernate or migrate to escape these adverse conditions. The meadow jumping mouse hibernates in the banks of cranberry bogs. Most reptiles and amphibians are dormant during the winter in the Northeast, since they are cold-blooded and cannot function during extremely cold weather. Timber rattlesnakes in New Jersey's Pine Barrens hibernate in Atlantic white-cedar swamps. Turtles burrow in muddy bottoms of ponds and lakes, and diamond-back terrapins burrow in subtidal muds of coastal embayments. Frogs also hibernate underwater in the mud, while most salamanders burrow in the soil or hide beneath rotting logs, rocks, and other debris in moist shaded forests. It is interesting to note that the wood frog, the most northern-ranging frog, survives the bitter arctic cold by creating a type of antifreeze. It increases the amount of glucose (sugar) in its cells which lowers their freezing point. Wood frogs can withstand temperatures to $-42.8°F$ or below (even to $-53.6°F$ in Alaska) when up to 70 percent of their body fluids freeze. In Alaska, these frogs remain frozen from November through April—that's quite a feat! Scientists are studying the wood frog's adaptation in search of ways to preserve human donor organs for organ transplant.

Migratory birds avoid northern winters by flying south to wintering grounds with favorable conditions, returning to northern wetlands the following spring. Some northern birds find the Northeast coastline favorable in winter due to the presence of open water and available food. Black ducks, loons, mergansers, and brants are examples. Black ducks are a common sight in many salt marshes in winter, where they feed on snails. During winter storms, these ducks seek out hardwood swamps for refuge.

White-tailed deer frequent northern white-cedar swamps, Atlantic white-cedar swamps, and other evergreen forested wetlands for winter shelter and food. These swamps are called "deeryards" because deer are concentrated in them in winter. Moose can also be found in these areas.

Wetlands as Wildlife Habitat

Wetlands of all sizes, shapes, and types provide important wildlife habitats. Larger animals typically require lots of space, preferring larger wetland complexes, sometimes in combination with relatively unbroken tracts of forest. Migratory species need a series of wetlands along their travel routes for feeding and resting. Many species can make a home out of small wetlands. One of the smallest wetlands, a vernal pool a few feet wide, can provide homes for hundreds of fairy shrimp and serve as a breeding area for salamanders, wood frogs, and spring peepers. Wetland regulations tend to protect larger wetlands, especially those associated with streams, while small isolated wetlands are largely forgotten. Many of these wetlands are critical breeding grounds for amphibians. Others are vital for waterfowl, because they provide the earliest open-water habitat in spring. They also provide vast quantities of invertebrates for feeding and isolated habitats for breeding. Recent studies suggest that areas with numerous small scattered wetlands are important to local populations of turtles, small mammals, and small birds. Losses of these wetlands increases dispersal travel distances and may adversely affect survival of the animals. The nation's wildlife need a wide range of wetland types and sizes to survive, reproduce, and thrive.

Fish and Shellfish Habitat

Wetlands are critically important to fishes and other aquatic animals in many ways. Hydrophytic vegetation produces oxygen for the water, as do aquatic plants in an aquarium. Shallow-water marshes and aquatic beds provide habitats for aquatic invertebrates, an important part of the diet of many fishes and other wildlife. These wetlands also serve as spawning grounds and nursery areas for fishes. Young fish seek shelter among the plant leaves and stems, thereby avoiding predation. Freshwater fishes are almost entirely dependent on wetlands; common species in the Northeast are chain and grass pickerel, basses, crappies, bluegills, bullheads, shiners, and carp (Figure 6.3). Pickerel lay their eggs over vegetation, and the young seek out shallow-water marshes for protection from predators. Adults of anadromous alewife and blueback herring migrate from the open ocean into freshwater rivers to spawn. Spawning has been observed in tidal fresh marshes of the Delaware River drainage, and the young use these wetlands as nursery grounds.

Nearly 180 species use Great Lakes coastal wetlands for spawning, nursery grounds, and feeding areas. Permanent marsh residents include northern pike, muskellunge, smallmouth bass, largemouth bass, rock bass, walleye, yellow perch, white perch, freshwater drum, log perch, black bullhead, white crappie, gizzard shad, banded killifish, and various shiners and minnows. Most of these species spawn during high water periods in the spring. Lake sturgeon and channel catfish are among other species that migrate from deeper waters to use these marshes for spawning. The marshes produce an enormous supply of aquatic invertebrates—food for young fishes—using the marshes as nursery grounds.

About two-thirds of the major commercial fishes in U.S. waters depend on estuaries and salt marshes for spawning or nursery grounds. This number approaches 90 percent for Delaware Bay and Chesapeake Bay fisheries. Some of these wetland-dependent species are menhaden (pogies), bluefish, fluke, white perch, weakfish (sea trout), mullet, croaker, striped bass, and drum. Forage fishes like bay anchovies, killifishes, mummichogs, and Atlantic silversides are common in tidal creeks. Forty species were found spawning in New Jersey coastal embayments along the Atlantic coast, while 136 species used these estuaries as nursery grounds.

Commercially important invertebrates also depend on wetlands. Bay scallops require eelgrass beds for larval attachment and growth. Scallop larvae stay attached for about a month while growing, then become motile. Blue crabs and grass shrimp depend on the marsh-estuary complex for survival. They are common in tidal creeks and frequent salt marshes at high tide.

Amphibian Habitat

Frogs and salamanders are descendants of the first animals that emerged from water to live on land over 350 million years ago. Amphibians can live on land, but nearly all must return to water to breed. Most of the approximately 190 species of North American amphibians are wetland-dependent. Common frogs in the Northeast and Great Lakes region are bull frogs, green frogs, pickerel frogs, leopard frogs, chorus frogs, and wood frogs (Figure 6.4). Common wetland salamanders include spotted salamanders, bluespotted salamanders, Jefferson salamanders, red-spotted newts, dusky salamanders, and marbled salamanders.

The red-spotted newt has an unusual life

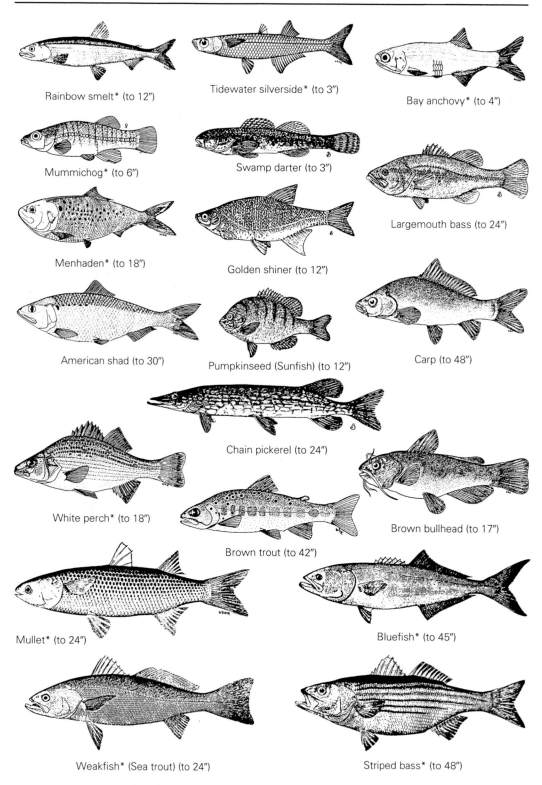

Rainbow smelt* (to 12")

Tidewater silverside* (to 3")

Bay anchovy* (to 4")

Mummichog* (to 6")

Swamp darter (to 3")

Largemouth bass (to 24")

Menhaden* (to 18")

Golden shiner (to 12")

American shad (to 30")

Pumpkinseed (Sunfish) (to 12")

Carp (to 48")

Chain pickerel (to 24")

White perch* (to 18")

Brown trout (to 42")

Brown bullhead (to 17")

Mullet* (to 24")

Bluefish* (to 45")

Weakfish* (Sea trout) (to 24")

Striped bass* (to 48")

Figure 6.3. Examples of fish found in and around coastal and inland wetlands. Coastal species are marked with an asterisk (*). White perch occur in both situations. Maximum lengths (for southern New England) are given in parentheses. (Sources: Whitworth et al. 1968; Thomson et al. 1971)

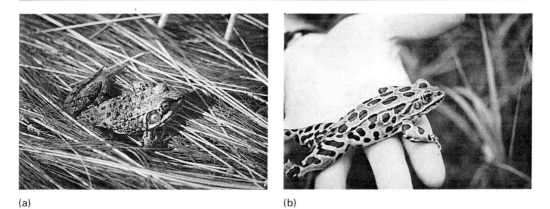

(a)   (b)

Figure 6.4. The green frog (a) is a familiar sight along pond shores and in marshes, while the leopard frog (b) can be found leaping across wet meadows.

history. It begins in water as an aquatic larva with gills. After about 3 months, it loses its gills and is transformed into a terrestrial salamander, the red eft. It lives on land until reaching sexual maturity, when it returns to the waters of lakes, ponds, and marshes as an aquatic adult without gills.

For breeding, wetland-dependent amphibians congregate in wetlands or vernal pools. Most salamanders are spring breeders, and many enter the edges of ice-covered ponds in February and March. Some amphibians, including the spotted salamander and wood frog, return every year to the same pools to breed. Other vernal-pool breeders are marbled salamander (the only fall breeder), Jefferson salamander, eastern spadefoot, chorus frogs, gray tree frog, spring peeper, and Pine Barrens tree frog. American and Fowler's toads breed in vernal pools, but also reproduce in ponds and lakes along with green frogs, bull frogs, and cricket frogs.

Hundreds of larvae may be produced in a single vernal pool only a few feet wide and long. In a one-acre pond in eastern Massachusetts, nearly 14,000 adult amphibians were counted, including about 8,000 spring peepers and over 4,500 wood frogs. A two-acre pond in western Massachusetts had an estimated 5,000 to 10,000 spotted salamanders and several times as many wood frogs and spring peepers during the breeding period. The animals are so abundant at this time that most researchers don't bother counting them. This reproductive strategy has been called "predator swamping" because more young are produced than predators can eat, assuring survival of the next

generation. (See Chapter 14 for best management practices on how to protect vernal pools and surrounding habitats.)

Reptile Habitat

Turtles, lizards, alligators, and snakes are the most familiar reptiles to most Americans. These cold-blooded animals are the modern-day relatives of dinosaurs that ruled the earth for over 200 million years (from 280 to 65 million years ago). They evolved from the amphibians by developing protective skin to prevent water loss and eggs with hard or leathery coatings, which allowed them to colonize and populate lands far removed from water. Unlike amphibians, reptiles do not have to return to water for breeding, although many reptiles are wetland and aquatic animals. Nearly all reptiles (except for sea snakes) have their young on land. Even aquatic turtles must leave water to lay their eggs on land.

In the Northeast, several turtles and a few snakes are swamp reptiles, while lizards are mostly terrestrial creatures. The only lizards that might be found along the edges of swamps are the five-lined skink and the northern fence lizard. The American alligator, the largest reptile in the United States, is wetland-dependent, but does not occur north of North Carolina.

The eastern painted turtle is the region's most common turtle, usually observed basking on logs, rocks, or the banks of ponds and sluggish rivers. The snapping turtle, the largest reptile in the Northeast (maximum size about 70 pounds and shell length of nearly 20 inches),

Figure 6.5. The spotted turtle is found in a variety of wetlands from marshes and wet meadows to forested wetlands and bogs. Though not endangered or threatened, it is listed as a species of concern by Massachusetts.

is more secretive. Quite abundant in bodies of fresh water, it has been seen in salt-marsh creeks on occasion. More typical of salt marshes, however, is the diamond-back terrapin. Spotted turtles can be found in wet meadows and swamps (Figure 6.5). Box turtles are principally terrestrial, but also frequent wetlands. The bog turtle, endangered in several states and proposed federally for threatened status, prefers wet meadows. Other turtles associated with Northeast wetlands are the eastern mud, eastern spiny softshell, red-bellied, stinkpot, and wood turtles.

Three snakes are common in Northeast wetlands: garter, ribbon, and water snakes (Figure 6.6). Timber rattlesnakes overwinter in Atlantic white-cedar swamps in New Jersey's Pine Barrens, and copperheads occur along the edges of coastal-plain swamps. Other less-common

snakes inhabiting wetlands in the region include black rat, eastern king, eastern worm, northern black racer, northern red-bellied, queen, brown, and smooth green snakes.

Bird Habitat

Roughly half of the 674 species of breeding birds in North America may have some dependency on aquatic habitats (including oceans and riparian habitats). About 26 percent are wetland obligate species and 5 percent have some association with wetlands (Dave Davis, pers. comm. 2004). The former group includes species like least bittern, green heron, snowy egret, mallard, snow goose, clapper rail, semipalmated plover, solitary sandpiper, common snipe, laughing gull, least tern, belted kingfisher, alder flycatcher, marsh wren, prothonotary warbler, and swamp sparrow, while the latter group includes palm warbler, hooded warbler, Acadian flycatcher, tree swallow, red-winged blackbird, veery, common yellowthroat, song sparrow, Canada goose, barred owl, and red-shouldered hawk. These numbers are conservative since they emphasize breeding and don't include upland birds that use wetlands for feeding on insects or nectar (e.g., ruby-throated hummingbird) or those using wetlands during migration or in winter (e.g., ring-necked pheasant).

Wetlands provide habitats for a diverse assemblage of avian species (Figure 6.7). The number of species found in wetlands along rivers tends to be higher than in upland habitats outside river valleys. Many birds, especially waterfowl, shorebirds, and colonial nesting birds, are wetland specialists, yet most birds are faculta-

Figure 6.6. Northern water snakes are often mistaken for poisonous water moccasins or cottonmouths, but the latter species do not occur in the Northeast.

Figure 6.7. Many birds besides waterfowl, shorebirds, and wading birds make use of wetlands, including songbirds like the yellow warbler. (Fred Knapp)

tive users of wetlands, showing no preference over upland habitats. Some species are wetland specialists only during the breeding season, such as the palm warbler, which breeds in northern New England bogs and is a generalist during migration. Avian breeding activity is greatest from May through August, when insects, other invertebrates, fruits, and seeds are readily available. The coming of winter, with its cold temperatures and low food supplies, forces most birds to migrate south to more favorable climates where food and water are accessible. Many birds use wetlands during migration for resting and feeding, even though they may not frequent wetlands at other times.

Salt marshes are nesting grounds for several birds, including laughing gulls, Forster's terns, sharp-tailed sparrows, clapper rails, black ducks, blue-winged teals, marsh hawks, and seaside sparrows. Wading birds such as herons, glossy ibises, and egrets feed in these marshes and nest in and adjacent to them. During fall and spring, migrating snow geese, wading birds, shorebirds, and peregrine falcons can be observed. Shorebirds are especially abundant on mud flats and pannes in the high marsh, where they feed on various invertebrates. Swallows and purple martins often can be seen flying over the marshes feeding on air-borne insects. Winter residents in Northeast salt marshes include black ducks, mallards, scaup, snow geese, brant, and Canada geese.

Tidal fresh marshes provide important nesting and feeding areas for many birds. The wild rice that dominates many marshes is prime food for red-winged blackbirds, grackles, and bobolinks, among others. Forty-eight species of birds have been seen nesting in New Jersey's tidal marshes. Typical nesting species include red-winged blackbirds, long-billed marsh wrens, least bitterns, clapper rails, American goldfinches, common yellowthroats, swamp sparrows, indigo buntings, yellow warblers, alder flycatchers, wood ducks, green-backed herons, and common moorhens.

Inland wetlands are used for nesting, feeding, and resting areas by hundreds of species, including year-round residents and migratory species. When most people think about wetlands and birds, they envision ducks, and when they think about duck production, the Prairie Potholes region should come to mind. This region, in the upper Midwest (the Dakotas and western Minnesota) and in adjacent Canada, is

pockmarked with millions of glacially scoured basins, ranging in size from less than a tenth of an acre to several hundred acres. The Pothole region accounts for only 10 percent of the continent's duck-breeding area, yet produces about half of the continent's ducks. Some of these ducks, like canvasbacks, make their way to the Northeast in winter. Wetlands in arid regions serve as stepping stones for birds migrating to northern breeding grounds. They are critical to the survival of these species.

Most Northeast waterfowl migrate within the "Atlantic Flyway" along the Atlantic Ocean from Canada to the Gulf of Mexico. Midwestern waterfowl tend to travel within the "Mississippi Flyway" (Figure 6.8). Ducks and geese migrate along these corridors, spending winters in more southerly locations, then flying north to summer breeding areas. In fall and spring, many ducks, geese, shorebirds, and others can be seen on their annual migrations.

Not all waterfowl migrate out of the region. Several species, including mallard, black duck, and the ubiquitous Canada goose, are year-round residents, while some of their members migrate within the region. Many black ducks overwinter in Northeast coastal marshes and go north or inland to breed.

The wood duck, America's most beautiful duck, is another familiar waterfowl in the region. Wood ducks are often seen in mated pairs in woodland streams. Like woodpeckers, they make their nests in cavities of dead trees in or near open water. They also nest in man-made nesting boxes erected in marshes.

Wading birds, shorebirds, some raptors, and a large variety of songbirds are also wetland species. Typical marsh birds include green-backed heron, great blue heron, spotted sandpiper, Virginia rail, sora, red-winged blackbird, swamp sparrow, and long-billed marsh wren. Woodcock feed on earthworms in alder swamps. Barred owl (swamp owl), northern harrier (marsh hawk), red-shouldered hawk (swamp hawk), and osprey (fish hawk) are typical raptors feeding on wetland animals. Veeries and yellowthroats use forested wetlands and shrub swamps, respectively, for nesting and feeding. Coastal-plain hardwood swamps are utilized by nearly fifty breeding species, which is more than nest in either Atlantic white-cedar swamps or mixed hardwood-cedar swamps. Among the important wetland breeders are great crested flycatcher, pine warbler, towhee,

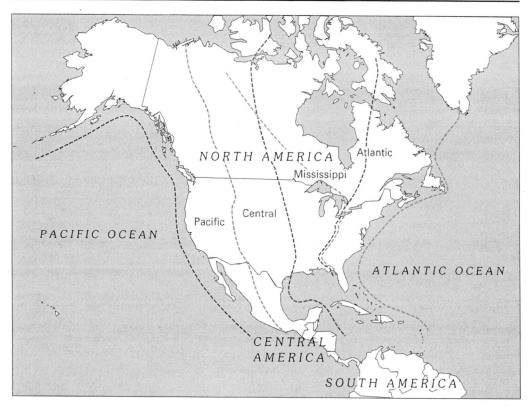

Figure 6.8. Major flyways in the North and Central Americas. Migratory birds overwinter in southern regions and migrate to northern breeding grounds along these corridors. There is some migration from west to east and back by some species. (Source: Stewart 1996)

chickadee, titmouse, prothonotary warbler, scarlet tanager, vireos, acadian flycatcher, ovenbird, black and white warbler, gray catbird, common yellowthroat, brown creeper, hooded warbler, and black-throated green warbler. Other breeding birds in cedar swamps are eastern wood pewee, wood thrush, parula warbler, yellow warbler, redstart, and song sparrow.

Mammal Habitat

Most mammals are terrestrial animals, yet a few species, such as river otter, beaver, muskrat, and star-nosed mole, spend virtually their entire lives in wetlands. Other mammals, such as raccoon, oppossum, striped skunk, cottontail rabbit, and snowshoe hare, visit wetlands in search of food, since an abundance of fruits, nuts, insects, and other food items can be found. It is not surprising to see many terrestrial animals making forays into marshes and swamps. Red and gray foxes often prowl the edges of marshes in search of food. In summer, bats (including the

eastern pipestrelle, little brown bat, and silver-haired bat) can be seen flying at dusk over marshes and swamps, feeding on flying insects. Some mammals are nocturnal, including southern flying squirrels, many rodents, mink, long-tailed weasels, and raccoons.

Muskrats and beavers are the most familar wetland mammals. Muskrats are much more common and wide-ranging, inhabiting both coastal (brackish) and inland marshes. Their lodges are often visible in cattail marshes. Beavers are found in woodlands, usually in more mountainous areas, but they also live in the coastal plain, where, feeding selectively on hardwoods, they help maintain Atlantic white-cedar swamps. Beaver ponds and their stick-built lodges and dams are common sights through much of New England. These ponds are valuable habitats for myriads of creatures like wood ducks (which construct nests in cavities of dead trees), great blue herons (which build nests of sticks in the treetops), amphibians, turtles, and aquatic invertebrates.

Figure 6.9. White-tailed deer frequent wetlands throughout the Northeast. (Jim Hudgins)

Larger animals also visit swamps, and in some areas, wetlands form a critical component of their habitat (Figure 6.9). White-tailed deer are so plentiful in some forested wetlands that a distinctive browse line can be seen in the forest. (Their dependency on evergreen swamps in winter has already been noted.) In some mountainous areas of the Northeast, black bears are common denizens of swamps, where they feed on skunk cabbage in spring and blueberries and other plants in summer. In the past, bears inhabited coastal-plain swamps, but they have been extirpated from most areas. Where bears still exist, wetlands are critical to their survival.

Additional Readings

Benyus, J. M. 1989. *Northwoods Wildlife: A Watcher's Guide to Habitats.* NorthWood Press, Inc., Minocqua, WI.

Brooks, R. P., D. A. Devlin, and J. Hassinger. 1993. *Wetlands and Wildlife.* Pennsylvania State University, College of Agricultural Sciences, School of Forest Resources, University Park, PA.

Caduto, M. J. 1985. *Pond and Brook: A Guide to Nature Study in Freshwater Environments.* Prentice-Hall, Inc., Englewood Cliffs, NJ.

DeGraaf, R. M., and D. D. Rudis. 1986. *New England Wildlife: Habitat, Natural History, and Distribution.* USDA Forest Service, Northeastern Forest Experiment Station, Broomall, PA. General Technical Report NE-108.

Majumdar, S. K., R. P. Brooks, F. J. Brenner, and R.W. Tiner, Jr. (editors). 1989. *Wetlands Ecology and Conservation: Emphasis in Pennsylvania.* Pennsylvania Academy of Sciences, Lafayette College, Easton, PA (see Chapters 13–19).

Niering, W. A. 1984. *Wetlands.* Alfred A. Knopf, Inc., New York.

Weller, M. W. 1981. *Freshwater Wetlands: Ecology and Management.* University of Minnesota Press, Minneapolis.

CHAPTER 7

Swampland at Work
Wetland Functions and Values

Wetlands are important to people as well as to fish, wildlife, and specially adapted plants, providing highly valued goods and services for society. After centuries of abuse and neglect, marshes, swamps, and other wetlands are now recognized as one of America's most valuable natural resources. Their location along the nation's waterways, their frequent prolonged wetness, and other conditions endow wetlands with many qualities that other lands do not possess. Wetlands benefit the general public by protecting property from floods, stabilizing shorelines, helping cleanse the nation's waters, reducing siltation in navigable waterways, and providing critical fish and wildlife habitat, among other things. This is why laws have been passed to protect wetlands or control their development on both public lands and private property, whereas most upland habitats have not been afforded similar protection (Chapter 9).

In general terms, functions are physical, chemical, and biological processes and their influences on plants, animals, and hydrology. Major wetland functions include water storage, maintenance of high water tables, nutrient cycling, sediment retention, accumulation of organic matter, and maintenance of plant and animal communities (Table 7.1). These functions are not necessarily performed continuously throughout the year, but most operate on a frequent and recurring basis. Moreover, not all wetlands provide all functions. The fact that each wetland has a limited capacity to perform certain functions must also be appreciated. Some functions are truly unique to wetlands, such as creating distinctive plant communities and providing habitat for wetland-dependent species, whereas other functions may be performed by uplands at times. For example, upland portions of floodplains temporarily store water and trap sediments during extraordinary flood events.

Several properties or characteristics of wetlands allow them to perform a range of functions. Landscape position, climate, and landform (geomorphology) greatly affect wetland functions and vegetation and soil development, which in turn also influence wetland functions. Wetlands located along rivers and streams are in prime landscape positions to execute many functions. The depressional nature of many wetlands also promotes certain functions that wetlands on slopes cannot accomplish.

Functions are value neutral, meaning that they take place whether or not they are considered important by society. In contrast, values represent people's opinions and are subject to change over time, even within short periods. A hunting-and-gathering society should have a different view of wetlands or the natural environment in general than an agrarian or industrial society. A whole host of variables enter the picture when speaking of values, including cultural backgrounds, education, scientific knowledge and certainty, landownership, and regional differences. American public opinion toward wetlands has changed markedly in the past thirty years, from a view of wetlands as wastelands to considering wetlands a valuable natural resource (see Chapter 9).

Wetlands largely operate as a holistic, integrated system within a watershed, waterfowl flyway, or ecoregion. In other words, individual wetlands work together to provide valuable functions. While it is true that any wetland will produce some benefits whether or not other wetlands exist, the real strength of wetlands is in their collective capacity to perform watershed- and region-scale functions. A small depressional wetland stores a little water, but a network of small wetlands can store great volumes of water and in the process provide significant flood protection for downstream property owners. The presence of many small wetlands in an area may be vital for maintaining local populations of certain plant or animal species. The value of the system or complex of wetlands is greater than the sum of its individual parts. Each wetland can

Table 7.1. Major wetland functions and some of their values.

Function	Value
Water storage	Flood- and storm-damage protection, water source during dry seasons, groundwater recharge, fish and shellfish habitat, water source for fish and wildlife, recreational boating, fishing, shellfishing, waterfowl hunting, livestock watering, ice skating, nature photography, aesthetic appreciation
Slow water release	Flood-damage protection, maintenance of stream flows, maintenance of fresh and saltwater balance in estuaries, linkages within watersheds for wildlife and water-based processes, nutrient transport, recreational boating
Nutrient retention and cycling	Water-quality renovation, peat deposits, increases in plant productivity and aquatic productivity, decreases in eutrophication, pollutant abatement, global cycling of nitrogen, sulfur, methane, and carbon dioxide, reduction of harmful sulfates, production of methane to maintain Earth's protective ozone layer, mining (peat)
Sediment retention	Water-quality renovation, reduction of sedimentation of waterways, pollution abatement (contaminant retention)
Provision of substrate for plant colonization	Shoreline stabilization, reduction of flood crests and water's erosive potential, plant-biomass productivity, peat deposits, organic export, fish and wildlife habitat (specialized animals, including rare and endangered species), aquatic productivity, trapping, hunting, fishing, nature observation, production of timber and other natural commodities, livestock grazing, scientific study, environmental education, nature photography, aesthetic appreciation

be viewed as a link in a chain, and the strength of the chain (translated into the ability to perform a given regional, flyway, or watershed function) is only as strong as its weakest link. Removing links without question weakens the chain, and thereby reduces the functional capacity of the system. This functional interdependence is often easily overlooked when one is viewing an individual wetland.

Water-Quality Protection

While the thought of drinking swamp water might make most people cringe, they'd be surprised to learn that water from Atlantic white-cedar swamps was a valuable resource for mariners of the past. Before sailing back to Europe, sea captains would have their crew fill wooden barrels with cedar swamp water. This tea-colored water (the color was due to tannic acid from plant decomposition) stayed fresh longer than other water, so sailors could drink it long after they departed land on their ocean journeys. Of course, not all swamp water is potable. Marshes typically release organic matter seasonally, which adds particulates to water, giving it a light brownish coloration. Yet these particulates provide food for microscopic and

other small aquatic organisms that are at the bottom of the food chain and help make wetlands the productive environments they are.

Nutrient Recycling and Pollution Abatement

While most of today's wetlands do not produce spring water, they help improve water quality by cycling various elements and performing other functions. Wetland plants remove nutrients, especially nitrogen and phosphorus, from flooding waters and help prevent eutrophication (over-enrichment) of natural bodies of water. Some of the plant uptake is seasonally released back to the water when wetlands are reflooded after the herbs have died back and leaves have fallen from woody species. Much of the biomass of herbaceous plants, however, is in their underground parts (roots and rhizomes), where longer-term storage can occur. Some wetland trees have demonstrated higher heavy-metal concentrations in their tissues than nonwetland trees. These mechanisms help wetlands protect the quality of existing waters and water supplies, a major reason that wetlands are protected by regulations.

Wetlands are important recyclers because anaerobic conditions promote biochemical processes that change the nature of certain elements (e.g., nitrogen, iron, manganese, sulfur, and carbon). All of these elements are converted from oxidized forms to reduced forms under these low-oxygen conditions. Wetland soils are sinks for many elements and heavy metals. Some of these elements are transformed to useable materials. For example, nitrogen is a vital, often growth-limiting, element for plants. Alders and sweet gale are nitrogen-fixing plants; they possess bacteria that convert nitrogen gas to ammonia which is then transformed into nitrate that can be taken up by plants or transported by groundwater (Figure 7.1). One of the more important transformations performed by wetlands is called "denitrification." In this process, anaerobic soil microbes convert nitrate to various nitrogen oxides and eventually nitrogen gas that is released to the atmosphere completing the nitrogen cycle. Nitrate removal occurs year-round in many wetlands, even in southern New England where more than 80 percent of nitrate was removed by streamside wetlands during both the growing season and dormant season. The combination of water, soil, plants, and microbes creates a biologically active substrate.

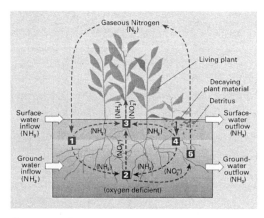

Figure 7.1. Cycling of nitrogen in wetlands. (Source: Carter 1996)

Explanation:
1. Bacteria change gaseous nitrogen (N_2) to ammonia (NH_3).
2. Bacteria change ammonia (NH_3) to nitrate (NO_3) (another form of nitrogen that the plant can use).
3. Plant roots absorb ammonia and (or) nitrate formed in processes 1 and 2 and incorporate nitrogen into the plant proteins and nucleic acids that nourish the plant.
4. Nitrogen compounds of decaying plants are broken down by bacteria and release ammonia through steps 2 and 3.
5. Bacteria change nitrate to gaseous nitrogen.

Prolonged anoxia and soil saturation cause a buildup of organic matter in the soils of many wetlands; such wetlands are carbon sinks. Although wetlands may represent about 2–5 percent of the Earth's land mass, they contain somewhere between 18–30 percent of the world's soil-stored carbon. Northern bogs and fens (peatlands) and temperate marshes and swamps on muck soils have much higher amounts of carbon than their mineral soil counterparts (e.g., floodplain forests, pine flatwoods, or wet meadows). Changes in carbon storage due to natural processes (e.g., fire and droughts) and human actions (e.g., wetland drainage, timber harvest, soil disturbance, and burning of fossil fuels) release carbon dioxide to the atmosphere, contributing to global warming ("the greenhouse effect"). Wetland restoration including afforestation (e.g., restoration of wet forests on agricultural land) increases carbon storage and thereby helps reduce the amount of carbon dioxide in the atmosphere.

Recent studies have found that forested wetlands along rivers and streams perform significant water-quality functions. This is especially true of wetlands in headwater positions (in the upper parts of drainage basins), where there is much contact between wetland vegetation and soils and the water, facilitating water-quality renovation. A forested wetland buffer strip of 300 feet can provide enormous water-quality benefits, while its canopy cover helps moderate water temperatures for cold-water species like trout. A 100-foot vegetated buffer strip along streams also yields substantial improvement in water quality. Protecting existing wetlands and restoring lost wetlands in key locations in watersheds may significantly enhance water quality.

With increased discharges of various pollutants into our nation's waterways, the ability of wetlands to renovate water is a very important function, especially considering that an estimated 40 percent of America's fresh water is unusable. Water-quality renovation is performed by most wetlands, especially those bordering farmland, urban development, and other sources of water pollution. It must be recognized, however, that each wetland has a limited assimilation capacity and can become saturated with pollutants after several years.

Tinicum Marsh, a 512-acre tidal fresh marsh just south of Philadelphia, is one of the world's best-known examples of significant water-quality renovation. Three sewage-treatment plants discharge treated sewage into marsh waters. On a daily basis, the marsh removes 4.9 tons of phosphorus, 4.3 tons of ammonia, and 138 pounds of nitrate from the water, while adding 20 tons of oxygen to the water. Without this marsh, aquatic life in Darby Creek, a tributary of the Delaware River, would be seriously depleted due to nutrient loading and increased oxygen demands. Neighboring tidal fresh marshes along the Delaware River have been found to significantly reduce nutrients and heavy-metal loadings from urban runoff. Wetlands downstream from urban areas elsewhere undoubtedly play a similar role.

Water-quality renovation can, however, lead to changes in wetland plant communities, as observed in New Jersey Pine Barren swamps. Nutrient-enriched runoff from expanding urban and suburban development has modified local soil-water chemistry. The acidic soils that once favored native Pine Barrens vegetation have become more neutral in pH, leading to a decrease in nutrient-poor species and an increase in nutrient-rich species. Similar results have also been observed in wetlands that have received direct wastewater discharge from sewage-treatment systems.

For control of point-source pollution, it is better to construct and manage artificial wetlands specially designed for this purpose than to discharge into natural wetlands. European countries have been using constructed wetlands for wastewater treatment and more recently for treating storm-water runoff. The Max Planck Institute of Germany has a patent to create biofiltration systems in which bulrush serves as the primary waste-removal agent. Nearly two hundred such systems are in operation in Germany. Various localities have been testing constructed wetlands for wastewater treatment in the United States (see Hammer 1989). Common reed, cattails, bulrushes, and reed canary grass are among the more frequently used species.

Trapping Sediments

Many wetlands form in depositional environments where water-borne sediments drop out of suspension, thereby lowering turbidity of the nation's waters. This function is important for maintaining aquatic life as well as reducing siltation of ports, harbors, rivers, and reservoirs. Removal of sediment is also a key factor in water-quality renovation because sediments often transport adsorbed nutrients, pesticides, heavy metals, and other polluting toxins.

Depressional wetlands retain all the incoming sediments. A classic study of Wisconsin wetlands found that watersheds with 40 percent coverage by lakes and wetlands had 90 percent less sediment in the water than watersheds with no lakes or wetlands. The Army Corps of Engineers, recognizing the value of marshes for reducing turbidity of dredged-material runoff and for removing contaminants, built artificial marshes in dredged spoil impoundments and found that after passing through about 2,000 feet of marsh grasses, the effluent had a turbidity similar to that of the adjacent river.

Floodwater Storage

A March or April visit to many Northeast wetlands provides first-hand evidence of their flood-storage function. By temporarily storing floodwater and slowly releasing it, wetlands protect downstream property owners from flood damage. This function is often called "flood desynchronization." When floodwater encounters resistance from wetland vegetation, its velocity is decreased, lowering the height of the crests and reducing the water's erosive potential. This process results in smaller, less damaging floods and encourages the deposition of water-borne sediments.

Approximately 134 million acres of land in the coterminous United States have severe flooding problems. Most of this land (about 93 million acres) is farmland, while only about 3 million acres are urban. Most of the flood-prone farmland is former wetland, and whenever there is a serious flood that wipes out crops, the government usually pays flood damages. Each year floods cause about $3 billion in damage and a loss of about two hundred lives. Protecting wetlands can help reduce future losses, and restoring lost wetlands can increase the flood-storage capacity of watersheds.

The great Midwest flood of 1993 (Missouri and Upper Mississippi Rivers), the most disastrous flood in recent U.S. history, caused about $20 billion in property damage and the loss of at least thirty-eight lives. Channelization, diking and leveeing floodplains, and wetland drainage—projects intended to move water downstream faster—served to raise the toll. These projects encouraged development on floodplains, which increased the amount of property at risk from this type of low-frequency flood (occurring once in one or two centuries). While wetlands cannot store all the water from such events, they can help lessen damage. Wetland destruction has been significant in this region: almost 20 million acres of wetlands were drained for agriculture. It was estimated that if these wetlands were in their natural condition, they would have been able to hold enough water to fill a tank a thousand football fields in area and 4½ miles deep (30 million acre-feet of storage). An estimated 13 million acres of wetlands would have kept the Mississippi River within its banks. Current findings are that pothole wetlands would be most effective in reducing damage from floods with a frequency of twenty-five years or less. The federal government has initiated an Emergency Wetlands Reserve program in this area to provide natural flood storage through wetlands restoration.

Flood-storage potential of wetlands is especially important near urban areas. Urban development with its tremendous increase in impervious surfaces (roads, driveways, parking lots, and buildings) and lawns, coupled with the loss of wetlands and forests, causes more water to run off the land during and immediately after rainstorms. This raises flood heights, increases the frequency of destructive floods, and escalates the cost of damage (Figure 7.2). The Passaic River basin in New Jersey and New York provides an excellent example of this. Between 1940 and 1978, nearly half of the wetlands in the central portion of the basin were lost to housing, reservoir construction, and industrial and commercial development. Much of the development probably took place in the 1960s, during a period of extended droughts when wetland regulations did not exist. A 1984 flood killed three people, forced the evacuation of nine thousand residents, and caused more than $400 million in property damage. Flooding remains a serious problem for the basin, and the government is spending billions of dollars to remedy the situation. Options under consideration include a $2.5 billion buyout of flood-prone properties and a $1.9 billion flood tunnel (based on Corps of Engineers estimates). This illustrates why development should not be allowed on floodplains.

In New England, the Corps of Engineers has recognized the value of wetlands as a means of preventing future flooding problems. In the Charles River watershed of eastern Massachusetts (the Boston area), the Corps found that if 40 percent of the wetlands were destroyed, annual flood damages would increase by at least $3 million (mid-1970s dollars). Loss of all basin wetlands would cause annual flood damages of $17 million. The Corps concluded that wetland protection through "natural valley storage" was the least-costly solution to future flooding problems. The Corps has now acquired about 8,500 acres of Charles River wetlands, its target for preservation and flood protection.

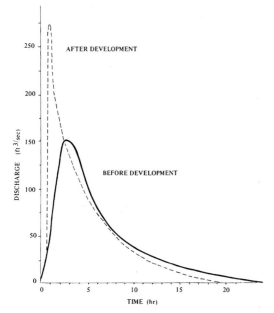

Figure 7.2. Urban development with its impervious surfaces increases surface-water runoff into rivers. This raises flood peaks and can lead to more flood damage. Downstream floodplain wetlands are important water-storage basins. (Source: Tiner 1985; redrawn from Fusillo 1981)

Shoreline Stabilization

Many wetlands are situated between watercourses and uplands where they buffer the uplands from water-generated erosion. Wetland vegetation, especially woody plants, increases the durability of the soil through binding with roots. Trees stabilize river and stream banks worldwide, and some treed banks have not been eroded in two hundred years. Plants also help dampen wave energy through friction, which also reduces current velocity and floodwater's erosive potential. The role of wetlands in re-

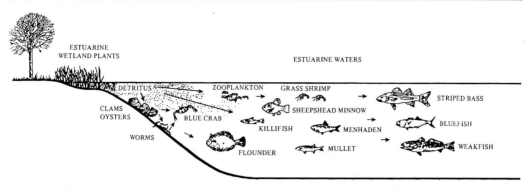

Figure 7.3. Coastal wetlands are the basis for supporting many marine and estuarine animals that provide food for recreationally and commercially important fishes.

ducing shoreline erosion is so highly regarded that various state governments are encouraging the establishment of marsh vegetation to stabilize shorelines, especially in coastal areas. Maryland and Delaware have initiated programs that subsidize such nonstructural techniques, planting smooth cordgrass and other marsh grasses.

Water Supply

Although many laws protecting wetlands cite the benefit of wetlands in recharging water supplies, most wetlands are groundwater discharge sites. Indeed, they are good indicators of water sources. Because of this, many towns have established public wells in wetlands. Such use may cause significant adverse effects if water tables are lowered to the point that wetland hydrology is significantly disrupted (Chapter 8). In agricultural areas, many farm ponds have been constructed in wetlands to provide irrigation water or livestock watering holes. Some wetlands do recharge groundwater, at least seasonally, such as those occurring on sandy soils and on floodplains. Depressional wetlands probably recharge groundwater at their upper edges when maximum flooding occurs, as do prairie pothole wetlands in the Midwest.

Aquatic Productivity

Some types of wetlands are among the most productive natural ecosystems in the world. Marshes and tropical floodplain forests produce as much biomass (organic matter, leaves, stems, fruits, etc.) as the Midwest's best cornfields. Wetland plants are particularly efficient converters of solar energy through photosynthesis.

Thus tons of organic matter are produced by each acre of tidal marsh, for example.

While the seeds and plant parts of hydrophytes provide food for many animals, the major value of plant productivity is realized when the biomass is broken down into fine particles of organic matter called "detritus." Detritus is the foundation of a food web that is best documented for salt and brackish tidal marshes (Figure 7.3). This web begins with detritus-eating invertebrates (zooplankton, clams, worms, shrimp, and crabs) and forage fishes like killifish, menhaden (pogies), bay anchovies, silversides, and mullet. Such "detritivores" are common in salt-marsh creeks and adjacent coastal waters, where they are eaten by larger predaceous species like bluefish, sea trout (weakfish), and striped bass. These and other predaceous fishes serve as food for people. Commercial and sport fisheries, therefore, depend on wetlands. Tidal marshes can be considered aquatic farmlands, producing tons of organic matter that is eventually transformed into seafood for our tables and food for many aquatic organisms. The majority of freshwater aquatic animals also depend either directly or indirectly on food produced in wetlands.

Seasonal woodland ponds called "vernal pools" are also extremely productive habitats, but in other ways. Their nutrients are supplied mainly by leaf litter from surrounding trees. This organic detritus forms the base of a food chain that includes aquatic invertebrates and larval amphibians, and their predators such as green frogs and water snakes. Organic matter from this ecosystem is exported to adjacent upland forests when juveniles migrate from the pools to spend their adult lives in these forests and as vernal pool predators move to other locations.

Provision of Wildlife Habitat

As we have seen in Chapter 6, marshes, swamps, bogs, and other wetlands provide homes for many species, breeding grounds for others, and important feeding and resting areas along migration routes. Without question, they are vital habitats for the nation's wildlife. They are the only habitat for about 8 percent of the vascular plants in the United States and provide habitat for as much as 33 percent of the nation's vascular flora. An estimated 43 percent of the nation's rare and endangered species (plants and animals) depend on wetlands for at least part of their life cycle. All of this is especially noteworthy since wetlands represent only about 6 percent of the land area in the coterminous United States; a surprisingly high percentage of the nation's plant and animal species lives in a disproportionately small percentage of the country's land area. And it is not only the pristine wetlands that are important for wildlife. In many cases, urban wetlands represent the last vestiges of natural habitat in these highly developed landscapes. They serve as refugia for plants and animals that were once more abundant in these locales. The Hackensack Meadowlands, one of the largest wetlands in northern New Jersey, is surrounded by cities, yet over three hundred species of birds, including sixty-five nesting species, have been observed in this ecosystem. In agricultural regions of once-forested landscapes, the presence of remnant wetland forests provides refugia for woodland wildlife. In the Southeast, for example, remaining coastal populations of black bear survive in forested wetlands. Pothole wetlands in the Upper Midwest (the Dakotas and western Minnesota) remain the nation's most productive waterfowl breeding grounds in spite of being surrounded by cropland. Many wetlands and their wildlife seem to be quite tolerant of human disturbance, although they have experienced significant changes in community structure and population size.

Local Climate Regulator

The presence of wetlands can affect air temperatures and humidity of local areas. For a first hand example of this, visit an Atlantic white cedar swamp on a hot, summer day. Instantly you'll notice a change in temperature with cooler, damper air. On a larger scale, a study of South Florida recently found that as the Everglades wetlands were drained and converted to cropland, colder air moved south. This has increased the frequency of crop freezes causing hundreds of millions of dollars damage to sugar cane and vegetable crops. A freeze on January 19, 1997, ruined $300 million worth of crops. The wetlands don't freeze over because their wet soils retain more heat. They also release more water vapor to the air, making the atmosphere more humid and keeping it warmer at night. Where wetlands dominate the landscape, they undoubtedly have a significant effect on local climates.

Databanks of the Earth's History

Anaerobic conditions responsible for the formation of wetlands have endowed them with exceptional preservation qualities. Many wetlands are storehouses of ancient pollen, fossils, and human artifacts. Some European bogs have produced artifacts dating back to 800 B.C. The remains of extinct mammals such as woolly mammoth and mastodon have been preserved in bogs. Well-preserved bodies of people from the Iron Age have been recovered from Scandinavian bogs.

The history of changes in vegetation is also recorded in many wetlands. Through the study of pollen (palynology) in different depths of peat-bog soils and lake-bottom deposits, scientists are able to reconstruct vegetation patterns. Since certain plants are indicative of wetter or drier climates, past climates can also be inferred. For example, pollen from spruce and fir suggests a cold, wet climate, whereas beech pollen tends to indicate a moist, warm climate, pine pollen a dry, cool one, and hemlock pollen a moist one (either cool or warm). A change in the type of peat from peat moss to woody peat implies a shift from a cool, moist climate to a warmer, drier climate. The pollen record is as important to the study of vegetation and climate change as fossils are to the study of plant and animal evolution.

Reaping Nature's Wet Bounty

Wetlands produce raw materials and other commodities that have served human needs since the beginning of time. It may be surprising to learn that the so-called cradle of civilization—the Tigris-Euphrates Delta in southern Iraq—is an enormous wetland. The productivity of this region made it an attractive area to settle (see Maxwell 1966). Other present-day cultures that developed around wetlands include the Gaguju of Australia, the Aka pygmies of central Africa, the Camarguais of southern France, and various groups in Southeast Asia. Like these peoples, Native Americans found wetlands attractive places to live. The archeologist George Nicholas, who has extensively studied the interaction of early northeastern peoples and wetlands, believes that most Americans would probably be astonished to learn about this history because people tend to impose their cultural geography upon past landscapes and their value system on other cultures. Besides food, water, and fiber, swamps and marshes offered Native Americans convenient places to find refuge from attacks by warring tribes or colonists. The rest of this chapter gives examples of how wetlands in their natural state have benefited people. Of course, they have also been used in ways that have altered their character or eliminated many of the natural functions they perform (e.g., farming, mining, disposal of dredged material, and development). These uses are discussed in the next chapter.

Precolonial Human Uses

Native Americans visited wetlands to harvest fish and shellfish, hunt waterfowl and other game, and pick wild berries and other edible plants. Many hydrophytes were used medicinally by these peoples as well as by early European settlers (Table 7.2). Wood from wetland species was used for making dwelling structures, arrows, and eating ware, as well as for firewood. Black ash was used in basket making. Old garbage dumps, called "shell middens," at campsites reveal evidence of the food Native Americans ate. Along the coast, most of these middens are dominated by shellfish such as oysters. There is a large shell midden at Tuckerton, New Jersey, and another at Tottenville, Staten Island, where oyster shells are twice as large as today's.

Wetlands along rivers have always been desirable areas for settlement due to the availability of food and the convenience of water-borne transportation. In precolonial times, millions of alewife and shad migrated upstream from late March through May in rivers like the Delaware and Connecticut. These anadromous fish were so abundant that they could be caught by hand. Because they provided Native Americans with a surplus of food, fish were smoked and dried for later consumption. Shellfish such as oysters, clams, and freshwater mussels were harvested.

When the anadromous fish runs were over in June, many Native Americans moved to camps near lakes, streams, or the sea to harvest berries, nuts, and herbs and to hunt marsh birds, waterfowl, passenger pigeons, frogs, and turtles. During the molting season, flightless waterfowl were easy prey that could be caught by hand. Wetlands also provided important food plants, including cattails, blueberries, cranberries, and tubers of duck potato (big-leaved arrowhead). Cattail was one of the more prized species, because its parts could be eaten year-round. Young shoots were consumed as green herbs, immature green spikes were roasted, and roots were eaten directly or dried and ground into flour enriched with pollen. It was such a versatile food that it was a staple for many Native American tribes, especially the "Cattail-eater" Northern Paiutes of Nevada.

The Lenape of New Jersey began planting crops about 1000 A.D., thousands of years after some Southeast and South American peoples had been farming corn, beans, and squash. The Lenape practice was more gardening than farming, on floodplains and near small streams, usually in natural openings or openings made by burning down trees. This type of agriculture did not have a great impact on the natural environment. The Lenape didn't clear the forests for at least two reasons: 1) their stone tools were not capable of felling massive forests, and 2) their population was in balance with the resources available from the natural environment, so they did not need to intensively manage the land.

Postcolonial Human Uses

Natural products were as important to European settlers as they were to Native Americans. The animals inhabiting wetlands were trapped, hunted, or fished for food and other uses. Timber, bog iron, and other materials were extracted from wetlands. Shakers, a well-known religious group, established their communal villages near black-ash swamps, where they could harvest the wood for basket making and

Table 7.2. Some wetland plants that have been used medicinally.

Plant	Examples of Uses
Sweet Flag	Root chewed for treating mouth sores and relieving stomachaches; also dried and powdered, then mixed with water for various ailments (potentially carcinogenic)
Boneset	Tea from dry leaves and tops used as cold and flu ("breakbone fever") remedy to break fevers
Joe-Pye-weed	Root simmered or boiled in water to make tonic for breaking fevers
Water Plantain	Fresh leaves applied to skin to draw out infection and relieve burns
False Hellebore	Roots boiled in milk used to rid oneself of lice; tincture for depressing heart action, and internal remedy for arthritis; also used to treat gout
Jack-in-the-Pulpit	Dried and grated root applied to skin to relieve headache
Skunk Cabbage	Dried and powdered root used externally as styptic (to stop bleeding)
Water Lily	Roots dried and powdered, drunk with water to relieve bellyache; also used as eye-wash, gargle for sore throats, and cure for baldness
Goldthread	Root chewed to relieve toothache and treat mouth sores
Peppermint	Crushed leaves applied to skin to relieve headache and pains, also used to prevent seasickness; tea from leaves and tops used for relief from stomach gas, headache, and heartburn
Blue Flag	Tea from root drunk cold for stomach and intestinal problems
Jewelweed	Raw juice used to counteract poison ivy and to remove warts, corns, and ringworms; entire plant boiled and applied to treat poison ivy
Sweet Gale	Juice from boiled leaves used to treat poison ivy
Winterberry	Bark used to make an internal and external tonic for treating fevers, gangrene, jaundice, and sores; berries used to rid children of worms
Silky Dogwood	Bark simmered in water makes tonic to induce vomiting
Elderberry	Ointment from leaves and flowers applied to wounds; berry tea is a laxative; inner bark used for relieving constipation; berries used to treat gout, infected lymph glands, and syphilis
Wax Myrtle	Bark chewed for toothache relief; tea from dried and powdered root bark used to treat diarrhea and dysentery and as mouthwash for sores and bleeding gums
Northern White Cedar	Ointment from fresh leaves and bear fat used to treat rheumatism; tincture from young twigs, leaves, and flowers used to treat venereal disease
Eastern Hemlock	Bark boiled then mashed, applied to heal wounds and reduce swelling
Alder	Chewed bark put on cuts and bruises to relieve pain
Cranberry	Berries used for treating scurvy and reducing fevers

other crafts. Some people still earn part of their wages from trapping wetland animals and harvesting natural products. Until the 1950s, many people living in the New Jersey Pine Barrens were tied to traditional ways of life, reaping the seasonal bounties of nature: fish, shellfish, ducks, rails, muskrats, snapping turtles, peat moss, salt hay, and timber, for example (see Moonsammy et al. 1987).

Trapping, Hunting, and Fishing

During the colonial period, beaver trapping provided a great source of income for those willing to take up the difficult life of a fur trapper. In

1703, one beaver pelt could buy a yard and a half of cotton cloth, five pecks of corn, two pints of gunpowder, one pint of shot, two axes, two hoes, six knives, six combs, ten pounds of pork, and one shirt. With value like that, it is little wonder that the beaver was extirpated from much of the Northeast. The beaver trade was such an important part of Canada's history that the beaver appears on the back of the five-cent coin. Today, there is not much demand for pelts, and prices are low, ranging between ten and thirty-five dollars. This low demand plus reintroductions have probably aided their reestablishment in many areas. Muskrat dominates the fur harvest in most Northeast states. Besides their fur value, muskrats are served as delicacies at some restaurants on the Delmarva Peninsula.

During the 1800s, waterfowl hunting provided food for tables and feathers for the millinery trade. This claimed a heavy toll on bird populations. The pressure was so intense and the decline in waterfowl populations so precipitous that market hunting was banned in 1900 by the Lacey Act (prohibiting interstate trade of illegally killed game). States began to set daily bag limits for many species, and hunting was restricted to sport and regulated by the government. Today, sportsmen across the country spend over $600 million each year to hunt waterfowl.

The millinery trade created a demand for bird feathers to adorn hats and bonnets. An estimated 5 million birds of about fifty species (including snowy egret, great egret, and terns) were killed each year. By the 1890s, many of these species were on the brink of extinction. The National Audubon Society was established by women concerned about the plight of these species. It stimulated interest in bird conservation by increasing public awareness. In 1918, Congress passed the Migratory Bird Treaty Act regulating hunting of migratory species.

Fish and shellfish populations followed a similar path of exploitation in many areas. Until the 1900s, the Delaware River provided a wealth of sturgeon, shad, and other fishes, with annual harvests of 14 to 20 million shad alone. In the early 1900s, a rapid decline in fish populations occurred, due in part to overfishing but mostly to environmental degradation from water pollution, dredging at or near spawning grounds, and sedimentation from upland development. In other rivers of the Northeast, fishing

interests lost out to manufacturing interests, as the construction of dams to provide waterpower for mills cut off migration of anadromous species. Today, there are serious efforts in many states to restore such historic fish runs.

Raritan Bay along the south shore of Staten Island had valuable fisheries for oysters, eels, and shad in the 1600s and 1700s. Its oysters were in such heavy demand that laws restricting harvests were passed in the early 1700s by both New York and New Jersey to prevent overfishing. These laws controlled harvests, but not external forces that adversely affected the beds. Sedimentation from forest clearing for agriculture had largely covered the beds with silt by the 1800s. The beds were reseeded, but by 1940, the oyster industry was dead due to pollution: industrial wastes, oil pollution, and disposal of treated and untreated sewage. The Bay's other fisheries were similarly affected.

Wetlands are vital to about three-quarters of America's fish production, which is worth $111 billion and provides jobs for 1.5 million people. Some East Coast states have higher dependencies, with 98 percent of Delaware's commercial species being estuarine-dependent (needing the salt marsh–estuary system for survival). Chesapeake Bay's commercial fishery is similarly dependent. Maryland ranks fifteenth nationally in commercial fish landings, with an annual value of over $53 million. Its blue-crab industry alone is worth over $20 million. New Jersey's recreational fishing, commercial fishing, and shellfish industries each yield about $160 to $220 million annually.

Commercial fishing in the Great Lakes was once a major economic enterprise. Overfishing, habitat destruction, and pollution especially toxic contamination of fish tissue led to fishery closures in most areas in the United States. Most of the U.S. harvest now goes into animal feed. Canada still maintains a significant commercial fishery in the Great Lakes. Their Lake Erie fishery landed about $59 million worth of fish in 1991; this amounts to about two-thirds of Canada's total Great Lakes harvest. Today, the Great Lakes are more important for sport fishing. In the 1960s, nonnative sport fishes such as coho, chinook, and Atlantic salmon were introduced to feed on alewife that increased greatly when the sea lamprey devastated lake trout populations (after entering the Lakes from the St. Lawrence River via locks and dams). Native sport fishes include walleye, lake trout, brook trout, yellow perch, white bass, and northern pike.

Hay Production

Early settlers cut salt hay from marshes for winter fodder, and some wetlands were grazed by cattle, sheep, and horses. As the sea level rose, salt-hay marshes became wetter, necessitating some type of water control. A combination of dikes, ditches, and tide gates was constructed for this purpose. Some farmers now manage by flooding the marshes for a couple of months in the spring to maintain the salt-marsh species and control undesirable weeds, and then keeping the tides out for the rest of the year. Salt hay is harvested from midsummer into winter. Some marshes have been harvested for two hundred years.

Salt hay grass (*Spartina patens*), salt grass (*Distichlis spicata*), and black grass (*Juncus gerardii*) are the principal species in upper high marshes (see Chapter 5). In colonial times, black grass was the preferred winter fodder. Other historic uses of salt hay include mulch (especially for strawberries and sweet potatoes), packing material for glassware and pottery, rope for casting iron pipe, and butcher's wrapping paper. Today, it is prized by suburban and urban gardeners as a weed-free mulch. A bale of salt hay sells for around ten dollars, compared to two to three dollars for a bale of upland mulching hay.

Although the soils were too wet to farm, inland wet meadows were pastured or hayed, and many have been used as such since the 1600s. Most wet meadows probably were created from forested wetlands—usually red-maple swamps—that were cut over and stumped to produce hay. In some cases, dams were built to provide enough water to create sedge meadows. Annual mowing or grazing is required to keep woody plants in check. With recent abandonment of agriculture in many areas, wet meadows are reverting to forested wetlands.

Peat-Moss Harvest

Peat moss has been used since colonial times for soil enrichment, and it remains valuable for urban and suburban gardens. Adding peat moss to gardens helps improve soil moisture. In New England and elsewhere, farmers collected peat moss and added it to poor glacial soils. Today

Figure 7.4. Peat moss possesses unique qualities that make it important for wetland establishment as well as for human uses.

horticultural peat moss is mined from bogs (Chapter 8).

Peat mosses (genus *Sphagnum*) have an uncanny ability to absorb and retain water. As noted before, they can retain up to twenty-five times their weight, like super sponges (Figure 7.4). This absorptive capacity and the plants' aseptic properties also led Native American women to use dried moss during menstruation and as diapers for babies and toddlers, and armies of several nations used it as sterile surgical dressing. Before the creation of styrofoam, peat was widely used for holding flowers in arrangements. In the New Jersey Pine Barrens, a few local woodsmen may still gather peat the traditional way by raking it from the swamps.

Forestry

Wetland forests provide millions of acres of timber for firewood, newsprint, and building lumber. Atlantic white cedar and loblolly pine are perhaps the two most heavily utilized commercial trees in the Northeast, while much red maple is cut locally for firewood. On the Delmarva Peninsula, pine silviculture is a major industry, in which managed loblolly-pine forests produce fiber for newsprint (Chapter 8).

Atlantic white cedar is the most profitable timber product from Northeast wetlands. Its many uses include boat building, posts for docks and bulkheads, cooperage (tubs and tanks), woodenware, church organ pipes, fencing, houses, corduroy roads through swamps, hope chests, decoys, and model boats. Cedar has even been cut for Christmas trees. In the 1600s and 1700s, cedar cutting was so heavy that Benjamin Franklin recommended planting red cedars and employing other forestry techniques to supply the nation's demand for cedar. In the 1800s and early 1900s, buried cedar logs were mined from cutover cedar swamps. Today, many acres of Atlantic white cedar are managed for timber, and efforts are under way to restore some former cedar swamps.

Wild-Plant Collection and Harvesting

The beauty of some wetland plants has led to their collection and, in some cases, to their cultivation. White water lily, sweet pepperbush, sweetbay, and rosebay rhododendron are examples of cultivated species. People collect the twigs of pussy willows and common winterberry with its bright red berries in early spring and fall, respectively, for dried arrangements. Flowering stalks of common reed are also gathered for similar use, a practice that might aid in the spread of this invasive species. Flower stalks of sea lavender and the fertile spikes of narrow-leaved cattail are still gathered for florists. Beautiful wreaths of sea lavender are sold in some places, like Downeast Maine. This plant has been harvested to the point that some communities prohibit picking it. Overharvesting of some evergreen plants for Christmas greens led some states to outlaw their collection. Among the wetland evergreens, laurel-leaved greenbrier, inkberry, and American holly were much sought after.

Many edible wetland plants have provided sustenance for people since early times (Table 7.3). Highbush blueberries and cranberries are in such demand that they have been cultivated (Chapter 8), but these and other berries are still harvested in the wild as well. Tea can be made from the leaves of several species and from rose hips. Tubers of aquatic plants like white water lily and spatterdock can be used as a potato substitute, while the young leaves of pickerelweed and marsh orach and young stalks of glassworts can be added to salads. Cattail, cherished by Native Americans as a food plant, has also been used as fiber for rush seats and stuffing for mattresses. Bright yellow muffins can be made by adding cattail pollen to your favorite muffin recipe. Also its immature male spikes can be boiled, buttered, and eaten like corn-on-the-cob.

Honey

Some wetland plants are prized sources of nectar for producing honey. In the Northeast, the fragrant sweet pepperbush of coastal-plain swamps provides the primary source of nectar for the honey bearing its name. Interestingly, the leaves of this plant have also been used by southern fisherman to clean their hands. The grocery shelf may have tupelo honey derived from the black gum (swamp tupelo) and water tupelo of southern river swamps. Purple loosestrife was introduced into this country for the bee industry. Although this plant is viewed as an invasive pest, beekeepers feel that the August bloom of this plant helps sustain their bees and increases honey yield per hive. All the same, the rate at which purple loosestrife is replacing native plant communities and their wildlife is a great cause for alarm.

Places of Refuge

Just as Native Americans sought refuge in the swamps from warring tribes and later from colonists, so did others. One famous Revolutionary War hero, Francis Marion of South Carolina, rode with his men into the Santee River swamps and hid from the British he had recently attacked. It's hard to follow tracks in a flooded swamp, and the presence of alligators and rumors of quicksand kept the British away. The successful retreat earned Marion the nickname "Swamp Fox." Later in the 1800s, swamps from Florida to Pennsylvania served as important links on the Underground Railroad. These seemingly inhospitable places were havens for freedom-seeking slaves running away from Southern plantations and heading north. Harriet Tubman, a Maryland slave, escaped to freedom and then returned on numerous occasions to guide hundreds of other escaped slaves to freedom, hiding in swamps by day and traveling by night to reach friendly homes and eventually northern cities. The infamous John Wilkes Booth eluded federal officers after assassinating Abraham Lincoln by hiding in the Zekiah Swamp on Maryland's lower Western Shore.

Enjoying the Wetlands

Activities that do not consume resources, such as hiking, canoeing, nature photography, bird watching, and other kinds of nature observation, are being enjoyed by an increasing number of people in and around wetlands (Figure 7.5). The flat terrain of most wetlands makes for easy hiking. Many wetlands are easily traversed in

Table 7.3. Examples of edible wetland plants.

Edible Part	Plant
Berries/fruits	Blueberry, Cranberry, Dangleberry, Huckleberry, Shadbush, Blackberry, Chokeberry, Wild Raisin, May Apple, Paw Paw, Riverbank Grape, Bunchberry, Creeping Snowberry, Partridgeberry, Northern Fly Honeysuckle
Berries (wine)	Blueberry, Elderberry
Berries (spice)	Spicebush
Fruits (jelly)	May Apple, Riverbank Grape, Elderberry, Blackberry, Northern Fly Honeysuckle, Cranberry, Wild Raisin, Chokeberry, Blueberry, Huckleberry, Shadbush
Flowers (fried in fritters)	Elderberry
Flowers (candied)	Violet
Flower buds (capers)	Marsh Marigold
Flower buds/spikes (boiled, buttered)	Water Lily, Cattail
Nuts	Shellbark Hickory
Tubers (raw)	Cattail
Tubers (cooked)	Cattail, Pickerelweed, Spatterdock, Water Lily, Arrowhead, Spring Beauty, Silverweed, Water Parsnip, Trout Lily, Turk's-Cap Lily, Water Shield, Soft-stemmed Bulrush
Tubers (dried, ground into flour)	Cattail, Water Lily, Wild Calla, Spatterdock, Water Shield, Common Reed, Soft-stemmed Bulrush, Giant Bur-reed
Tubers (candied)	Sweet Flag
Young leaves/fleshy stems (raw)	Marsh Orach, Glasswort, Seaside Plantain, Ostrich Fern, Violet, Wintergreen, Sweet Flag, Water Shield, Pickerelweed, Common Greenbrier, Cattail, Soft-stemmed Bulrush
Young shoots (boiled)	Wood Nettle
Young stems (pickled)	Glasswort
Leaves (boiled)	Ostrich Fern, Wood Nettle, Common Greenbrier, Water Lily, Seaside Plantain, Sea Blite, Marsh Marigold, Trout Lily, Violet, Pickerelweed, Riverbank Grape
Leaves (tea)	Wild Mint, Peppermint, Spearmint, Labrador Tea, Sweet Gale, Spicebush, Wintergreen, Creeping Snowberry, Shrubby Cinquefoil, Violet, Eastern Hemlock, Wood Nettle
Young shoots (boiled)	Jewelweed, Soft-stemmed Bulrush
Seeds (grain)	Wild Rice

late summer or early fall, when water levels are at their lowest. Some suggested places to visit are listed in Appendix B.

Additional Readings

Hammer, D. A. (editor). 1989. *Constructed Wetlands for Wastewater Treatment: Municipal,* *Industrial and Agricultural.* Lewis Publishers, Chelsea, MI.

Majumdar, S. K., R. P. Brooks, F. J. Brenner, and R. W. Tiner, Jr. (editors). 1989. *Wetlands Ecology and Conservation: Emphasis in Pennsylvania.* The Pennsylvania Academy of Sciences, Easton, PA.

Marshall, C. H., R. A. Pielka, Sr., and L. T. Steyaert. 2003. Wetland: crop freezes and

Figure 7.5. Nature observation in and around wetlands can be enjoyed by people of all ages. These kinds of outings are great for families. (Barbara Tiner)

land-use change in Florida. *Nature* 426: 29–30.

Maxwell, G. 1966. *People of the Reeds.* Pyramid Books, Harper & Row, New York.

Mitsch, W. J., and J. G. Gosselink. 1986. *Wetlands.* Van Nostrand Reinhold Company, Inc., New York.

Moonsammy, R. Z., D. S. Cohen, and L. E. Williams (editors). 1987. *Pinelands Folklife.* Rutgers University Press, New Brunswick, NJ.

Peterson, L. A. 1977. *A Field Guide to Edible Wild Plants of Eastern and Central North America.* Houghton Mifflin Co., Boston.

CHAPTER 8

Swampland Now and Then
Wetland Status and Trends

The nation's wetland resources have changed greatly over the past two hundred years. Most of the change has been the consequence of human activities, seeking to make these lands more useful to society. While wetlands in more exposed or vulnerable locations are in constant flux due to natural forces, the majority of wetlands remain relatively stable. Now that fire is largely suppressed and the beaver population is lower than presettlement numbers, the natural changes tend to be more gradual—for example, in response to climatic shifts. Where change is rapid, it is usually brought on by fire, beaver construction, or human actions, and such changes are often reflected in the vegetation (see Chapter 5). Changing human attitudes toward wetlands have stimulated an interest in wetland protection, conservation, and restoration which has significantly decreased wetland losses through human actions.

Wetland Status

While it is impossible to determine the exact acreage of wetlands at a given point in time due to changing uses and natural processes, data from recent wetland inventories conducted by government agencies provide a reasonable estimate of the extent of wetlands today. The U.S. Fish and Wildlife Service has been mapping the nation's wetlands for nearly three decades through its National Wetlands Inventory program. This survey, based largely on interpretation of aerial photographs, provides a standard basis for assessing the status of wetlands across the country.

National Status

In the 1990s, the United States had an estimated 292 million acres of wetlands (Table 8.1). This represents about 12.5 percent of the nation's land surface or an area about ten times the size of New York or Ohio (442,000 square miles). Sixty percent of the nation's wetlands are in Alaska, covering approximately 48 percent of the state. Twenty-three states have more than one million acres of wetlands, and Florida, Minnesota, and Louisiana have more than 10 million acres each. In the coterminous United States (lower 48 states) only 6 percent of the land is wetland. Most occur along the Atlantic-Gulf Coastal Plain, Mississippi Valley, and the northern woods (Maine to Minnesota). Fifty-one percent of the inland wetlands are forested types, 25 percent are marshes, wet meadows, fens, and wet prairies, 18 percent are shrub swamps, and 6 percent are ponds. Coastal wetlands comprise about 5 percent of the wetlands in the lower 48 states. Roughly three-quarters of these wetlands are salt and brackish marshes, while 11 percent are nontidal tidal flats and beaches. Coastal wetlands are most extensive along the Atlantic-Gulf Coastal Plain.

Northeast Status

In the Northeast, wetlands are most abundant along the coastal plain and in glaciated sections due to a combination of favorable climates and landscape features conducive to wetland formation (Chapter 3). Maine's glaciated terrain, plus about 50 inches of annual rainfall, relatively low temperatures (and thus low evapotranspiration rates), high beaver populations, and an irregular coastline subjected to high tidal ranges (with resulting extensive intertidal flats), has resulted in an estimated 5.2 million acres of wetlands. Maine has the most wetland acreage of any northeastern state, with almost five times as much as Virginia and New York. This acreage accounts for more than 25 percent of the state's land area (Table 8.1). States with more than 10 percent of their land surface occupied by wetlands include Delaware,

Table 8.1. Estimated extent of wetlands in the U.S. by state.

State	Area of Wetland (acres)	Proportion of State	National Rank
Alabama	2,651,000	8.2%	14
Alaska	174,684,000	47.8%	1
Arizona	600,000	0.8%	33
Arkansas	3,573,000	10.7%	13
California	462,000	0.5%	38
Colorado	1,000,000	1.5%	23
Connecticut	173,000	5.5%	47
Delaware	223,000	18.0%	45
Florida	11,403,000	32.9%	2
Georgia	6,520,000	17.5%	5
Hawaii	52,000	1.2%	50
Idaho	386,000	0.7%	43
Illinois	1,254,000	3.5%	19
Indiana	810,000	3.5%	28
Iowa	692,000	1.9%	30
Kansas	435,000	0.8%	40
Kentucky	388,000	1.5%	42
Louisiana	10,313,000	36.2%	4
Maine	5,199,000	26.2%	9
Maryland	598,000	9.5%	34
Massachusetts	590,000	11.8%	35
Michigan	5,583,000	15.3%	7
Minnesota	10,607,000	20.8%	3
Mississippi	4,365,000	14.4%	11
Missouri	643,000	1.5%	31
Montana	840,000	0.9%	27
Nebraska	1,906,000	3.9%	17
Nevada	457,000	0.6%	39
New Hampshire	200,000	3.5%	46
New Jersey	916,000	19.2%	26
New Mexico	482,000	0.6%	37
New York	1,025,000	3.4%	22
North Carolina	5,048,000	16.1%	10
North Dakota	2,490,000	5.6%	15
Ohio	719,000	2.7%	29

(continued)

Table 8.1. (continued)

State	Area of Wetland (acres)	Proportion of State	National Rank
Oklahoma	950,000	2.2%	24
Oregon	1,394,000	2.3%	18
Pennsylvania	404,000	1.4%	41
Rhode Island	65,000	9.7%	48
South Carolina	4,091,000	21.2%	12
South Dakota	2,240,000	4.6%	16
Tennessee	632,000	2.4%	32
Texas	6,412,000	3.8%	6
Utah	558,000	1.1%	36
Vermont	243,000	4.1%	44
Virginia	1,075,000	4.2%	21
Washington	938,000	2.2%	25
West Virginia	57,000	0.4%	49
Wisconsin	5,331,000	15.3%	8
Wyoming	1,250,000	2.0%	20
All States	282,926,000	12.5%	
Coterminous U.S.	108,242,000	5.7%	

Note: Acreages have been rounded-off to nearest thousand. (Source: Tiner 1999 from numerous references)

Maryland, Massachusetts, New Jersey, and Rhode Island. Mountainous states like West Virginia and Pennsylvania have relatively small portions of their landmass occupied by wetlands. Coastal wetlands in the Northeast are most abundant in three states (New Jersey, Maryland, and Virginia), and make up more than 30 percent of the wetlands in the former two states (Table 8.2).

Great Lakes Status

The most extensive wetlands occur in the glaciated portions of the Great Lakes region (Minnesota, Wisconsin, and Michigan). Wetlands are also particularly abundant along the Ohio and Missouri Rivers in southern Illinois and Indiana. Minnesota has the largest wetland extent with over 10.6 million acres (Table 8.1). Michigan and Wisconsin have nearly equal amounts of wetland, with 5.6 and 5.3 million acres, respectively.

Wetland Trends

Wetlands have been greatly altered since colonial times. Current estimates suggest that the coterminous 48 states have lost over 50 percent of their original wetlands. Today, about 100 million acres remain. The total wetland loss since the 1700s amounts to an area the size of the state of California. These habitats have been largely converted to agriculture, as well as being filled for urban and suburban development, impounded for reservoirs, and modified in other ways (see the following section). Some states have lost almost all their wetlands, with California and Iowa being prominent examples. States that have had over half of their original wetlands destroyed include Michigan, North Dakota, and Minnesota.

National Trends

In the 1600s, the continental United States possessed an estimated 221 million acres of wet-

Plate 1a. Buttressed trunk (water gum) and cypress knees.
Plate 1b. Fluted trunk (American elm). (Geoff Knapp)
Plate 1c. Adventitious roots (common reed).
Plate 1d. Water roots and hypertrophied lenticels (willow).
Plate 1e. Oxidized root channels.
Plate 1f. Aerenchyma tissue. (U.S. Army Corps of Engineers)

Plate 2. New England salt marsh in the fall.
Plate 3. Black-needlerush brackish marsh (lower Delmarva Peninsula).
Plate 4. Brackish marsh (North River, Massachusetts).
Plate 5. Tidal saltwater swamp (Delmarva Peninsula). (Salt-marsh species with loblolly pine.)
Plate 6. Cattail marsh (Sebago Lake, Maine).
Plate 7. Wet savanna (southern New Jersey).

8.

9.

10.

11.

12.

13.

Plate 8. Tussock-sedge meadow (northern New Jersey).
Plate 9. Wet meadow (western Maryland).
Plate 10. Fen, with larch swamp in background (eastern Maine).
Plate 11. Lake-fill shrub bog (Adirondacks, New York). (Bill Zinni)
Plate 12. Black spruce forested bog (eastern Maine).
Plate 13. Shrub swamp (western Maryland).

14.

15.

16.

17.

18.

20.

19.

Plate 14. Red-maple swamp (central Pennsylvania)
Plate 15. Skunk cabbage in spring, the dominant groundcover in many hardwood swamps.
Plate 16. Coastal-plain hardwood swamp (Cape May County, New Jersey).
Plate 17. Floodplain forested wetland (South Branch of Raritan River, New Jersey).
Plate 18. Northern-white-cedar swamp (eastern Maine).
Plate 19. Atlantic-white-cedar swamp (southern New Jersey).
Plate 20. Hemlock swamp (western New York).

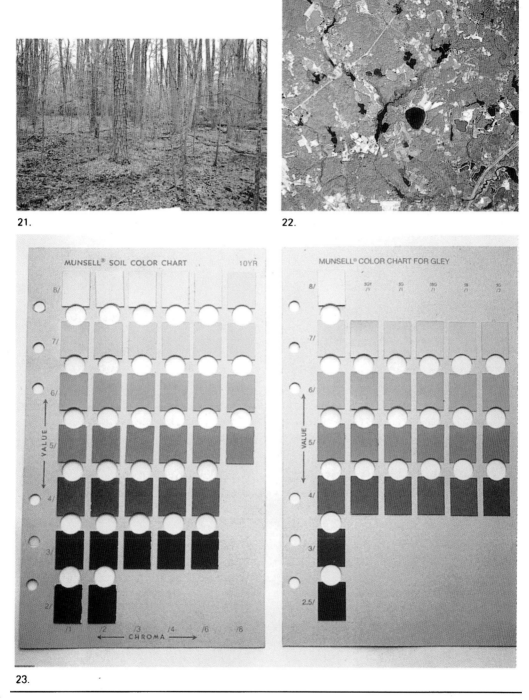

21.

22.

23.

Plate 21. Wet loblolly pine flatwood in late winter (Coastal Plain).

Plate 22. Well-manicured cranberry bogs (smooth red patches) are easily seen in this aerial view of southeastern Massachusetts. Deciduous forested wetlands (dark gray areas along streams) are also evident.

Plate 23. Example of Munsell color charts (10YR page, left; gley page, right). On the 10YR page and other hue pages, low-chroma colors are the left two columns. The gley page contains only low-chroma colors. Courtesy of Kollmorgen Corporation, Munsell Color. Reprinted by permission.

24.

25.

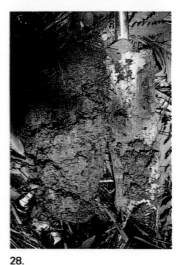

26.

27.

28.

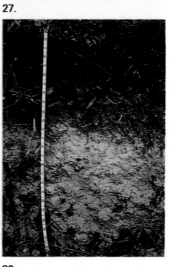

29.

30.

Plate 24. Muck, characteristic of marshes and the wettest swamps.
Plate 25. Peat, typical of New England shrub bogs.
Plate 26. Hydric nonsandy soil (very poorly drained), characteristic of swamps and wet meadows.
Plate 27. Hydric nonsandy soil (poorly drained), from New England wet meadow.
Plate 28. Hydric nonsandy soil (bluish gray subsoil).
Plate 29. Hydric nonsandy soil (poorly drained), from a mid-Atlantic wet meadow.
Plate 30. Hydric nonsandy soil (poorly drained), from a Coastal-Plain forested wetland.

31.

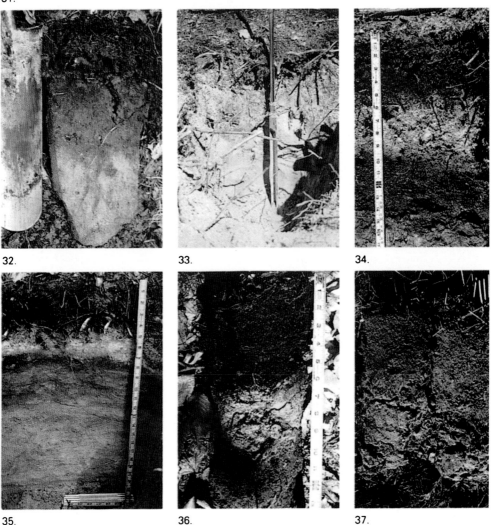

32.   33.   34.

35.   36.   37.

Plate 31. Nonhydric nonsandy soil (somewhat poorly drained).
Plate 32. Hydric sandy soil (very poorly drained), from an Atlantic-white-cedar swamp.
Plate 33. Nonhydric sandy soil, typical of many Coastal-Plain upland forests.
Plate 34. Hydric spodosol (poorly drained), from a mixed forested-shrub bog in eastern Maine.
Plate 35. Hydric spodosol (poorly drained), from a forested wetland in southwestern Maine.
Plate 36. Nonhydric spodosol (moderately well drained), typical of northern New England evergreen forests.
Plate 37. Hydric (poorly drained, on right) and nonhydric (well-drained, on left) mineral soils found within a few feet of one another in an overgrown field in New York.

38a.

38b.

38c.

38d.

38e.

38f.

38g.

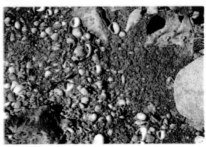

38h.

38i.

Plate 38a. Water-stained leaves.
Plate 38b. Silt line on trunks (9 feet above the ground).
Plate 38c. Water mark on rock (evidence of high tides).
Plate 38d. Tidal wrack (dead stems) left in high marsh by flood tide.
Plate 38e. Water-borne debris on floodplain. (Pile indicates direction from which flood waters came.)
Plate 38f. Sediment deposits.
Plate 38g. Algal mats.
Plate 38h. Remains of aquatic invertebrates.
Plate 38i. Scars of hypertrophied lenticels.

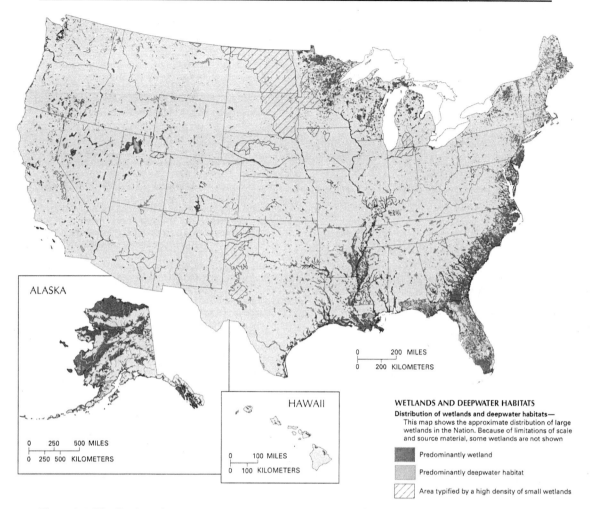

Figure 8.1. Distribution of wetlands and inland waters in the United States. (Source: Fretwell et al. 1996)

lands. With European settlement, westward expansion, and the later emergence of an industrialized society, many wetlands were converted to various human uses. Wetland alteration tended to follow this pattern of settlement (Figure 8.2). From the 1600s to 1800, most wetland alteration took place in the original thirteen states, mainly from Massachusetts through South Carolina. In the 1800s, the U.S. population began a westward movement into recently acquired lands (e.g., following the Louisiana Purchase in 1803). From 1800–1860, more wetlands were lost in the Midwest and South due to the development of drainage tile in the 1800s and passage of the Swamp Land Acts of the mid-1800s, which promoted wetland drainage. From 1860–1950, the federal government subsidized wetland alteration through agricultural

and public works programs, and by the 1950s, over 110 million acres of wetlands—half of the original wetlands—had been destroyed.

From the 1950s to the 1970s, the coterminous U.S. experienced a net loss of 9 million acres of wetland—an area twice the size of New Jersey. An average of 458,000 acres (440,000 of inland wetlands and 18,000 of saline tidal wetlands) were lost annually during this period, equaling an area about half the size of Rhode Island. Agriculture was responsible for 87 percent of these losses, with urban development accounting for 8 percent and other development 6 percent. Nearly 6 million acres of forested wetland, 2.7 million acres of marshes and wet meadows, and 400,000 acres of shrub swamp were converted to cropland. Most of these losses took place in the Mississippi Valley

Table 8.2. Tidal wetland acreage in northeastern states.

State	Tidal Wetlands	% of State's Wetlands
Connecticut	20,400*	11.8
Delaware	98,000	43.9
Maine	110,700	5.4
Maryland	280,000	46.8
Massachusetts	79,000	13.4
New Hampshire	8,700	3.0
New Jersey	297,000*	32.4
New York	72,000	7.0
Pennsylvania	200	<0.1
Rhode Island	8,000	12.3
Virginia	240,000	22.3

*excludes tidal forested swamps
Note: Acreages are rounded off the nearest hundred. Estimates are based on largely on results from the U.S. Fish and Wildlife Service's National Wetlands Inventory mapping and data reported by Tiner (1999) and recently amended for New Hampshire and Maine.

(bottomland forests), the Dakotas (pothole marshes and meadows), North Carolina (pocosins), Florida (Everglades), and Nebraska (Sandhills and Rainwater Basin). Coastal wetland losses were heaviest along the Gulf Coast (Florida, Louisiana, and Texas). Dredging and filling operations destroyed coastal wetlands in Florida, Texas, New Jersey, and California.

With the passage of state wetland laws and increased federal regulations in the decade that followed, annual losses dropped to 290,000 acres. An estimated 3.4 million acres of forested wetlands were lost from the mid-1970s to the mid-1980s, with 2.1 million acres converted to nonwetlands and most of the rest converted to other wetlands. Some of the latter change was simply the result of timber harvest. Freshwater marshes and meadows increased by 220,000 acres, yet 375,000 acres were converted to farmland, 38,000 acres to urban uses, and 64,000 acres to ponds.

Wetland conservation and protection initiatives (such as strengthened regulations, elimination of drainage subsidies, and wetland restoration programs) were likely responsible for a further decline in the annual loss rate between the mid-1980s to the mid-1990s. An estimated average of 58,500 acres were lost each year during this time period (57,500 acres of inland wetland and 1,000 acres of tidal wetland)—an 80 percent drop since the 1980s. Urban development accounted for 30 percent of the total losses, agriculture 26 percent, 23 percent silviculture, and 21 percent rural development. For coastal wetlands, 75 percent of the losses involved filling to create upland, while 14 percent became freshwater wetlands and 12 percent open water.

Regional Trends

In the Northeast, the most alteration has probably occurred in coastal-plain states (i.e., Maryland, Delaware, and Virginia) where farming and forestry are major land uses (Table 8.3). Virginia continues to endure heavy losses, especially in the Norfolk-Hampton area. New York has also experienced significant wetland conversions to agriculture as forested wetlands and marshes have been converted to mucklands. Coastal wetlands around the world have been exploited for port development, navigation uses, and residential housing, and the Northeast is no exception. Connecticut has only about half its original coastal marshes remaining. Throughout the region, coastal wetlands were being destroyed at alarming rates prior to the passage of state tidal-wetland protection laws. In the 1950s and 1960s, New Jersey was losing about 3,000 acres of coastal marsh per year, and New York and Maryland about 1,000 acres annually. These loss rates and the acknowledged value of the wetlands for marine fisheries prompted legislative action to protect them.

Agriculture has played the major role in wetland conversion in the Midwest, spurred on by the Swamp Land Acts of the mid-1800s (Figure 8.3). Through these Acts, the federal government transferred millions of acres to several states including all Great Lakes states (Illinois—1.46 million acres; Indiana—1.26 million acres; Michigan—5.68 million; Minnesota—4.7 million; Ohio—26,372 acres; Wisconsin—3.36 million). States receiving these land grants were supposed to reclaim these swamplands for agriculture and other development. Illinois may have converted up to 90 percent of its original wetland to other uses. All that remains of the Great Kankakee Swamp that originally contained over one million acres is a few pieces of swamp along the Kankakee River. Indiana lost 85 percent of its wetlands to cropland. Today

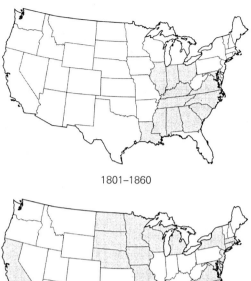

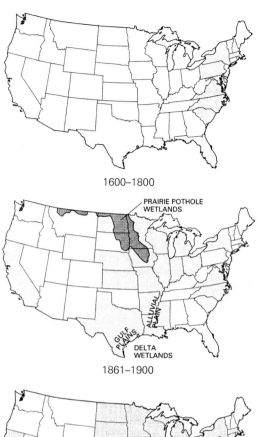

Figure 8.2. Pattern of wetland exploitation in the coterminous United States from 1600–1990. Maps show states with significant wetland losses during various time periods. (Source: Dahl and Allord 1996)

about 1–3 percent of its remaining wetlands are converted annually, still mostly to farm uses. Most of Michigan's lost wetlands were drained for cropland uses prior to 1930, yet from 1934–1940, significant drainage was initiated to reduce the risk of malaria. Most of the drainage took place in the southern third of the state. Industrialization in the Saginaw River and from the St. Clair River to Lake Erie has filled much wetland acreage. More than 70 percent of Minnesota's prairie wetlands were drained by the 1980s. Early efforts to drain bogs in northern Minnesota largely failed due to the lack of suitable topography, but many remain degraded and some are actively mined for peat, harvested for timber, or cultivated. Only about 10 percent

of Ohio's wetlands remain. Its wetlands were mostly concentrated in two areas: the Black Swamp (northwest Ohio to eastern Indiana) and along the shores of Lake Erie. In the 1700s, the Black Swamp was about the size of Connecticut (120 miles long and 40 miles wide). Between 1859–1885, over 3 million acres were drained and some of it became the state's most productive farmland. Today, only 5 percent of the swamp remains. Ohio's coastal marshes have been impounded, filled, and converted to farmland; today only 10 percent of these wetlands exist. Like other Midwest states, Wisconsin's heaviest wetland losses were caused by agriculture. Some of this loss is attributed to cranberry cultivation. In the southeastern part of the

Table 8.3. Examples of wetland trends in the Northeast, Great Lakes region, and for the coterminous United States (lower 48).

Geographic Area	Wetland Type	Acreage Lost (percentage loss; dates)
Coterminous United States	All wetlands	120 million (50%; 1700s–mid-1970s)
	All wetlands	9 million (8%; mid-1950s–mid-1970s)
	All wetlands	2.6 million (3%; mid-1970s–mid-1980s)
	All wetlands	644,000 (0.6%; 1986–1997)
Connecticut	Inland wetlands	150,000 (50%; 1700s–1990)
	Tidal Wetlands	10,000 (45%; 1884–1980)
New Jersey	All wetlands	200,000 (20%; 1900s–mid-1970s)
	Tidal wetlands	61,000 (25%; 1953–1973)
	Northern tidal wetlands	23,000 (75%; 1925–mid-1970s)
Delaware	All wetlands	270,000 (40–50%; 1700s–1980s)
	Tidal wetlands	8,000 acres (9%; 1938–1973)
	Tidal wetlands	7,500 acres (9%; 1954–1971)
	Tidal wetlands	120 (0.1%; 1973–1979)*
Maryland	All wetlands	600,000 (45–60%; 1700s–1990)
	Inland vegetated wetlands	15,000 (6%; 1955–1978)
	Coastal marshes	10,400 (8%; 1955–1978)
	Inland wetlands	4,300 (1.4%; 1982–1989)
	Coastal wetlands	600 (0.5%; 1982–1989)*
Great Lakes Basin	Coastal wetlands	Not available (70%; 1700s–1980s)
Ohio	All wetlands	4.5 million (90%; 1700s–1980s)
	Coastal wetlands (southwest Lake Erie)	951,000 (96%; 1850–1993)
Michigan	All wetlands	5.6 million (50%; 1700s–1980s)
	Coastal wetlands	Not available (70%; 1700s–1980s)
Indiana	All wetlands	4.8 million (85%; 1700s–1980s)
Illinois	All wetlands	7.0 million (85%; 1700s–1980s)
	Great Kankakee Swamp	>1.0 million (99%; 1830s–1990s)
Wisconsin	All wetlands	4.5 million (46%; 1780s–1980s)
Minnesota	All wetlands	4.0–7.0 million (27–39%; 1700s–1980s)
	Prairie wetlands	Not available (>70%; 1700s–1980s)

*the tremendous decline in tidal wetland loss is due to passage of state tidal wetland protection laws.
Note: Numbers have been rounded off. (Data from U.S. Fish and Wildlife Service wetland trends studies and state reports and other sources including Fretwell et al. 1996.)

state, development is the major threat. Permitted losses amounted to nearly 12,000 acres statewide from 1982–1990 for an average annual loss of over 1,300 acres.

Causes of Wetland Loss and Degradation

The notion that wetlands were not as valuable to society in their natural state as they would be with alteration (wetlands as wastelands) has largely caused the demise of wetlands worldwide. In the name of "land reclamation" (an ironic term for areas that were never dry land), people have filled wetlands for a host of uses in the civilized world. Wetlands have also been dredged, impounded, drained, and mined for various purposes (Table 8.4). Many of these uses have benefited human societies, but often at the expense of significant and, in many instances, irreparable environmental degradation. The effects of wetland destruction are obvious when wetland functions and values are lost. Ac-

Table 8.4. Major causes of wetland loss and alteration in the Northeast and Great Lakes region.

Human Threats

1. Drainage for farming, forestry, and mosquito control.
2. Dredging and stream channelization for navigation, flood protection, coastal housing developments, and reservoir maintenance and to facilitate drainage.
3. Filling for disposal of dredged material and solid wastes, road and highway construction, and commercial, industrial, and residential real estate (buildings).
4. Construction of dikes, dams, levees, and sea walls for flood protection, water supply, irrigation, storm protection, and wildlife habitat management.
5. Discharge of pesticides, herbicides, industrial wastewater, and municipal wastewater.
6. Runoff from agricultural, urban, and suburban areas.
7. Mining of peat, sand and gravel, and coal.
8. Extraction of groundwater and river water for domestic, agricultural, and industrial usage.

Natural Threats

1. Subsidence due to rising sea level.
2. Droughts.
3. Hurricanes and other storms.
4. Erosion.
5. Animal actions (flooding induced by beaver; muskrat and snow goose eat-outs; insect infestation).
6. Invasive species.
7. Climatic change due to global warming (induced at least in part by human actions).

tions that degrade wetlands may eliminate some functions and significantly impair or diminish others. Natural processes are also operating to reshape or destroy wetlands. For example, sea-level rise is permanently flooding some salt marshes in Chesapeake Bay, and marsh and upland islands have been eroding away. The following sections highlight some of the human-induced causes of wetland loss and degradation in the Northeast, with reference to other regions. Natural factors affecting wetlands have been discussed earlier (Chapter 3).

Wetland Destruction

The unique properties and the waterside location of many wetlands destined them for human use. The fertility of their soils made floodplain wetlands, prairie pothole wetlands, and mucky swamps highly desirable sites for cultivation. The rich peat deposits of bogs attracted mining interests. Wetlands adjacent to navigable waters targeted them for port development, navigational uses, and the establishment of vacation communities. Other wetlands simply were located in expanding urban areas or in areas of sand and gravel or coal deposits where they were either filled for development or exploited for minerals.

Drainage for Farming

Cultures the world over have utilized fertile floodplain wetlands to produce crops ranging from rice (the staple for most of the world's human population) to corn and soybeans. Some farming can be accomplished by simply removing existing vegetation and planting desired crops, whereas most farming in wetlands requires drainage. Land drainage for crop production has been practiced for at least two thousand years. In the second century B.C., the Roman statesman Cato drained wet spots in fields to improve crop yield. In 1748, the Connecticut clergyman Jaret Elliot promoted wetland drainage and the use of muck as fertilizer in his essays on field husbandry.

Beginning in the late 1800s, the emergence of soil-drainage science, industrial production of tile drains, and mechanization of ditch-digging faciliated drainage projects throughout the world. In the United States, these innovations, coupled with the passage of the Swamp and Overflow Lands Acts (1849, 1850, and 1860), accelerated wetland drainage. Through the swampland acts, the federal government gave fifteen states authority to claim federally owned wetlands, provided the lands were drained and put into agricultural production. By 1905, a total of 82 million acres had been claimed, with half of the total in Florida (20 million), Louisiana (12 million), and Arkansas (9 million). Northeastern states were not included in these acts, since the federal government never had any significant landownership in the original thirteen states.

Land-grant colleges taught swamp draining as part of their curriculum, and agricultural drainage is still a fundamental course. Books like *Productive Soils* (Weir 1920) and *Text-book of Land Drainage* (Jeffrey 1921) were standard authorities used in many schools. Wilbert Weir's book was a general work on soils that included a chapter on land drainage and irrigation in which Weir noted that in many instances, only a spade was needed to dig a shallow trench to a nearby roadside ditch for drainage. Such ditches have facilitated wetland drainage in many areas of the country, especially in the Midwest prairies, where millions of potholes have been drained. Joseph Jeffrey reported that a combination of ditches and pumps was used to drain a 40,000-acre Louisiana tract, including part of New Orleans, and the Owosso Sugar Company's 10,000-acre farm near Saginaw, Michigan. Another book, *Engineering for Land Drainage: A Manual for the Reclamation of Lands Injured by Water* (Elliott 1912), included chapters on levee drainage systems for river bottomlands, reclamation of tidal lands, and drainage of peat and mucklands, among others. John Elliott called John Johnston of Geneva, New York, the "Father of tile-drainage in the United States," because from 1835 to 1851 he had laid sixteen miles of tile. Elliott also noted that New York City's Central Park, a swampland considered a menace to public health, was drained in 1858 under the direction of Colonel George Waring, who later wrote *Draining for Profit and for Health*. This was the largest drainage project in the country at that time. Elliott's book describes a drained tidal marsh in Delaware City, Delaware, that had been producing corn, wheat, and grass since the 1850s.

Across the country, drainage districts were established to organize local landowners to accomplish wetland drainage. Such organizations were needed, since draining one parcel upstream would undoubtedly increase water problems downstream and farmers had to cooperate in order for all to benefit from drainage. The passage of state drainage laws provided the legal authority to form such bodies, to collect assessments to reduce the costs of drainage to individual landowners, to pay for any inadvertent damage resulting from drainage projects, to establish the perpetual right of members to use ditches or drains, and to sell bonds to secure needed monies to fund drainage projects. New Jersey passed the first drainage law as early as 1772. Delaware probably has the most active and best organized wetland-drainage program in the Northeast today. Drainage laws in other states produced similar cooperative ventures. For example, Ohio laws in the mid-1800s facilitated drainage of the Black Swamp by 1900. Drainage districts established in Illinois drained nearly 5 million acres between 1870 and 1930.

Besides giving authority over wetlands to fifteen states, the federal government supported wetland conversion to farmland nationwide for more than a hundred years. The Department of Agriculture's Division of Drainage Investigations conducted research to determine the best methods for drainage, and the Soil Conservation Service offered technical assistance, including the preparation of plans, to farmers seeking to drain wetlands. The swampbuster provision of the 1985 Food Security Act now prohibits such drainage and encourages wetland restoration (see Wetland Protection).

Agricultural conversion has been responsible for the greatest losses of the nation's wetlands, even in recent years. From the mid-1950s to the mid-1970s, drainage and cultivation of wetlands was responsible for almost 90 percent of the human-induced losses: a total of roughly 9 million acres, which represents an area twice the size of New Jersey. By 1978, only 5.2 million acres of bottomland hardwood forests remained in the lower Mississippi valley, representing only 45 percent of the bottomland forests (11.8 million acres) present in 1938. These wetlands were cleared and drained for growing corn, soybeans, cotton, and other crops.

Millions of wetland acres have been converted to productive farmland since colonial times (Figure 8.3). Great Lakes states have converted much more acreage to farmland than Northeast states, draining over 30 million acres. Drainage of the Great Black Swamp in northwestern Ohio alone converted over 3 million acres of wetland to farmland between 1859 and 1885. Crops such as cranberries, blueberries, corn, mint, onions, celery, carrots, potatoes, lettuce, and radishes are grown in these areas. Most of this acreage is effectively drained and is no longer wetland, yet there still remains considerable acreage that is wet enough to be considered farmed wetlands, namely cranberry bogs, mucklands, and partly drained areas. Corn may be the most widely planted crop in the region's former wetlands, especially along floodplains. Sod farms have also been established on drained wetlands. From Delaware and Maryland south, many forested wetlands on the coastal plain have been cleared, drained, and leveled for agriculture (e.g., soybeans and corn), similar to what was done in the Mississippi valley.

When one is thinking of wetland crops, cranberries and blueberries are probably the

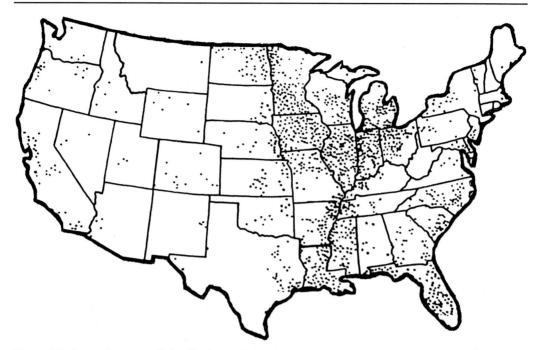

Figure 8.3. General extent and distribution of drained agricultural land in the coterminous United States, 1985. (Source: Dahl 1990)

first plants to come to mind since their berries can be harvested in natural wetlands. Increased demand for these berries led to their cultivation. Over 11,000 acres of Massachusetts wetlands are managed cranberry bogs (Plate 22). Nearly 10,000 acres of New Jersey's former wet pitch-pine lowlands and hardwood swamps now produce blueberries, while about 3,000 acres of former Atlantic white-cedar swamps yield cranberries.

Cranberry cultivation began in Dennis, Massachusetts, in 1810. Growing cranberries has traditionally involved removing natural vegetation (mainly Atlantic-white-cedar-swamp species), leveling the area by removing the upper 2 to 4 inches of turf, sanding the bog with 3 to 4 inches of sand, constructing a series of dikes and ditches to manage water levels, and often building a pond to hold water for irrigation and cold-season flooding of the bog. More recently, many cranberry bogs have been expanded by excavating sandy uplands down to the water table. Cranberries are worth almost $100 million annually. Massachusetts has always led the nation in cranberry production and acres under cultivation: slightly less than half of the nation's cranberry crop comes from this state. Wisconsin is the second-leading state, pro-

ducing less than one-third of the nation's crop. New Jersey is third-ranked, accounting for less than 10 percent of production (see Eck 1990.)

In 1916, Elizabeth White (a cranberry grower) and Dr. Frederick Coville of the U.S. Department of Agriculture began cultivating blueberries in Whitesbog, New Jersey. Blueberries are propagated from "whips" collected in spring before budbreak, which are planted in a peat-sand mix then transplanted to the fields in the fall. New Jersey's blueberry crop generates an annual income of nearly $40 million, with a net return of nearly $4,000 per acre. Nationally, New Jersey produces about one-quarter of the highbush blueberry crop and ranks second in blueberry production (behind Michigan's 15,000 acres and annual income of $50–60 million). (For additional information on blueberry cultivation, see Eck 1988 and Eck and Childers 1989.)

In northern New Jersey and several regions in New York, thousands of acres of wetlands, many on former glacial lake beds, are now farmed for truck crops such as carrots, lettuce, and onions (Figure 8.4). The rich organic soil of these mucklands is perfect for growing these crops. Due to mucklands' severe wetness, water must be pumped out to control water tables.

Figure 8.4. Many wetlands with organic soils in northern New Jersey and several regions of New York have been converted to mucklands. Nearly 40,000 acres of mucklands exist in New York.

These farmed wetlands are among the most productive and profitable agricultural lands in the Northeast.

Although the impacts from agriculture have been orders of magnitude greater than those from other causes of wetland loss, they are not irreversible. Drained wetlands are among the easiest and most cost-effective sites for wetland restoration (see Wetland Protection at the end of this chapter). Decreased market demand, inadequate drainage, and changes in life-styles have led to the closure of many farms in the region. With abandonment, many formerly drained acres are reverting to wetlands. In other cases, farmers who have found that much of their drained land is marginally productive at best are now restoring wetlands through government-subsidized programs, such as the Fish and Wildlife Service's Partners for Fish and Wildlife program and the Department of Agriculture's Wetlands Reserve program.

Dredging and Channelization

Dredging involves the excavation of hydric soils or bottom substrates of waterways and the depositing of this material in other wetlands or deeper water. It is usually performed to maintain or improve navigation or to create ports, marinas, and other developments such as waterside housing (Figure 8.5). Until recently, excavated material was simply disposed of on marsh surfaces or in impounded marshes. Under new environmental laws, dredged material must now be disposed of on upland sites or barged to deepwater disposal sites as the capacity of previously diked sites is exhausted.

Dredging activities have filled thousands—perhaps millions—of acres of marshland along the nation's coastal and inland rivers. Significant losses of wetlands have occurred along the Atlantic Intracoastal Waterway, a system of man-made canals and natural waterways extending from New Jersey south to Florida to facilitate commercial and recreational boating. Periodic dredging is required to maintain this system. River dredging and spoil disposal has extirpated or seriously depleted populations of numerous wetland plant species in some river systems, such as the Delaware River. Today dredging is strictly regulated under various federal and state water-resource laws (see Wetland Protection).

(a)

(b)

Figure 8.5. Dredge-and-fill operations and filling of coastal marshes from the upland side have destroyed many tidal wetlands in port cities and other coastal locations. (a) The Hackensack Meadowlands, one of the largest remaining tidal wetlands in northern New Jersey. Most of the tidal wetlands in the New York City–Newark metropolitan area have been filled for urban, industrial, and port development and sanitary landfills, while the remaining acreage is seriously degraded by pollution. Despite this, these wetlands are still important for wildlife. (Robin Burr) (b) Aerial view of a dredge-and-fill development in progress on the south shore of Long Island. Such developments were common along the East Coast in the 1950s and 1960s, prior to passage of laws to protect tidal wetlands. (Al Dole)

Channelization is a type of dredging to widen or deepen an existing channel. Sometimes a project is designed to improve a navigable waterway, but in most cases it is an expanded form of ditching intended to provide improved drainage upstream or flood protection downstream. Until recently, the Department of Agriculture's Small Watersheds Program promoted such drainage improvements in many areas. Small streams in bottomland forests were channelized to facilitate drainage of agricultural fields, often on former wetlands upstream. The improved surface-water runoff often resulted in additional wetland drainage upstream as well as adjacent to the channel. This domino effect has been observed in many parts of the country, especially in the Southeast. Levees formed by the excavated material may also alter the local hydrology of adjacent wetlands, with varied effects. Some levees prevent seasonal overflows that are important in maintaining wetlands, while others have an impounding effect, where runoff from neighboring uplands is substantial.

Filling

Wetlands along navigable waterways have also been filled for port and urban development. Many of the nation's leading coastal cities, including New York, Boston, Philadelphia, Washington, Baltimore, Norfolk, Tampa, and New Orleans, were built partly on wetlands. This is also true of some of the world's leading port cities, with Venice and Amsterdam being among the more notable examples.

The New York City–Newark area has probably experienced the heaviest urban-wetland losses in the Northeast. In the late 1800s, large tracts of the Hackensack Meadowlands were drained and then filled for the construction of railroad terminals and factories. More recently, the Meadowlands have been filled for a sports complex, a section of the New Jersey Turnpike, and other developments (see Figure 8.5a). At one point, filling of the Meadowlands was occurring at a rate of 30, 000 tons per week.

From the 1950s to the early 1970s, many tidal marshes were filled for residential developments (Figure 8.5b). A combination of dredging and filling was employed to convert hundreds of acres of marsh to lagoon-type housing communities. Parts of the marshes were dredged to create canal systems, while the dredged material was deposited on the adjacent marsh, creating dry land for home construction. Examples of these dredge-and-fill developments can be seen on the south shore of Long Island, in many New Jersey shore communities (e.g., Beach Haven, Forked River, Tuckerton, Sea Isle City, Seaside Heights, and Toms River), and in the vicinity of Ocean City, Maryland. Since the early 1970s, coastal-wetland losses have been greatly reduced by state and federal regulations.

Inland marshes and bogs were filled for urban and residential development. In some communities, wetlands were the last pieces of undeveloped real estate. Many wetlands were zoned for industrial development. Prior to wetland protection, filling was encouraged by municipalities that hoped to benefit through taxes on new developments while making productive use of "wasteland."

Many wetland losses are attributable to roads, railroads, and airports. The direct impacts of transportation corridors on wetlands can be readily seen along many major highways as well as secondary roads throughout the Northeast. Although the loss and fragmentation of wetlands by highway development is obvious, the more significant impact of these roads is subtle: the increase in development of lands once outside the mainstream transportation network and now linked to major metropolitan areas through interstate highways. This indirect impact of the new roads accelerates urban and suburban development in wetlands as well as in forests and farmlands.

The broad, flat stretches characteristic of large wetlands have made them optimal locations for airports. Logan (Boston), Newark, Kennedy, Philadelphia, and Washington National Airports are all built on filled tidal wetlands and bodies of water. A proposal to build an airport in the Great Swamp in Somerset and Morris Counties, New Jersey, rallied local citizens, leading to the establishment of the Great Swamp National Wildlife Refuge and the Somerset Environmental Center. The same thing happened in the Florida Everglades, where an airport was proposed and eventually Everglades National Park was established.

Low spots on the landscape have always been convenient places to dispose of civilization's waste products. Many sanitary landfills and hazardous-materials disposal sites are located in former wetlands. Thus many of Staten Island's original wetlands have been filled with

New York City's refuse. Such uses have created significant groundwater and surface-water pollution. Wetland and other environmental laws have virtually stopped this type of use, although pollution from existing sites continues unless remediated.

Impoundment

Reservoir construction along rivers affects wetlands in two ways. Some wetlands in the reservoir area are filled by the dam embankment itself, while most on the valley floor are permanently inundated. The change in river hydrology adversely affects downstream wetlands and their ecology by disrupting the natural seasonal flow patterns (producing higher off-season floods and lower dry-season flows) and by reducing the amount of sedimentation for maintaining these dynamic systems. Such impacts are perhaps best demonstrated by the Everglades, where the entire ecology of the system has been imbalanced by heavy water withdrawals for agriculture and urban uses. Artificial impoundments in wetlands may have similar impacts. In coastal regions, reservoirs can greatly alter salinity levels in tidal rivers by reducing freshwater flows. For example, construction of the Ordell Reservoir increased saltwater intrusion in the Hackensack River in northeastern New Jersey, changing the ecology of that portion of the river and its bordering wetlands.

Thousands of ponds have been created in wetlands in the Northeast through excavation, diking of streams, or a combination of these methods. Most ponds are built in uplands by digging down to the local water table. The majority of ponds are probably on farms, where they are used for livestock watering and perhaps for irrigation during droughts. In northern New Jersey and the Poconos of Pennsylvania, recreational lakes and ponds have been created by impounding wetlands in mountain valleys.

Wetlands have been impounded by federal and state agencies for waterfowl-management purposes. Many coastal wetlands in federal refuges and state fish-and-wildlife-management areas are diked. Water levels are strictly managed to promote valuable waterfowl food plants like bulrushes and widgeon-grass or to provide freshwater-wetland habitats along the coast. While these modifications benefit waterfowl, shorebirds, and other birds, use by estuarine organisms (fishes and marine invertebrates) is either severely limited or eliminated by the dikes and water-control structures.

Mining

An abundance of valuable soil and minerals lies hidden in or beneath many wetlands. In some places, these materials are extracted by mining operations. Peat deposits up to 50 feet deep or more are associated with many northern wetlands. Peat harvest involves drainage, clearing, and excavation of the peat with heavy machinery that destroys the bog, leaving an excavated depression. The United States is the third leading nation in peat resources with an estimated 124 million acres; the former U.S.S.R. ranks first with 371 million acres and Canada is second with 272 million acres (Canadian Sphagnum Peat Moss Association, www.peatmoss.com). In 2003, three states were responsible for 85 percent of the United States' peat production: Florida, Michigan, and Minnesota. Michigan leads the nation in peat harvest, producing "black peat" (muck) for potting soils. Minnesota harvests "brown peat" (peat moss) that is used for horticultural purposes. Most of this peat moss comes from Canada. In the Northeast, peat moss is harvested in New England (especially Maine), New York, northern New Jersey, the Poconos, and western Maryland.

Coal strata and sand and gravel deposits are buried beneath wetlands in certain areas. Strip mining for coal has reshaped entire landscapes, while the effects of sand and gravel pits are usually more localized. Both activities can adversely affect both wetlands and uplands. Coal-mining impacts are most common in West Virginia, Pennsylvania, and western Maryland. Sand and gravel mining takes place in many areas across the region, especially on the coastal plain and where glaciation has left sizable deposits. Mining has inadvertently created some wetlands where excavation exposes underlying water tables. Some abandoned sand and gravel pits harbor unique plant communities and provide valuable breeding grounds for salamanders.

An unusual type of mining affecting wetlands in the 1700s and 1800s involved bog iron. A form of limonite, this mineral was extracted from a subsoil layer of cemented sands (orstein) glued together by an iron-aluminum-organic complex. Interactions between the decaying leaf litter and the wet sandy soils (spodosols) in certain evergreen forested wetlands cause the formation of bog iron. Many Northeast swamps were sources of bog iron during colonial times. The New Jersey Pine Barrens are one of the more notable areas where bog iron was mined and smelted. The Batsto furnace was built in 1766 along Batsto Creek. At that time, iron production required charcoal for running the furnaces (provided by the abundant pitch pines), sand for castings and insulation (from coastal-plain soils), lime for smelting (obtained from oyster shells), and water for power and transportation (from Batsto Creek and the Mullica River). The availability of these resources in and around the Pine Barrens plus the proximity of Philadelphia made Batsto an excellent location for this industry. The bog-iron ore was excavated from the swamps and floated down streams to Batsto, where it was crushed by water-driven hammers and then processed into usable iron. Some items produced were cast-iron pots, pans, Dutch ovens, stoves, and firebacks (for fireplaces), wrought iron (for many uses, including horseshoes), iron rods, hammers and nails, and even cannon balls for the Continental Army. The Batsto furnace flourished until the mid-1800s, when the discovery of coal near iron-ore deposits in Pennsylvania and new technology for smelting iron made it obsolete. (See Pierce 1966 for more information on this subject.)

Wetland Degradation

Many existing wetlands have been degraded in various ways that have impaired their functions. Activities that destroy some wetlands, such as draining and impounding, have also adversely affected remaining wetlands. Other factors that reduce wetland quality or impair wetland functions include pollution, certain kinds of land uses adjacent to wetlands, introduction of exotics or invasive species, forestry practices that lower habitat diversity for many wildlife species, heavy groundwater withdrawals, and other actions that alter natural hydrologic regimes.

Pollution

Pollution poses a major threat to many bodies of water as nutrient inputs from cities, agricultural fields, residential lawns, and failing septic systems cause eutrophication. Direct discharges of storm water from urban areas, parking lots, roads, and other impervious surfaces have degraded many wetlands. Leachates from landfills and hazardous-waste disposal sites also

pose problems for some wetlands. In the coal regions of the Appalachians, acidic mine drainage pollutes streams and associated wetlands.

The federal government has made substantial progress in improving water quality since the 1970s through the Clean Water Act. Discharges of pollutants from industrial and municipal sources and wetland alteration have been regulated, with the net public benefit of improved quality in many bodies of water. As discharges are successfully controlled, more attention needs to be given to control nonpoint sources such as runoff from the land. Stormwater runoff from urban areas and agricultural lands poses particular challenges for today's water-quality managers.

Changing land-use patterns have an enormous impact on water quality. Historically, the conversion of forests to farmland exposed soils to erosion, and much of the nation's topsoil has been deposited in wetlands and waterways. Ironically, this activity helped form some wetlands along major rivers. The expansion of impervious (paved) surfaces due to urbanization usually increases runoff, with more sedimentation and higher nutrient loadings in streams. This adversely affects wetland and aquatic plants and animals, often favoring less desirable invasive plants and decreasing plant and animal diversity. Plants like purple loosestrife and common reed have become characteristic plants of many urban wetlands.

The ecologist Joan Ehrenfeld of Rutgers University recorded dramatic changes in the species composition of wetlands in the New Jersey Pine Barrens due to urbanization. About 25 percent of the species typical of pristine wetlands disappeared from wetlands in developed watersheds. Elevated pH and increased flooding from urban developments threatened acid-loving native species.

Forestry Management

Harvesting trees is not in itself particularly harmful if done in a manner that promotes the continued growth of native forests and minimizes water-quality degradation. In the short term, cutting and removal of trees clearly changes a forested wetland to an emergent or shrub wetland, but eventually trees and a forested wetland return (Figure 8.6).

Depending on the swamp type, the harvest procedures employed, and the forestry management plan, the impacts can be more significant,

Figure 8.6. Timber harvest of wet flatwoods on Maryland's Eastern Shore.

leading to the elimination of certain forested wetland communities. For example, many Atlantic white cedars were harvested without regard to regeneration, and consequently, most former cedar swamps are now overgrown with red maple. Select or small-tract cutting practices favor replacement by hardwood swamps rather than white-cedar regeneration. Forestry management can, however, reverse this past trend. In some places, cedars are now harvested in a way that encourages their natural regeneration, and special efforts are being made to restore former cedar swamps.

Some forestry practices require minor drainage to improve stand yield, while others completely change the composition of wetland forests. On the coastal plain from Delaware south, many forested wetlands have been transformed into loblolly-pine plantations for pulpwood production. These intensively managed monoculture systems lack the diversity of natural forests and alter the value of such forests for use by wildlife.

Exotic and Invasive Species

Over a hundred exotic plants now grow in Northeast wetlands (Table 8.5; Figure 8.7).

Three of the more widespread and familiar invasive species are common reed, purple loosestrife, and multiflora rose (see Species 7, 84, and 94 in Chapter 10). Detailed descriptions and photos of many invasive species can be found online at: http://invasives.eeb.uconn.edu/ipane/ (the Invasive Plant Atlas of New England).

Along the coast, salt marshes have been crossed by roads and railroads. In many cases, the culverts maintaining a hydrologic connection between the divided marshes are not adequate to pass the full spring tide. This flow restriction has changed the salt balance in the upstream marshes, transforming salt marshes to brackish marshes and even fresh marshes where the tides have been completely cut off. Such conditions favor the establishment and spread of common reed, a tall plumelike grass that changes a marsh from an open low grassland to an almost thicketlike tall-grass monoculture. Common reed has been rapidly spreading in the Northeast since the 1950s, leading to speculation that it is an exotic plant. However, remains of its rhizomes have been found in 3,000-year-old salt-marsh peat in Connecticut, and common reed has also been found in Native American mats about 1,000 years old. Common reed

Table 8.5. Some common invasive species in Northeast and Great Lakes wetlands.

Life Form	Common Name	Scientific Name
Aquatic	Brazilian Waterweed	*Egeria densa*
	Eurasian Water Milfoil	*Myriophyllum spicatum* (Figure 5.10)
	Hydrilla	*Hydrilla verticillata*
	Curly Pondweed	*Potamogeton crispus*
	Water Chestnut	*Trapa natans*
Herb	Garlic Mustard	*Alliaria petiolata* (Figure 8.7)
	Common Ragweed	*Ambrosia artemisiifolia*
	Canada Thistle	*Cirsium arvense*
	Bull Thistle	*Cirsium vulgare*
	Barnyard Grass	*Echinochloa crusgalli* (33b)
	Quack Grass	*Elytrigia (Agropyron) repens*
	Japanese Stilt Grass	*Eulalia (Microstegium) viminea* (35b)
	Ground Ivy	*Glechoma hederacea* (Figure 8.7)
	Tall Pepperweed	*Lepidium latifolium*
	Moneywort	*Lysimachia nummularia* (Figure 8.7)
	Purple Loosestrife	*Lythrum salicaria* (84)
	Yellow Iris	*Iris pseudacorus* (61b)
	Reed Canary Grass	*Phalaris arundinacea* (36)
	Common Reed	*Phragmites australis* (7)
	Mile-a-minute Weed	*Polygonum perfoliatum* (Figure 8.7)
	Japanese Knotweed	*Polygonum cuspidatum* (Figure 8.7)
	Bittersweet Nightshade	*Solanum dulcamara* (Figure 8.7)
	Blue Cattail	*Typha* × *glauca* (48b)
Shrub/Tree	Japanese Barberry	*Berberis thunbergii*
	Bell's Honeysuckle	*Lonicera* × *bella*
	Morrow's Honeysuckle	*Lonicera morrowii*
	Tartarian Honeysuckle	*Lonicera tatarica*
	Common Buckthorn	*Rhamnus cathartica* (Figure 8.7)
	Glossy Buckthorn	*Frangula alnus* (161b; Figure 8.7)
	Multiflora Rose	*Rosa multiflora* (94b)
	White Willow	*Salix alba*
	Weeping Willow	*Salix babylonica* (153b)
	Crack Willow	*Salix fragilis* (153b)
Woody Vine	Japanese honeysuckle	*Lonicera japonica* (133)

Note: Numbers listed in parentheses refer to species number for plants described and illustrated in Part II of this book; those with a letter following the number are described as similar species.

is a native species, yet it has recently been behaving like an invasive exotic. This suggests a possible introduced strain from Europe or elsewhere or a genetic mutation, but to date its status is undetermined.

Because of its invasive nature and attraction to bare soil, the presence of common reed usually indicates a disturbance of some kind such as degraded water quality or marsh filling. It is frequently observed in roadside ditches and has even been found growing in shallow water as a floating mat. In this country, there is almost universal interest in controlling this species. In coastal areas where salinities are 18 parts per thousand or greater, common reed may be reduced by simply returning full tidal flow to a re-

stricted marsh (provided there has not been a lot of development in low-lying areas or land subsidence in the marsh). A variety of other techniques including herbicide treatment have been used to manage this species.

Purple loosestrife, a Eurasian species introduced, as noted already, for beekeepers and horticulture, is an invader of freshwater marshes and meadows, replacing native species. Its prolific flowers bloom from late July to mid-August, giving the marshes a sensational pinkish purple hue. Once established, it is extremely difficult to eradicate since each plant produces a million or more seeds. Attempts are being made to control the spread of purple loosestrife through the introduction of root-boring and leaf-eating insects

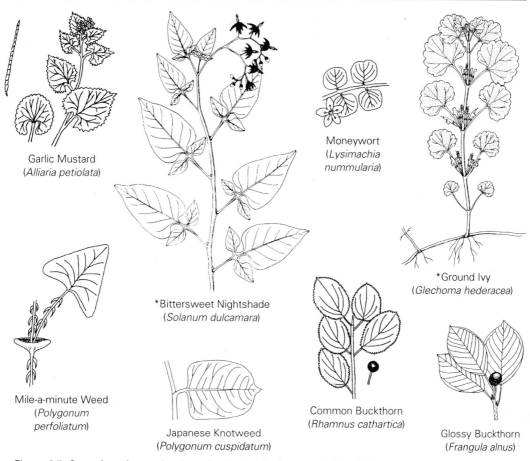

Garlic Mustard
(*Alliaria petiolata*)

Moneywort
(*Lysimachia nummularia*)

*Ground Ivy
(*Glechoma hederacea*)

*Bittersweet Nightshade
(*Solanum dulcamara*)

Mile-a-minute Weed
(*Polygonum perfoliatum*)

Japanese Knotweed
(*Polygonum cuspidatum*)

Common Buckthorn
(*Rhamnus cathartica*)

Glossy Buckthorn
(*Frangula alnus*)

Figure 8.7. Some invasive wetland species in the Northeast and Great Lakes region. (* denotes illustrations by Abigail Rorer; the rest are by the author.)

specific to the plant. More than twenty states are now using this type of biological control.

Multiflora rose is a small white rose that forms impenetrable thorny thickets. Introduced as a hedgerow plant by the USDA Natural Resources Conservation Service, it has proved to be very effective for this, but has escaped and is now common in many wetlands, especially in recently abandoned farmed wetlands and adjacent meadows.

Others exotics are common or becoming so in the region's wetlands. Japanese honeysuckle, a fragrant flowering vine, forms the ground-cover in many wetlands from Delaware south. European or glossy buckthorn is now a dominant understory shrub in many red-maple swamps of southern New England. Similarly, common buckthorn, a European introduction with fine-toothed leaves and short thorns at the end of the branches, has become a dominant

small tree of both wetlands and uplands in western New York, Pennsylvania, and the Great Lakes states. Other less-threatening introductions have been here since colonial times and have stable populations. Most people probably would not recognized them as exotics. Such species include common glasswort and salt-marsh fleabane in salt marshes, and yellow iris, slender blue flag, true forget-me-not, nodding beggar-ticks, pipeworts, water pepper, big duckweed, spearmint, and peppermint in freshwater marshes.

Hydrologic Manipulation

Alteration of the hydrology of watersheds and withdrawals of groundwater have perhaps the most insidious adverse impacts on wetlands. While the visual impacts of dredging, filling, and impounding wetlands are striking, the impact of hydrologic alteration often goes unnoticed, except for the channels and ditches formed or where the wetland is converted to specific uses such as cropland. The effect on river flows may be most pronounced in summer during normal low-water periods. For example, public and industrial wells in and adjacent to Ipswich River wetlands in Massachusetts cause this section of the river to dry up in summer. Gauging the effects of groundwater withdrawals and alterations of river and stream flows on wetlands requires a rigorous assessment of hydrology and a comparison with the hydrology of similar unaffected wetlands.

Most wetland plant species can grow in drier, mesic conditions and will survive and likely reproduce under the altered hydrologic regime. The vegetation remains largely unchanged, providing no indication of significant damage. If the wetland has organic soils, the ground will subside over time as the muck or peat dries up and disintegrates. Exposed tree or shrub roots well above the ground surface are evidence of these impacts. In any case, the former wetland continues to provide habitat for some wildlife, but the natural seasonal cycle of wetness and dryness that produced a rich plant and animal community and provided valuable functions such as water filtration is lost or significantly modified.

When subject to heavy groundwater withdrawals, some wetlands may change from being groundwater discharge sites to groundwater recharge sites. At first glance, this might be viewed as a positive—adding to groundwater supplies—but it is more a sign of system failure. The capacity of the natural aquifer to supply water for human use without disrupting natural ecosystems has been exceeded. Groundwater flow paths reverse direction, and more surface water enters the groundwater, thereby taking surface water away from rivers and streams, with potential adverse impacts on aquatic species. While there may be no immediate signs of such impacts, the long-term effects could be disastrous.

In the western United States due to lower annual precipitation, higher temperatures, and intensive agricultural uses, there is a long history of water shortages. The East is blessed with abundant water supplies and plenty of rain (40 to 50 inches on average, compared to 20 inches or less for western states). But the expanding population and increasing recreational (e.g., ski resorts and golf courses) and industrial uses for water are beginning to exhaust the natural capacity to satisfy eastern water needs. Major volumes of water are being diverted from rivers and streams, and groundwater is being withdrawn at rates that are altering the hydrology of wetlands at withdrawal sites and downstream. The salt balance in some estuaries is being changed. In the future, water laws may play a more prominent role in the eastern courts, but probably not until municipalities fail to fulfill the public's water needs in spite of upstream diversions and significant groundwater withdrawals that have dewatered many wetlands.

Wetlands Still At Risk

While there is a high degree of wetland protection through federal and state laws across the country (Chapter 9), not all activities that negatively impact wetlands are regulated nor are all wetland types covered by the laws. The better-protected wetlands include tidal marshes, mangrove swamps, floodplain wetlands, lakeside wetlands, and wetlands along rivers and streams or connected to such waters. In general, the wetter the wetland and the more obvious its connection to navigable waters, the more likely it is to be protected to some degree. There are, however, certain wetlands that remain highly threatened by development. Isolated wetlands including pothole marshes, kettle-hole bogs (Figure 8.8), playas, dune swale wetlands, Carolina bays, woodland vernal pools, West Coast vernal pools, sinkhole wetlands, Great Basin terminal salt-flat wetlands, and various types of ponds are more vulnerable to development since a 2001 Supreme Court decision (the SWANCC decision) ruled that the Corps could not regulate an isolated water or wetland solely on its value to migratory birds. Small isolated wetlands like woodland vernal pools and many Carolina bays, however, probably were not effectively regulated prior to this decision. A special edition of *Wetlands* (the journal of the Society of Wetland Scientists) focused on "isolated wetlands" with attention placed on potential implications of this court ruling (*Wetlands* Vol. 23, No. 3, September 2003); see also a U.S. Fish and Wildlife Service report, *Geographically Isolated Wetlands: A Preliminary Assessment of their Characteristics and Status in Selected Areas of the United States* (posted on the web at: http://wetlands.fws.gov).

Although connected to navigable waters by

Figure 8.8. Kettle-hole bog, one of many types of isolated wetlands.

ditches and culverts, some wetlands may not be regulated in certain Corps districts. Head-water wetlands connected to streams by seepage slopes may also be viewed as outside of federal jurisdiction in certain districts. Other wetlands at risk are drier wetlands such as flatwood wet-lands on poorly drained soils along the coastal plain and on glaciolacustrine plains (e.g., Great Lakes plain) that may not meet all criteria ac-cording to the 1987 Corps of Engineers wetland delineation manual. The National Academy of Sciences pointed out several deficiencies in this manual that limited its ability to accurately iden-tify wetlands. Wetlands that are dry for most of the growing season seem to be the ones that may be excluded from jurisdiction following current guidelines; they typically lack the required evi-dence of hydrology at these times. However, if one was to evaluate this type of wetland in the early part of the growing season, the water table should be close enough to the surface to war-rant a wetland finding. Wetlands in urban areas tend to be under continued pressure for devel-opment as they often remain the last parcels of undeveloped real estate.

Additional Readings

Canadian Wildlife Service. 2002. *Where Land Meets Water: Understanding Wetlands of the Great Lakes.* Environment Canada, Downs-view, ON.

Dahl, T. E. 2000. *Status and Trends of Wetlands in the Conterminous United States 1986 to 1997.* U.S. Fish and Wildlife Service, Wash-ington, DC.

Dahl, T. E. 1990. *Wetlands: Losses in the United States, 1780s to 1980s.* U.S. Fish and Wild-life Service, Washington, DC. Report to Congress.

Dahl, T. E., and G. J. Allord. 1996. History of wet-lands in the conterminous United States. In:

J. D. Fretwell, J. S. Williams, and P. J. Redman (compilers), *National Water Summary on Wetland Resources*. U.S. Geological Survey, Water-Supply Paper 2425, pp. 19–26.

Dahl, T. E., and C. E. Johnson. 1991. *Status and Trends of Wetlands in the Conterminous United States: Mid-1970s to Mid-1980s*. U.S. Fish and Wildlife Service, Washington, DC.

Eck, P. 1990. *The American Cranberry*. Rutgers University Press, New Brunswick, NJ.

Eck, P. 1988. *Blueberry Science*. Rutgers University Press, New Brunswick, NJ.

Eck, P., and N. F. Childers (editors). 1989. *Blueberry Culture*. Rutgers University Press, New Brunwick, NJ.

Fretwell, J. D., J. S. Williams, and P. J. Redman (compilers). 1997. *National Water Summary on Wetland Resources*. U.S. Geological Survey, Water-Supply Paper 2425.

Herdendorf, C. E., S. M. Hartley, and M. D. Barnes (editors). 1981. *Fish and Wildlife Resources of the Great Lakes Coastal Wetlands within the United States*, Volume 1: Overview. U.S. Fish and Wildlife Service, Biological Services Program, Washington, DC. FWS/OBS-81/02-v1.

Herdendorf, C. E., C. N. Raphael, and E. Jaworski. 1986. *The Ecology of Lake St. Clair Wetlands: A Community Profile*. U.S. Fish and Wildlife Service, National Wetlands Research Center, Washington, DC. Biological Report 85(7.7).

Pierce, A. 1966. *Iron in the Pines*. Rutgers University Press, New Brunswick, NJ.

Southern Environmental Law Center. 2003. Georgia wetlands at risk: three case studies. Posted at: http://www.selcga.org.

Tiner, R. W. 1984. *Wetlands of the United States: Current Status and Recent Trends*. U.S. Fish and Wildlife Service, Washington, DC.

Tiner, R. W., H. C. Bergquist, G. P. DeAlessio, and M. J. Starr. 2002. *Geographically Isolated Wetlands: A Preliminary Assessment of their Characteristics and Status in Selected Areas of the United States*. U.S. Fish and Wildlife Service, Northeast Region, Hadley, MA. (posted at: http://wetlands.fws.gov)

Williams, M. (editor). 1990. *Wetlands: A Threatened Landscape*. Basil Blackwell, Cambridge, MA.

CHAPTER 9

Swampland—Wasteland or Watery Wealth
Wetland Conservation, Management, and Restoration

Their location, biological characteristics, and physiochemical properties have provided wetlands with many functions that are valuable to society (see Chapters 6 and 7). People have used wetlands to satisfy many of society's needs such as food, fiber, raw materials, and space (places to live and work), often in ways that have eliminated or seriously degraded their natural functions (see Chapter 8). Depending on a society's viewpoint, the value of wetlands may be more in their natural state or in an altered state. Wetlands are usually altered because of a need for land for either development or agriculture. So perspectives on wetlands vary in time and by culture. Important factors include the type of society (hunter-gatherer, agrarian, and industrial/commercial), religious beliefs, dependency on natural products harvested from wetlands, medical needs (i.e., to treat diseases borne by wetland organisms), state of scientific knowledge of wetlands, the level of public awareness of the functions and values of wetlands, the abundance of wetlands in a given area, the amount of "original" wetlands remaining, and the degree of threat to existing wetlands.

Changing Attitudes toward Wetlands

For much of America's history, wetlands were viewed by the general public mainly as wastelands and even as public health hazards that produced mosquitoes, biting flies, chiggers, snakes, and alligators and diseases such as malaria, cholera, typhoid, or, most recently, West Nile virus. Even today wetland terms are typically used to describe bad experiences, such as "bogged down," "swamped," or "mired." The use of these terms in literature and everyday speech perpetuates the notion that wetlands are awful and treacherous places—places to avoid or better yet, to drain, fill, and otherwise eliminate. Fortunately, public perceptions about wet-

lands have changed drastically over the past thirty years. For a detailed account of the human history of U.S. wetlands, see Ann Vileisis' *Discovering the Unknown Landscape: A History of America's Wetlands.*

Wetlands—Suppliers of Life's Basic Needs

Early human societies in North America and elsewhere were based on hunting and gathering. Inhabitants of these cultures fished, hunted game, and gathered nuts, fruits, and other foodstuffs to provide raw materials essential to life (such as food, shelter, and clothing). The environment was viewed as the provider—the sustainer of life. In coastal areas, middens (local garbage dumps) left by Native Americans reveal that they harvested waterfowl, shellfish, turtles, and fish from coastal wetlands. Inland tribes also made good use of the bounty of wetlands. Wetland food for these tribes included wild rice, cranberries, blueberries, cattails and other edible plants, fish, turtles, waterfowl, deer, moose, and other animals. Native Americans used wetland grasses, sedges, and twigs for basket making and the hides and feathers of various animals for clothing and adornment; some tribes used wetland plants as building materials. Subsistence cultures like these are the most dependent on wetlands and on the natural environment overall. As a society, they probably have the greatest appreciation for the natural wealth of these areas. These types of cultures still exist in other parts of the world.

Early cultures in the Americas also planted maize and other crops in gardens on rich floodplain soils. More advanced Native American civilizations in Central and South America, with cities of their own, converted substantial acreage of wetlands to cropland.

It is hard for us today to imagine how plentiful fish and wildlife were in these early times. Accounts from early explorers provide a vivid picture of this abundance. From these writings, we know, for example, that lobsters were so plentiful that they could be harvested by hand from shallow water and were used for fertilizing cropland.

Wetlands—Potentially Rich Farmland

European colonists brought their Old World perceptions about wetlands and land in general to the New World. For example, European wetlands were viewed as breeding grounds for diseases like malaria. The colonists also came from societies where land could be bought and sold. In the past, most people considered land left idle to be wasteland that should be "reclaimed" and used more productively. For wetlands, "better" uses could be derived through draining, filling, or other significant modifications. The combination of utilitarian thinking and a negative public attitude towards wetlands is most likely what directed government to treat wetlands the way it did from the 1700s through most of the 1900s. For over two hundred years, all levels of government were responsible for destroying wetlands and the public showed little concern about this despoliation. Wetland drainage for crop production was a national goal. Filling of marshes was as commonplace as harvesting timber. George Washington attempted to drain the Great Dismal Swamp in southeastern Virginia, believing it would better serve the nation by being converted to farmland.

Despite this, not all wetlands were considered worthless from an agricultural perspective. The earliest settlements in America grew up around coastal marshes. From the beginning of the country's settlement, salt marshes were recognized as valuable lands for the production of salt hay. Winter hay from the marshes was the determining factor in how many livestock a farmer could support. Colonists grazed their animals in the marshes harvested salt hay for winter fodder, and used cordgrass ("thatch") for roofing material. In 1879, Reverend Allen Brown, a New Jersey historian, referred to salt marshes as "natural privileges" since they produced annual hay crops without any cultivation (Figure 9.1). Wet meadows have been used as pastures since colonial times; many would have been farmed except they were too wet to plow.

As the population of colonial America grew, fewer people became dependent on coastal wetlands. Inland areas were cleared and upland hay was produced, diminishing the demand for salt hay. With the development of agriculture, animal husbandry, and forestry, people began to convert natural communities into managed ecosystems (farms, rangeland, and forests) to produce food and fiber. The environment was viewed as a resource to be utilized and managed. In many places, marshes became more important as sites for cropland, pasturage, or real estate.

Wetlands—For Real Estate

As the nation moved from an agrarian-based society to an industrial-based society from the late 1700s to the mid-1800s, the value of wetlands in the eyes of most Americans declined further. Cities sprung from small harbor towns and more commercial land was needed. Nearby wetlands were filled for real estate. As mentioned in Chapter 8, a large portion of most port cities on the East Coast was built on former coastal wetlands and shallow water habitat (e.g., Boston, New York City, and Washington, D.C.). The invention of the tractor in the late 1800s, and later the automobile and truck, made it much easier to fill wetlands and the nature of the land could be changed more rapidly. Roads and railroads could be more easily built through wetlands hastening their conversion for commercial and residential development. Many marshes and swamps in and around urban areas were filled for development (Figure 9.2). This rapid growth took place in the Northeast well into the twentieth century. With this marked increase in development, certain individuals and groups became concerned with the loss of wildlife habitat. The federal Duck Stamp program was started in 1934 to secure funds to save important waterfowl wetlands from destruction. The National Wildlife Refuge System was established around this time to save important breeding, overwintering, and migration areas for North American waterfowl. The notion of wetland preservation in the United States was thus introduced more than sixty years ago to benefit wetland wildlife.

Wetlands—Threatened Natural Resources

In the mid-twentieth century, the general public's attitude towards wetlands started to change.

Figure 9.1. Salt hay harvested from mid- to late summer was piled on "straddles" (cedar posts driven into the marsh) to prevent tidal flooding. This allowed hay to dry before removing it from the marsh by horse-drawn sleds or boat. (U.S. Fish and Wildlife Service, photo by W. French)

In the 1950s, the U.S. Fish and Wildlife Service began publishing reports on the threatened status of wetlands, revealing how wetland destruction was adversely affecting the nation's wildlife, especially waterfowl. In the 1960s, scientists studying coastal marshes, especially from University of Georgia's Marine Institute at Sapelo Island, began to make the public more aware of the vital link between these wetlands and nearshore marine fisheries. This linkage was not only critical to supporting commercial fisheries that provided seafood for human consumption and the livelihoods for many citizens in coastal communities, but was also critically important for maintaining recreational fisheries and local tourism. New information on coastal marshes coupled with the sight of bulldozers filling in these aesthetically pleasing vistas quickly changed the public's sentiment toward these habitats. Coastal marshes were no longer viewed as wastelands and as sites for landfills or development, but were now considered valuable natural resources worthy of protection and management. With public support, the government then began to protect wetlands.

In 1962, the Massachusetts legislature passed the first state wetland law in the country, recognizing the many societal benefits provided by salt marshes and giving marshes protected status. A few years later, it passed a similar law to control the use of inland wetlands. By the end of the 1970s, most coastal states in the United States had implemented some measure of protection for tidal wetlands. Most states in the Northeast also began to appreciate the values of their inland marshes and swamps, including water-quality renovation, shoreline stabilization, floodwater storage, and fish and wildlife habitat. During the 1970s and 1980s, the passage and implementation of federal and state laws, especially the Clean Water Act and various state wetland-protection laws produced a tremendous improvement in the status of America's wetlands.

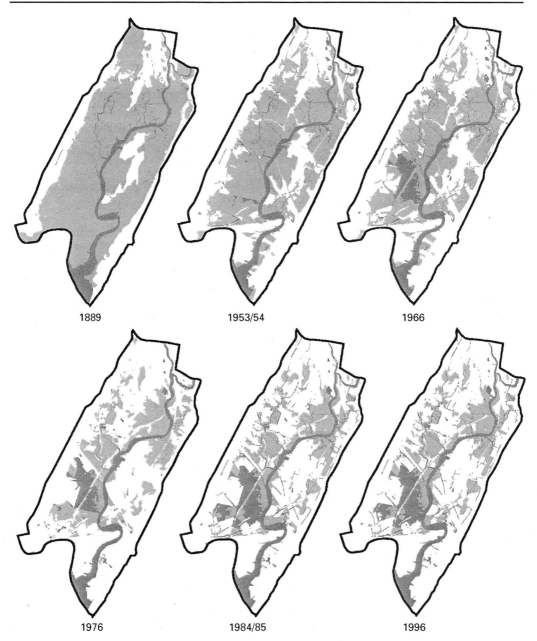

Figure 9.2. Pattern of wetland development in the Hackensack Meadowlands, northern New Jersey (1889–1995). Gray areas represent wetland. (Source: Tiner et al. 2002)

Wetlands—The Conflict between Public Benefits and Private Property

Today, wetlands are regarded as one of the nation's most valuable natural resources. The federal government and many states are protecting, acquiring, and even restoring wetlands of all kinds. The functions of wetlands are better understood and their values more widely appreciated than at any time in the nation's history.

Most people now recognize that land doesn't have to be farmed or developed to be valuable; idle land is not necessarily wasted land, especially if it is providing a vital public service. Moreover, as natural habitats disappear, the remaining lands become more valuable as open space.

Because environmental laws and wetland regulations apply to wetlands on both public and private lands, certain uses of privately owned wetlands now come under the jurisdiction of the U.S. Army Corps of Engineers. The scope of federal jurisdiction increased in 1989 when the federal government adopted standardized techniques for wetland delineation across the country. Previously, each Corps district decided what it would regulate in accordance with the Clean Water Act and there was no national consistency. As a result of the expansion in the federal wetland permit programs, some groups, industries, and individuals have tried and are still trying to get the government to roll back legislation designed to protect wetlands. Operating under the banner of private-property rights, some of these aggrieved parties have distributed misinformation about wetland regulation in an attempt to create a groundswell of opposition. For example, in 1989 an article in *Forbes* magazine profiled the plight of landowners facing increased federal wetland regulation. The spokesperson for a private property advocacy group claimed that the government's new wetland delineation procedures increased the extent of land subject to regulation in Dorchester County, Maryland, from 84,000 acres (prior to adoption of the federal interagency manual in 1989) to 259,000 acres (after 1989). The first number was actually the extent of tidal wetlands regulated by the state of Maryland, while the second was the acreage of hydric soil map units in the state that includes many acres of drained areas (that are no longer wetlands) as well as some upland soils that are not wetlands. While the federal government did increase the scope of wetland regulation in 1989, these statistics clearly overstated the case since the federal government was already regulating both tidal and nontidal wetlands before 1989. This type of misinformation, plus exaggerated accounts of the federal government picking on defenseless citizens (who actually had repeatedly violated federal law and not heeding numerous warnings, were brought to trial), helped galvanize a segment of the American public into a fairly significant private-property rights movement. This vocal body has had some success at getting the government to relax certain environmental-protection requirements.

There has been a noticeable decrease in federal wetland regulation since 1991 when agencies were forced by Congress to abandon use of the 1989 interagency technical manual for identifying wetlands. The federal government now uses procedures developed back in 1987 that provide less comprehensive wetland coverage. The change in federal practices also caused several states to conform their delineation procedures to those of the federal government. The result is that thousands of wetland acres no longer require a government permit for alteration, so they are being filled and converted to housing developments, industrial sites, and other uses without public review.

In 2001, a Supreme Court decision on a federal regulatory case in Illinois (called the "SWANCC decision"—Solid Waste Agency of Northern Cook County) ruled that the Army Corps of Engineers overstepped its regulatory authority when it took jurisdiction over certain ponds ("isolated waters") solely on their value to migratory birds. This led to a significant rollback in federal wetland regulations, since some Corps districts interpreted this decision to mean that they should not regulate any isolated wetland (including those connected to navigable waters by ditches). Other districts had a narrower interpretation of the court's ruling, limiting it to situations where the only interstate commerce connection was migratory bird habitat. A recent U.S. General Accounting Office (GAO) audit of the Corps Clean Water Act permitting program concluded that Corps districts differ in their interpretation and application of federal regulations regarding jurisdiction. In particular, it noted differences in handling of: 1) adjacent wetlands (criteria used to determine when a wetland is "adjacent" to navigable waters and can be regulated), 2) isolated wetlands (what criteria are used to determine whether or not to regulate this type of wetland), and 3) ditch and tile drain connections (whether wetlands connected to navigable waters by these structures are subject to jurisdiction). The GAO report also found that only three of the sixteen Corps districts evaluated had published guidelines that were available to the public, while the others relied on oral communication by staff. The GAO recommended that the Corps (in consultation with the EPA): 1) survey all district offices to determine if significant differences in jurisdictional interpretations exist, 2) determine whether and how to resolve these differences, and 3) require all thirty-eight districts to document their procedures and make this information available to the public.

Despite these setbacks, much progress had been made since the 1970s due to government environmental protection programs. The rivers and the air are cleaner, some endangered species have recovered to the point that they are no longer threatened by extinction (the bald eagle, our nation's symbol, is no longer endangered), wilderness areas have been set aside for all to enjoy, and wetlands are being better protected. Annual wetland losses have been cut at least in half. The United States has come a long way, but still has a great distance to go before citizens can afford to be complacent about environmental protection as evidenced by recent events. These significant gains did not just materialize out of thin air. They happened because the majority of the public wanted improvements in the environment, demanded that government do something, and saw to it that it happened. (See Appendix A for tips on how you can help conserve wetlands.)

Wetland Protection

During the last three decades, a great deal has been learned about wetland functions and wetlands are now regarded as among America's most valuable natural resources (Chapter 7). Development of most wetlands is regulated nationwide by the Army Corps of Engineers, with oversight from the Environmental Protection Agency. Most states in the Northeast and some in the Midwest have passed laws to protect wetlands from destruction and degradation. (For a more in-depth review of wetland regulations, see Kusler and Opheim 1996; for specific information on current regulations, contact the appropriate government agency, including EPA's Wetland Information Hotline 800-832-7828 or EPA's Water Channel on the Internet: http://www.epa.gov/owow.) Table 9.1 presents a timeline of some significant events affecting wetland protection nationally and in the Northeast and Great Lakes states.

Wetland Regulation

A combination of federal and state laws and regulations and local ordinances controls uses of wetlands in the Northeast and Midwest. At the federal level, the most significant wetland regulations are promulgated in accordance to Section 10 of the Rivers and Harbors Act of 1899 and Section 404 of the Clean Water Act. The former relates to wetlands along navigable waters, while the latter encompasses all waters of

the United States. The federal laws address wetlands indirectly—that is, wetlands are largely protected in an attempt to preserve water quality. In contrast, most state laws and local bylaws have been passed to explicitly protect wetlands for the myriad functions they perform.

Federal Wetland Regulations

Section 10 requires individuals seeking to dredge, deposit fill, or place structures in navigable waters (e.g., tidal waters, many nontidal rivers, and some lakes) to obtain a permit from the U.S. Army Corps of Engineers. These activities are regulated in wetlands from shallow water to the mean high tide level in coastal areas and to the ordinary highwater mark in inland waters.

Section 404 establishes a permitting program for the deposition of dredged and/or fill material in waters of the United States, including most wetlands (Figure 9.3). The Corps also administers this program, but with oversight from the U.S. Environmental Protection Agency. Regulated activities include filling for residential and commercial development, roads, dams, and airports. Drainage to convert wetlands to upland farms, orchards, or managed forests may be prohibited unless authorized by permit. Landowners must obtain a permit prior to engaging in regulated operations. Certain activities are exempt from these requirements, such as discharges from normal farming, forestry, and ranching operations, maintenance of existing ditches, ponds, and structures, and construction of temporary sediment basins, farm roads, forest roads, or temporary mining roads. The program only permits unavoidable impacts to regulated wetlands identified through the Corps of Engineers wetland delineation manual. Projects must go through a sequencing process that demonstrates first avoidance, and then minimization of wetland impacts prior to considering any wetland alteration. Most alterations require some type of compensation for or mitigation of unavoidable impacts, such as restoration or creation of wetlands and sometimes by wetland acquisition.

Section 401 of the Clean Water Act gives states the authority to deny water quality certification for federally permitted or licensed activities that involve a discharge into waters of the United States (including wetlands) if such activities fail to meet state water quality standards. A denial prevents the Corps from issuing a federal

Table 9.1. Timeline of significant events for wetland protection.

1849	First Swamp Lands Act enacted giving Louisiana 9.5 million acres of wetlands with intention that they be drained and "reclaimed"
1850	Swamp Land Act enacted giving twelve states over 50 million acres of wetlands to drain and reclaim (AL, AR, CA, FL, IL, IN, IA, MI, MS, MO, OH, and WI)
1860	Swamp Land Act gave two states (MN and OR) about 5 million acres of wetlands
1899	Rivers and Harbors Act enacted giving U.S. Army Corps of Engineers responsibility for regulating dredging, filling, and construction of structures in navigable waters
1903	President Theodore Roosevelt issued Executive Order to establish Pelican Island (FL) as the "first national bird reservation;" over fifty more reservations were authorized by 1909
1929	Migratory Bird Conservation Act authorized acquisition of wetlands for waterfowl habitat (no funding provided)
1934	Migratory Bird Hunting Stamp Act required hunters to purchase a duck stamp annually for waterfowl hunting and revenues were dedicated to purchase or lease waterfowl habitat (including wetlands)
	Fish and Wildlife Coordination Act required federal agencies (Agriculture and Commerce) to provide technical assistance to states and other federal agencies to protect and increase game and fur-bearing animals and to study the effects of pollution on wildlife resources
1940	U.S. Department of Agriculture authorized to provide technical assistance and cost sharing to farmers for draining wetlands under Agricultural Conservation Program (practice continued until mid-1970s); nearly 57 million acres of land (including wetlands) were drained
1946	Fish and Wildlife Coordination Act amended to require federal agencies to consult with the U.S. Fish and Wildlife Service (FWS) and state fish and wildlife agencies for projects requiring federal permit or license; objective was to prevent loss of and damage to wildlife
1954	FWS published *Wetlands of the United States* (Circular 39) to report on the current status of wetlands in the coterminous United States with emphasis on wetlands important to waterfowl
1958	Fish and Wildlife Coordination Act amended to require that fish and wildlife resources be given equal consideration when planning federal water resource development projects. Development agencies must consult with the U.S. Fish and Wildlife Service, National Marine Fisheries Service, and the state wildlife agency on the impact to fish and wildlife resources of proposed federal projects or projects requiring federal permits.
1960s	FWS published reports on threats to wetlands
1962	Congress passed law (87-732) prohibiting U.S. Department of Agriculture from providing technical assistance or funding to Minnesota, South Dakota, and North Dakota where the Secretary of Interior found that wildlife would be harmed by drainage
1963	Massachusetts passed first state wetland protection law in the nation to protect salt marshes (the Jones Act)
	Recreation Coordination and Development Act established Land and Water Conservation Fund that provided funds to states for outdoor recreation projects and to federal agencies for land acquisition (funds used to purchase wetlands)
1965	Massachusetts enacted first state law to protect freshwater (nontidal) wetlands (the Hatch Act)
1966	National Wildlife Refuge System Administration Act—organic act for the National Wildlife Refuge System (amended in 1997)
1967	New Hampshire passed law to protect coastal wetlands
1968	Land and Water Conservation Fund Act established a program to provide funding to states and local governments to acquire lands (including wetlands) for outdoor recreation and open space
	Rhode Island passed law to protect coastal wetlands

(continued)

Table 9.1. (continued)

1969	National Environmental Policy Act required federal agencies to prepare environmental impact statements for proposed projects
	Connecticut passed tidal wetlands protection law
1970	Maryland and New Jersey passed legislation to protect tidal wetlands
1971	Rhode Island passed freshwater wetland protection law
1972	Federal Water Pollution Control Act (later renamed Clean Water Act) amendments gave Corps of Engineers and Environmental Protection Agency authority to regulate deposition of fill material and other pollution in U.S. waters.
	Coastal Zone Management Act enacted providing federal funding to states to establish coastal zone management programs to protect and conserve coastal wetlands and other resources
	Endangered Species Act enacted to protect ecosystems that support endangered species; many such species require wetlands for survival
	Connecticut passed freshwater wetland protection law
1973	Delaware and New York enacted laws to protect coastal wetlands
1974	FWS initiated plans for a national wetlands inventory
1975	Court decision—*Natural Resources Defense Council v. Callaway* (in Florida)—determined that wetlands are part of the waters of the United States and that the Corps must regulate activities impacted these under Section 404 of the Clean Water Act
	Corps and EPA published proposed wetland definition in the Federal Register (July 25, 1975)
	FWS proposed initial wetland classification system (undergoes several iterations prior to final publication in 1979)
	New York passed the Freshwater Wetlands Act
1976	FWS used interim wetland classification to begin national wetlands inventory
1977	Corps and EPA finalized wetland definition based on review (July 19, 1977)
	President Carter issued Executive Orders 11990 (Protection of Wetlands) and 11988 (Floodplain Management) requiring all federal agencies to minimize wetland destruction or degradation and avoid development on floodplains when carrying out their duties, including providing technical assistance
	Clean Water Act amendments added provisions to delegate regulatory authority to states and directed the FWS to provide technical assistance to states in developing programs for regulating the discharge of dredge and fill materials into U.S. waters including wetlands. The FWS was also authorized to conduct a national inventory of wetlands.
1978	Pennsylvania's Dam Safety and Encroachments Act included provisions to regulate activities in wetlands (wetland regulations amended in 1991)
1979	FWS published final version of its wetland classification system *Classification of Wetlands and Deepwater Habitats of the United States* for conducting national wetlands inventory (later adopted as the national standard for wetland inventories and for reporting on the status of the nation's wetlands)
	Michigan passed state wetland protection law (Goemaere-Anderson Wetland Protection Act)
1981	Court decision—*Avoyelles Sportsman's League, Inc. v. Alexander Marsh* (in Louisiana)—established three criteria for wetland identification (hydrophytic vegetation, hydric soils, and wetland hydrology)
1982	Coastal Barrier Resources Act designated undeveloped barrier islands as part of the Coastal Barrier Resources System. Such areas are ineligible for federal financial assistance.
1984	FWS published national report *Wetlands of the United States Current Status and Recent Trends* documenting wetland status, trends, and values and reporting that over half of the wetlands in the lower 48 states had been lost
	Florida enacts law to regulate dredge and fill in wetlands

Table 9.1. (continued)

1985	U.S. Supreme Court decision—*United States v. Riverside Bayview Homes* (Ohio)—affirmed Corps jurisdiction in wetlands adjacent to navigable waters
	Food Security Act contained Swampbuster provision that encourages wetland conservation by denying federal agricultural subsidies to farmers who convert wetlands to cropland (commodity crops) after December 23, 1985
1986	Emergency Wetlands Resources Act authorized purchase of wetlands through the Land and Water Conservation Fund, required the Secretary of Interior to report to Congress at ten-year intervals on the status and trends of the nation's wetlands, and established completion dates for the National Wetlands Inventory (September 30, 1998, for lower 48 states; September 30, 1990, for Alaska and noncontiguous areas; funding not provided)
	North American Waterfowl Management Plan, signed by the United States and Canada, initiated joint ventures to restore and protect waterfowl habitat (Mexico became a signatory in 1994)
	Vermont passed the 1986 Act Relating to the Regulation of Wetlands
1987	Corps of Engineers published its wetland delineation manual as "guidance" for determining areas subject to jurisdiction under Section 404 of the Clean Water Act; use of the manual was discretionary
	New Jersey passed the Freshwater Wetlands Protection Act
1988	National Wetlands Policy Forum (comprising government, business, private citizens, academics, bipartisan representatives) recommended that the federal government adopt a policy of "no net loss of wetlands"
	Great Lakes Coastal Barrier Resources Act passed (similar to 1982 Act but specific for Great Lakes region)
	EPA published its wetland delineation manual for determining jurisdictional wetlands
	Maine's Natural Resources Protection Act included provisions for regulating uses of wetlands
1989	Interagency wetland delineation manual published; developed by four federal agencies (Corps of Engineers, EPA, FWS, and Soil Conservation Service) to serve as the technical standard for wetland identification (January 10, 1989)
	Corps and EPA issue memorandum of agreement adopting the interagency manual as the national standard for identifying and delineating wetlands subject to Clean Water Act jurisdiction (January 20, 1989); first mandated standard for delineation of federally regulated wetlands
	President Bush's administration announced policy of no net loss of wetlands
	North American Wetlands Conservation Act provided matching funds for organizations and individuals to acquire, restore, and enhance wetlands
	Maryland passed freshwater (nontidal) wetland protection law
	Illinois passed the Interagency Wetland Policy Act, promoting a goal of no net loss of wetlands due to state-funded projects
1990	Coastal Wetlands Planning, Protection, and Restoration Act established National Coastal Wetlands Conservation Grant Program providing funds for acquisition, restoration, and enhancement of these wetlands
	Vermont adopted rules for identifying and protecting wetlands
1991	EPA proposed revisions to interagency manual. Thousands of comments recommended rejecting this proposal since the requirements would significantly limit the number of wetlands subject to federal regulation.
	Minnesota passed the Wetland Conservation Act
1992	Energy and Water Development Appropriations Act prohibited the Corps from using the interagency manual. Corps adopted its 1987 wetland delineation manual as standard for making jurisdictional determinations.

(continued)

Table 9.1. (continued)

1993	National Research Council (NRC) created Committee on Wetland Characterization to conduct scientific review of wetland delineation practices used by the federal government and summarize current understanding of wetland functions
	President Clinton issued *Protecting America's Wetlands: A Fair, Flexible, and Effective Approach*, which reaffirmed President Bush's policy of no net wetland loss and laid out plans for a national wetland policy
1994	Memorandum of agreement making Soil Conservation Service (Natural Resources Conservation Service) the lead agency in making wetland determinations on farmland
1995	National Academy Press published *Wetland Characteristics and Boundaries* detailing the NRC findings on the science behind wetland delineation. It recommended that the federal government develop a new wetland delineation manual based on experiences learned since 1987.
2000	Corps published revisions of nationwide permits
	Estuaries and Clean Waters Act authorized development of a national estuary habitat restoration strategy and provision of grants for such restoration
2001	U.S. Supreme Court decision—*SWANCC v. U.S. Army Corps of Engineers*—invalidated Corps use of "migratory bird rule" for identifying certain "isolated" waters of the United States as jurisdictional waters, putting isolated wetlands at greater potential risk for alteration
	Wisconsin enacted legislation (Wisconsin Act 6) to protect isolated wetlands
	National Research Council published report on wetland mitigation concluding that lost wetlands are not being adequately replaced by mitigation
2004	President Bush announced program to increase wetland restoration and established goal of 3 million acres over the next five years

Note: Laws mentioned are federal laws unless stated otherwise. State laws outside the Northeast and Great Lakes region are not listed.

permit in accordance with the Rivers and Harbors Act and the Clean Water Act and the Federal Energy Regulatory Commission from granting federal licenses for hydropower projects. This can be a powerful water quality and wetland protection tool, especially for states without laws to protect wetlands.

Federal regulations have had a tremendous positive impact on wetland protection nationwide. From the mid-1950s to the mid-1970s (prior to the implementation of federal wetland regulations), wetland destruction was occurring at a rate of about 500,000 acres per year in the coterminous United States. Ten years after federal regulations went into effect, the annual rate was reduced to 290,000 acres. For the 1990s, the yearly decline was estimated at 58,500 acres.

Other Federal Wetland Conservation Policies

In 1985, Congress enacted provisions to help protect wetlands from agricultural drainage as part of the Food Security Act (also referred to as the Farm Bill). The provision known as "swampbuster" is intended to discourage the use of wetlands as farmland. If a farmer drained a wetland and produced a commodity crop like corn or soybeans after December 23, 1985, or converted a wetland for such purpose after November 28, 1990, he or she would be denied federal agricultural subsidies. The USDA Natural Resources Conservation Service (NRCS) is responsible for this program and conducts wetland delineations on farmland for this program. As of 2005, the NRCS coordinates these delineations with the Corps and advises farmers about the Clean Water Act.

There are many other programs that the federal government uses to promote wetland conservation (Table 9.1). For example, President Carter's Executive Orders 11990 and 11988 require that all federal agencies take steps to minimize wetland and floodplain alterations in conducting their agency's business. The Coastal Zone Management Act provides funding to states to develop programs to conserve coastal resources including wetlands. The Land and Water Conservation Fund Act makes

Figure 9.3. Federal laws and many state laws now prohibit the filling of most wetlands, yet illegal and unregulated filling still occurs. Drainage that converts wetlands to nonwetlands is not effectively regulated in most areas. Isolated wetlands (completely surrounded by dry land) may be the most vulnerable to destruction.

funding available to federal agencies, such as the U.S. Fish and Wildlife Service, to acquire wetlands and other habitats. The National Flood Insurance Program provides subsidized flood insurance to communities that have enacted local floodplain protection regulations. The U.S. Environmental Protection Agency grants funding to states to develop statewide wetland conservation plans.

State Wetland Regulations

About one-third of the nation's states have passed legislation to protect freshwater wetlands. The bulk of these laws are from the Northeast and Great Lakes region. Most coastal states in the Northeast have laws to protect tidal wetlands. However, Delaware, West Virginia, Ohio, Illinois, and Indiana have not passed special laws to protect freshwater wetlands; they tend to rely on the protection under authority of Section 401 water quality certification program of the Federal Clean Water Act.

The state laws recognize the values that wetlands provide to society and are intended to protect wetlands from development. They have been instrumental in turning the tide from wetland development. For example, before the passage of New Jersey's Wetlands Act of 1970, annual tidal marsh losses averaged 3,200 acres, but since then, yearly losses have been reduced to 50 acres or less. Similar laws in other states have had the same beneficial effect. Like the federal regulation, permits are required for most activities in wetlands. Regulated activities do vary from state to state, so consult the appropriate authority for specifics.

Wetland Acquisition

When it comes to wetland protection, acquisition may be preferred over regulation simply because regulations can be changed and what may be protected today might be abandoned to filling in the future. Acquisition involves securing wetlands by purchasing property or conservation easements on property, or by accepting land donations. Many wetlands are owned by public agencies or private nonprofit environmental organizations. Nevertheless, an estimated

75 percent of the nation's wetlands are privately owned, which seems to necessitate a largely regulatory approach to wetland protection in the United States.

Government agencies and nongovernment organizations manage many wetlands for varied purposes. Federal agencies such as the Fish and Wildlife Service, National Park Service, and Forest Service control uses of wetlands on National Wildlife Refuges, National Parks and Seashores, and National Forests, respectively. The National Wildlife Refuge System preserves and manages wetlands that are vitally important for migratory birds, especially waterfowl (Figure 9.4). To date, almost 100 million acres of land, mostly wetlands, are part of the over-500-refuge system. State parks, forests, and wildlife management areas have many protected wetlands, while county parks, town forests, and town conservation lands also encompass wetlands. Private nonprofit organizations including the Nature Conservancy, National Audubon Society, Massachusetts Audubon Society, Trustees of Reservations, and various land trusts have acquired many wetlands and adjacent uplands for preservation. There are numerous programs in place that provide funds for wetland acquisition (Table 9.1).

Wetland Management

To wildlife biologists, wetland management might address ways of modifying wetlands to improve waterfowl habitat (e.g., water-control devices, impoundment, controlled burning, and creation of nesting islands, and "green tree reservoirs") or habitat for other animals. It might also include control of exotic and invasive species to maintain natural plant communities. In a broader sense, wetland management includes various practices that may be employed to reduce adverse impacts from human activities around wetlands. For example, several activities are exempt from wetland permits, but state and federal laws may encourage the employment of "best management practices" (BMPs). Certain forestry operations, construction activities adjacent to wetlands and streams, and agricultural operations fall into in this category. While a few states (e.g., Maryland, Massachusetts, New Jersey, and Rhode Island) reg-

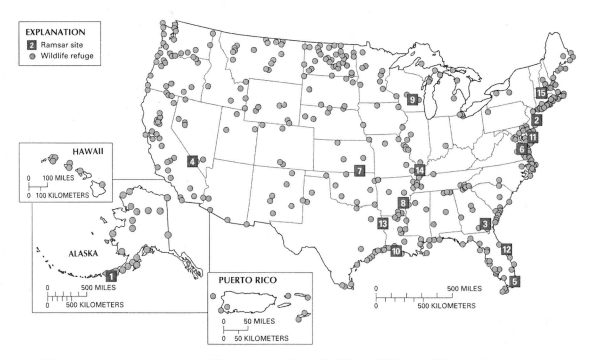

Figure 9.4. Location of national wildlife refuges in the United States. Wetlands of international importance (RAMSAR sites) are also shown. (Source: Stewart 1996)

ulate activities in the buffer zone around wetlands, the federal regulations do not. Knowledge of these BMPs provides an understanding of the major adverse impacts of these land uses and how to minimize their negative effects.

The main concerns about forestry are clearcutting, removal of riparian vegetation, soil disturbance (increases likelihood for soil erosion), and road building, especially stream crossings. There may also be issues related to harvesting certain species and the need to provide sufficient standing crop for regeneration (e.g., Atlantic white cedar). Table 9.2 outlines some forestry BMPs gathered from several state publications; contact your state forestry department for specific guidelines in your state.

Construction around wetlands and streams presents some of the same problems that forestry does, including soil disturbance and vegetation removal. Depending on the type of construction (e.g., roads or buildings), the eventual use of that land may pose threats to water quality or fish and wildlife habitat. Maintaining an adequate-sized vegetated buffer may reduce the impacts of such uses. Table 9.3 lists some recommended buffer widths to protect water quality and address most wildlife habitat concerns.

Agricultural practices include crop production and livestock grazing. When these uses take place in or immediately adjacent to waters and wetlands, the negative impacts are greatest (Figure 9.5). These impacts can be greatly reduced by maintaining or restoring vegetated buffers and establishing grassy drainageways. Vegetated buffers serve an added function as habitat for a variety of species such as nesting sites for many waterfowl. Runoff from agricultural lands likely contains agrochemicals (pesticides, herbicides, and nutrients) or livestock wastes that jeopardize water quality and fish habitat. Keeping livestock out of streams by fencing off riparian corridors and providing alternative watering sources are wise management practices that help maintain water quality and bank stability and protect aquatic life.

In Chapter 6, we learned that vernal pools are breeding grounds vital for the survival of many amphibians. They help sustain populations of these animals by producing hundreds and thousands of tadpoles and larval salamanders. Since these animals live as adults in neighboring woodlands up to three-quarters of a mile or more away from their natal pools, it is also important to manage these forests in ways that to continue supporting these populations. Table 9.4 offers some recommendations based largely on work in the Northeast.

Table 9.2. Some best management practices for forestry operations in wetlands and for dealing with stream crossings.

1. A harvest plan should be prepared addressing issues of environmentally sensitive features (e.g., streams, wetlands, and other waterbodies). In developing this plan, the forester should review maps and aerial photographs, conduct a field inspection of the entire site, and plan on using BMPs to address unavoidable situations.

2. Timing of timber harvest is an important consideration. Harvest should be done when soils are frozen or dry (when water tables are low). The wet season (spring and fall) and other wet soil conditions should be avoided. Protect vegetation in a streamside management zone (e.g., 25–200+ feet along each side of the streambed depending on slope and soil stability).

3. Construct roads on uplands wherever possible, following the natural topography and minimizing water crossings. Use soil stabilization techniques to reduce erosion. Keep piles of exposed materials away from areas of water flowage or where they may be washed into streams, wetlands, and other waterbodies.

4. If a road must cross a wetland, be sure to maintain the wetland's hydrology by providing proper-sized culverts to handle maximum flows (see #5 below). Construct ditches where necessary to route subsurface drainage to culverts (be sure not to create ditches that will lower the wetland's water table). Use mats or other temporary structures to reduce soil compaction and minimize rutting when equipment must be operated in the wetland. Use the minimum amount of fill necessary for the road.

5. When crossing a stream with a road, look for the narrowest point, areas with low, stable banks, a firm stream bottom, minimal surface runoff, and gentle slopes on approaches. Install appropriate culverts and

(continued)

Table 9.2. (continued)

bridges to maintain stream flow and unrestricted movement by aquatic life; recommended culvert pipe sizes vary due to upstream drainage area, e.g., for 10 acres, an 18″ pipe; 20 acres, a 20″ pipe; 50 acres, a 28″ pipe, and 100 acres, a 34″ pipe (as recommended by U.S. Forest Service; clay soils will need larger pipes). Limit construction to low water period or normal flow (for important fish streams avoid egg incubation period, e.g., October to April in New Hampshire). Culverts should not create pooling of water (scouring basins) on either side of the culvert.

6. Maintain a vegetated filter strip between the road and the stream. Width of strip varies with slope: a 0–10% slope should have a 50-foot buffer; an 11–20% slope, 51–70 feet; a 21–40% slope, 71–110 feet; and a 41–70% slope, 111–150 feet (as recommended by New York State).

7. Avoid routing surface water from roads directly into streams, wetlands, and other waterbodies; drain into a filter strip, other vegetated area, or sediment trap.

8. Use an existing landing if possible, unless it is near a stream or other waterbody in which case relocate the landing to a place with less potential for water quality impact if possible (i.e., at least 200 feet away from waterbodies and wetlands). If the landing must be constructed within 200 feet, use erosion control measures (e.g., straw bales or silt fences) to minimize erosion and divert water away from the landing by ditches.

9. Keep residue piles (e.g., slash, sawdust, or wood chips) away from places where they may wash into streams, wetlands, and other waterbodies.

10. Regrade and revegetate exposed areas (e.g., skid trails, landings, roadsides, and temporary roads) after harvesting is completed. Mulching is recommended to prevent erosion. Seeding with native species is preferred.

11. Utilize equipment with high flotation tires or low-ground pressure tracks if harvest must be done when soil is moist or if harvesting forested wetlands.

12. When spraying pesticides, keep at least 50 feet away from streams, marshes, and other waterbodies; do not apply when windy.

13. Do not clearcut up to stream or to the edges of open water, marshes, wet meadows, shrub swamps, shrub bogs, and similar wetlands. Do selective cutting within streamside management zones and forested buffers around these wetlands.

(Adapted from Alpaugh 1995; Cullen 1996; NYS Forestry 2000; and Rummer 2004)

Table 9.3. Recommended vegetated buffer widths to protect water quality and wildlife habitat in streams, wetlands, and riparian habitats.

Habitat	Function to be Protected	Recommended Buffer Width (Source)
Stream	Water Quality	100 ft for urban stream buffer wherever possible (Federal Interagency Stream Restoration Working Group 1998) 148 ft where sediment and adsorbed pollutants are major concerns (Desbonnet et al. 1994)
	Fish Habitat	98 ft forested buffer to maintain stream temperature (Castelle et al. 1994) (Note: a 16–33 ft buffer offers little protection for aquatic resources)
Wetland	Wildlife Habitat	50–100 ft to reduce human disturbance 200–300 ft to protect wetland-dependent species 100–200 ft to protect other wildlife (Castelle et al. 1994)
Bottomland Hardwood Forest	Bird Habitat	>1640 ft to maintain characteristic species (Kilgo et al. 1998)
Riparian Forest	Neotropical Migrant Birds	>328 ft to protect forest interior-nesting species (Keller et al. 1993)

Figure 9.5. A site where best management practices need to be employed to improve water quality.

Wetland Restoration, Creation, and Enhancement

Since 1989, the federal government and several states including Maryland, Massachusetts, and New York have adopted an environmental policy of "no net loss of wetlands." Through regulatory programs, efforts are being made to avoid or at least minimize losses, and to compensate for these losses through mitigation as a permit requirement. Mitigation allows for wetland replacement of some kind through creation or restoration, or in some cases by acquiring "unprotected" wetlands and adding them to a refuge, wildlife management area, or similar conservation area. These efforts have often been less than successful in establishing the desired wetlands. Moreover, regulatory programs do not protect all wetlands and certain wetlands continue to be converted to other uses with or without replacement through creation or restoration. So how can "no net loss of wetlands" be achieved?

Several government agencies have initiated proactive wetland restoration as a means to make up for these losses, and more significantly,

to simply restore wetlands because they are valuable natural resources. Wetland restoration is a vital component of the federal government's plan to improve the status of wetlands across the country. Through this effort, a net gain in wetlands may be achieved in the long-term, since wetland restoration can both increase wetland acreage and expand the functional capacity of damaged wetlands.

Wetland restoration can be separated into two basic types: 1) reestablishing former wetlands (type 1 restoration) and 2) repairing damaged wetlands (type 2 restoration). The prime objective of the former is to put a wetland back on the landscape and the precise wetland type may not be particularly important, especially if significant changes in land use or watershed hydrology have occurred. The latter restoration seeks to recover some lost or damaged functions (Table 9.5). It is important to distinguish between these types, because only type 1 projects lead to a net gain in wetland acreage, while type 2 projects result in an improvement in one or more functions. These differences are significant when the government attempts to assess whether "no net loss" is being achieved.

Table 9.4. Best management practices for forestry and construction activities near vernal pools.

Activity	Recommended Practices

Forestry	1. Do not disturb vernal pool
	2. Within 100 feet of the pool (the "vernal pool envelope"):
	  a. Maintain a minimum of 75% of existing canopy
	  b. Harvest only when soil is frozen or completely dry
	  c. Avoid using heavy equipment in this zone
	  d. Keep roads and landings out of this zone
	  e. Do not disturb fallen logs
	  f. Leave limbs, tops, and other slash from harvest
	  g. Avoid use of chemicals
	3. From 100 to 400 feet from pool (the "amphibian life zone"):
	  a. Maintain a minimum canopy cover >50% by trees 20–30 feet tall
	  b. Avoid harvest openings >1 acre
	  c. Avoid changes in forest type in this zone
	  d. Do not disturb fallen logs
	  e. Leave limbs, tops, and other slash from harvest
	  f. Harvest only when soil is frozen or completely dry
	  g. Minimize soil compaction and scarification (scraping of topsoil)
	  h. Minimize use of chemicals

Roads (including driveways)	1. Keep out of vernal pool and vernal pool envelope 2. Keep roads with 5 to 10 cars/hour 750 feet away from vernal pool 3. Eliminate curbing 4. Install box culverts (2 ft wide ×3 ft high) beneath roads in areas of known amphibian migration and use curbing to funnel amphibians to the culvert 5. Use shared driveways to cluster development, while keeping homes as far from pools as possible
Site Development	1. Limit disturbance to 25% of the zone from 100 to 400 feet from the pool 2. Cluster development (homes/buildings) 3. Avoid making ruts and other depressions; fill in perc test holes to grade 4. Use silt fencing and hay bales to control erosion and minimize use of such structures within 750 feet from pool 5. Install silt fencing around construction site to keep out amphibians 6. Limit lot clearing to <50% of lots that are ≥2 acres in size 7. Mark vernal pool protection area with granite monuments or stone cairns
Stormwater Management	1. Do not use vernal pools as detention basins 2. Locate detention bases >750 feet from pool 3. Minimize impervious surfaces to reduce runoff

(Adapted from Calhoun and Klemens 2002 and Calhoun and deMaynadier 2004)

It is also essential to distinguish wetland restoration from wetland enhancement and creation. Wetland enhancement involves changing the condition of an existing unaltered or relatively unaltered wetland to strengthen one or more functions usually at the expenses of others. An example would be impounding a wet meadow to create a marsh for the benefit of waterfowl at the expense of meadow species like leopard frogs. In some cases, there is a subtle difference between wetland enhancement and wetland restoration (type 2), yet the presence of a dike may be a useful indicator. In contrast, wetland creation is building a wetland where one never existed. This requires establishing wetland hydrology in a nonwetland area. It is usually done by excavating down to the local water table or by diking a drainageway and compacting soils or bringing in clay for an impervious substrate. Created wetlands are often built to treat stormwater runoff or to provide tertiary treatment of wastewater.

Type 1 wetland restoration is usually more successful than wetland creation in establishing wetlands for the following reasons: 1) the site is a former wetland, 2) it is located in a landscape position that can support desired functions, 3) it usually has drained or buried hydric soils, 4) it

Table 9.5. Types of wetland restoration projects.

Restoration Type	Conditions Necessitating Restoration	Restorative Action
Type 1	Filled	Remove fill to restore original elevations
	Effectively drained	Restore hydrology by blocking ditches, breaking tile drains, or using dikes and water-control structures
	Excavated	Restore elevations suitable for recolonization by wetland vegetation
	Impoundment	Remove or open dikes and install water-control structures to manage hydrology for wetland plant reestablishment
Type 2	Tidal restriction	Remove tide gates, enlarge culvert, install tide gates to improve tidal flushing, or open dikes to restore tidal flow
	Partly drained	Restore hydrology by filling in ditches, breaking tile drains, or by other means
	Partial filling	Remove fill to restore original elevations
	Exotic and invasive plants	Control exotic and invasive species by various means (using herbicides, pulling by hand, changing hydrology, biological control, etc.) and if necessary, plant desirable species
	Chemical contamination	Restore natural chemistry by eliminating the pollution source. Remove contaminated soils if necessary.
	Devoid of vegetation	Plant desirable species in suitable habitats

Note: Type 1 projects involve reestablishing a former wetland and Type 2 projects entail rehabilitating existing wetlands.

possesses a natural seedbank of hydrophytes that facilitate reestablishment of the wetland plant community, and 5) it has a modified hydrology that can usually be restored to some form of wetland hydrology. Evaluations of created wetlands intended to replace natural wetlands suggest a much lower success rate. Also it is unlikely that most created wetlands are functionally equivalent or similar to the natural wetlands they are intended to replace.

Proactive wetland restoration programs have been initiated by federal and state governments. The U.S. Fish and Wildlife Service's Partners for Fish and Wildlife Program has successfully restored thousands of wetland acres nationwide. This is a cost-sharing program that works with cooperating private landowners to restore wetlands and other natural habitats. In the Northeast, the program has been most active in New York and Pennsylvania where significant acreage of drained wetlands exists on farmland. These sites are perhaps the easiest and most cost-effective restoration projects, with costs usually in the $100–200 per acre range. The USDA Natural Resources Conservation Service administers the Wetland Reserve Program that also encourages farmers to restore wetlands on idle and marginally productive lands.

In coastal states, tidal-wetland restoration often gets more attention than inland-wetland restoration. Connecticut's wetland restoration program focuses solely on salt marsh restoration. Former mosquito control personnel are now actively engaged in this effort. When low-lying development is not present, salt marsh restoration may be simply accomplished by removing the tidal restriction, although concerns about possible subsidence may require a closer assessment of hydrologic conditions. In cases where development is in low-lying areas, special studies have to be conducted to determine the maximum level of tidal flooding that can be introduced without causing property damage. Such projects typically require installation of automated or self-regulating tidal gates.

Potential wetland restoration sites may be identified through a combination of photointerpretation, map interpretation (i.e., soil maps), and field inspections. The latter are always required for site assessment. Reference wetlands (unaltered wetlands or wetlands in the same

Table 9.6. Some basic steps for developing a wetland monitoring plan.

1. Establish and assess reference wetlands before initiating wetland restoration or creation to aid project design and development of performance standards

2. Before restoring or creating a wetland, develop a plan with measurable objectives; consider the following:
   a) Wetland type
   b) Size
   c) Hydrology
   d) Desired plant communities
   e) Desired functions
   f) Time for wetland vegetation establishment
   g) Risks (e.g., herbivory, invasive species, wet/dry years, and adjacent land use)

3. Develop performance standards for certain characteristics of wetlands (examples below):
   a) Wetland Properties
      (1) Vegetation (percent cover by certain species or types of species, reproductive success, no invasives)
      (2) Hydrology (hydrograph comparable to wetland of similar type, frequency and duration of surface water, seasonal high water table duration)
      (3) Soil (hydric soil properties, buildup of organic matter as appropriate for wetland type)
   b) Wetland Functions
      (1) Wildlife habitat (observations)
      (2) Water quality (sedimentation, riparian buffer)
      (3) Surface water detention (depth, duration, and frequency of inundation)
      (4) Shoreline stabilization (riparian vegetation)

4. Consider the sampling frequency (how often you need to sample a particular property or function to determine success)

5. Determine milestones to track project success (e.g., a forested wetland will take many years to establish and will go through several seral stages)

ecosystem) are often evaluated to gather data for project design and monitoring purposes. Watershed deficits such as flood storage, water quality, and fish and wildlife habitats may be evaluated so that wetland restoration can be targeted where it may improve watershed functions.

After a wetland restoration, creation, or enhancement project is built, a monitoring program should be employed to track project progress, to detect problems early on (e.g., lack of sufficient hydrology, too much water, overgrazing by muskrat or geese, or colonization and spread of invasive species), and to help develop adaptive management actions to get project on the proper trajectory (Table 9.6). Without monitoring, how can one determine project success?

Additional Readings

Biebighaser, T. R. 2003. *A Guide to Creating Vernal Pools.* USDA Forest Service, Morehead, KY.

Boyd, L. 2001. *Buffer Zones and Beyond: Wildlife Use of Wetland Buffer Zones and Their Protection under the Massachusetts Wetland Protection Act.* Department of Natural Resources Conservation, University of Massachusetts, Amherst.

Burke, D. G., E. J. Meyers, R. W. Tiner, Jr., and H. Groman. 1988. *Protecting Nontidal Wetlands.* American Planning Association, Chicago, IL. Planning Advisory Service Report 412/413.

Calhoun, A.J.K., and P. deMaynadier. 2004. *Forestry Habitat Management Guidelines for Vernal Pool Wildlife.* Metropolitan Conservation Alliance, Wildlife Conservation Society, Bronx, NY. MCA Technical Paper No. 6.

Calhoun, A.J.K., and M. W. Klemens. 2002. *Best Development Practices: Conserving Pool-breeding Amphibians in Residential and Commercial Developments in the Northeastern United States.* Metropolitan Conservation Alliance, Wildlife Conservation Society, Bronx, NY. MCA Technical Paper No. 5.

Castelle, A. J., A. W. Johnson, and C. Conolly. 1994. Wetland and stream buffer size requirements: a review. *J. Environ. Qual.* 23: 878–882.

Interagency Workgroup on Wetland Restoration. 2002. *An Introduction and User's Guide to Wetland Restoration, Creation, and En-*

hancement. National Oceanic and Atmospheric Administration, Environmental Protection Agency, Army Corps of Engineers, Fish and Wildlife Service, and Natural Resources Conservation Service, Washington, DC.

Kusler, J. A., and T. Opheim. 1996. *Our National Wetland Heritage.* Environmental Law Institute, Washington, DC.

Lewis, W. M. 2001. *Wetlands Explained: Wetland Science, Policy, and Politics in America.* Oxford University Press, New York.

Minkin, P., and R. Ladd. 2003. *Success of Corps-required Wetland Mitigation in New England.* U.S. Army Corps of Engineers, New England District Concord, MA. http://www.nae.usace.army.mil/reg/wholereport.pdf

Sibbing, J. M. 2004. *Nowhere Near No-net-loss.* National Wildlife Federation, Washington, DC. http://www.nwf.org

The Conservation Foundation. 1988. *Protecting America's Wetlands: An Action Agenda.* The Final Report of the National Wetlands Policy Forum. Washington, DC.

Tiner, R. W. 1995. Wetland restoration and creation. In *Encyclopedia of Environmental Biology,* ed. W. A. Nierenberg, 3: 517–534. Academic Press, San Diego.

U.S. Environmental Protection Agency. 1991. *The Federal Wetlands Protection Program in New England: A Guide to Section 404 for Citizens and States.* Region 1, Boston, MA.

Vileisis, A. 1997. *Discovering the Unknown Landscape: A History of America's Wetlands.* Island Press, Washington, DC.

Wenger, S. 1999. *A Review of the Scientific Literature on Riparian Buffer Width, Extent, and Vegetation.* Institute of Ecology, Office of Public Service and Outreach, University of Georgia, Athens.

Williams, T. 1995. The wetlands-protection farce. *Audubon,* March–April 1995: 30–36.

PART II

Wetland
Identification
Guide

CHAPTER 10

Recognizing Wetland Plants

Plants are conspicuous in most wetlands at all times of the year, with exceptions being non-vegetated wetlands (e.g., mud flats) and tidal-riverine wetlands and aquatic beds in winter. These latter wetlands are dominated by non-persistent plants, vegetation that dies back each fall and overwinters as underground rhizomes buried in the mud or as seeds. The rest of the wetlands have some type of vegetation present year-round, such as woody plants and persistent herbaceous species like cattail and common reed.

As noted already, many plant species are useful indicators of wetlands. In the past, analysis of vegetation was the typical method for wetland identification. Even today, there are many wetlands that can be easily recognized simply by observing vegetation. Only about a third of the nation's flowering plants ever occur in wetlands, so about two-thirds of U.S. plant species are eliminated from consideration. If the more common species in the Northeast are emphasized, fewer than three hundred species are encountered. Many plants are reasonably specific to certain types of habitats, so by knowing the wetland type, the choices are greatly narrowed.

Some plants are easily recognized by leaf shapes, leaf margins, flower types, and flower characteristics. Grasses and grasslike plants (sedges and rushes) are often difficult to identify; they are usually best left to experts, yet there are some species within these families that are easily recognized. By using the simple keys provided in this chapter along with the illustrations and brief species descriptions, you should be able to identify most of the Northeast's common wetland plants. Most of these species are common in wetlands of the Great Lakes region. Plants specific to midwest prairies are not included.

Wetland-Plant Lists

Through the years, botanists and ecologists have observed and recorded many species of plants growing in wetlands. Some of these plants grow only in wetlands, yet the majority of plants living in wetlands are more wide-ranging—they grow in both wetlands and uplands to varying degrees. The Fish and Wildlife Service, in cooperation with other federal agencies, has developed a list of plants that grow in the nation's wetlands, based on an exhaustive review of botanical manuals, with subsequent technical review by wetland experts and plant ecologists. Regional lists were prepared while producing the national list (Figure 10.1). The *National List of Plant Species That Occur in Wetlands: Northeast Region* (Reed 1988) contains over 2,500 species of vascular plants found in the region's wetlands. Each plant is listed by scientific name (according to the USDA Soil Conservation Service's *National List of Scientific Plant Names*) and by common names. In addition, the national and regional wetland-indicator status of each plant is given, along with information on habit and general distribution.

The wetland-indicator status represents frequency of occurrence—that is, how likely it is that a plant will be found in wetlands versus uplands. Four major categories are recognized: 1) obligate wetland (OBL), 2) facultative wetland (FACW), 3) facultative (FAC), and 4) facultative upland (FACU). OBL species almost always occur in wetlands (>99% probability). It is highly unlikely that these species would be found dominating a nonwetland or upland site, and thus they are the best plant indicators of wetland. FACW species usually occur in wetlands; they are more common in wetlands than uplands (67–99% probability). They are also good indicators of wetland, but can grow in uplands in considerable numbers. FAC species show no marked preference for wetlands or uplands (34–66% probability). FACU species are more commonly upland species, but they can be found in wetlands and sometimes may occur as dominant species (1–33% probability). Plant species not listed are strictly upland plants

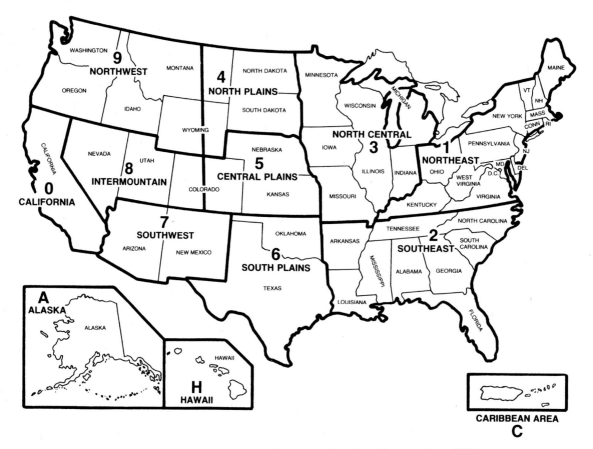

Figure 10.1. Map showing state coverage by regional wetland plant lists. (Source: Reed 1988)

(UPL). Areas dominated by UPL species should unquestionably be nonwetland.

There are many wetlands, especially floodplains and flatwoods, where FAC plants like sweet gum and ironwood predominate. By themselves, these plants may reveal very little about an area's status as a wetland, but by considering the presence, abundance, and distribution of all of the plants within an area—that is, the whole plant community—a better assessment can be made. In general, FAC species are nonindicators since they have nearly the same likelihood of occurring in wetlands as in uplands. In such cases, you'll usually need to consider the landscape position of the area and other factors like soils to separate wetlands from uplands.

In refining the wetland-plant list, the regional review panel assigned a plus sign (+) or a minus sign (−) to the three facultative categories to indicate whether a plant was on the "wetter" or "drier" side of the category's range.

For example, a FACW+ plant has a higher frequency of occurrence in wetlands (e.g., 90–99%) than a FACW− plant (e.g., 67–75%). Although the lists are not without their limitations, they are based on years of experience of plant and wetland ecologists, and they are generally pretty reliable.

A revised national list was published for review and comment in 1997. This list has generated considerable controversy because of possible implications to wetland jurisdictional calls by some of the cooperating agencies. As of January 2005, the final list has not been published. The draft list is, however, posted on the web at http://wetlands.fws.gov for reference in the meantime. For this list, some regions were further divided into subregions, in an attempt to extract the most specific information about a plant's habitat preferences across each region. For example, pitch pine in the New Jersey Pine Barrens is actually more of a FAC species than a FACU plant, being just a little more common

Table 10.1. Number of vascular plant species occurring in wetlands in the Northeast and North Central Regions.

Region	Indicator Status	Number of Species	% of Species Found in Wetlands	% of Region's Vascular Flora
Northeast	OBL	823	31	12
	FACW	593	22	9
	FAC	441	17	7
	FACU	571	21	8
	NI	252	9	4
	Subtotal	2,680	100	40
	UPL	4,047	—	60
	Total	6,727	40	100
North Central	OBL	669	30	10
	FACW	496	22	7
	FAC	431	19	7
	FACU	474	21	7
	NI	192	8	3
	Subtotal	2,262	100	34
	UPL	4,465	—	66
	Total	6,727	34	100

Note: Species are listed by wetland indicator status defined by Reed (1988); species with + and − signs have been placed in the general category (e.g., FAC− and FAC+ into FAC). Upland species (UPL) and species occurring in wetlands but not assigned an indicator status (NI) are also listed. Percentages are rounded off to equal 100 percent.

on uplands, so it is designated as FAC− in the coastal plain. This updated list better represents a given species's affinity for wetlands in different areas than does the previous list.

For the Northeast and North Central (Great Lakes) regions, 34–40 percent of the region's vascular plant species are commonly found in wetlands (Table 10.1). The majority of each region's vascular flora grows strictly in uplands, yet there are over 2,250 species that grow in wetlands more than one percent of the time. About 30 percent of these species are obligate hydrophytes—restricted to wetlands. From this categorization, it is clear that the majority of plants growing in wetlands can also grow in uplands. This fact has made the question "What is a wetland plant?" more difficult to answer than it would be if such species grew only in water and wetlands.

The varied distribution of plants can be explained largely by competition. Although most plants can grow in well-drained soils, those that are found there are present because they are the best competitors. Other plants have to find alternative places to grow. The rest of the land has soils with some limitations: too wet, too dry, too shallow, too rocky, too sandy, too salty, for example. Plants will adapt to these conditions in order to have a place to grow and reproduce. Some plant species exhibit an amazing degree of ecological plasticity, being found in a wide array of habitats. Pitch pine exemplifies this adaptive behavior, growing under a range of wet to dry conditions, from seasonal ponds and organic soils to the driest sandy sites in the Pine Barrens (Figure 10.2). As a result, its presence does not reveal much about a particular habitat. In contrast, the OBL species, such as Atlantic white cedar, skunk cabbage, cattails, and smooth cordgrass, are excellent indicators

Figure 10.2. Pitch pine is a remarkable plant, capable of growing in diverse habitats ranging from seasonal ponds and Atlantic-white-cedar swamps to sand dunes. It dominates both wet and dry sites, and thus the presence of this plant provides no clue as to whether an area is a wetland, in spite of its predominance in many New Jersey Pine Barrens wetlands.

of wetlands, since they are exclusive wetland species.

The wetland-indicator status of plant species is important for determining the likely presence of hydrophytic vegetation. The assemblage of different species in an area with a distinctive landscape, soils, and hydrology form a plant community. When more than 50 percent of the dominant species of a plant community (considering all types: tree, sapling, shrub, herb, and woody vine) consists of OBL, FACW, and FAC species, the plant community has a good probability of being wetland and meets the current federal criterion for being a positive indicator of hydrophytic vegetation for wetland identification. (Note: The Corps wetland delineation manual treats FAC− species like upland plants—nonwetland indicators.) The more OBL and FACW species, the greater the likelihood of wetland. The most obvious wetlands are dominated

by OBL species. Thus knowing the indicator status of the dominant plants provides a sense of whether the area is likely to be wetland or not. Many wetlands, however, are dominated by FAC species. The presence of hydric soils and signs of wetland hydrology in certain landscape positions help identify these plants as hydrophytes and the communities as wetlands (Chapter 13).

How to Identify Common Hydrophytes

The following sections deal with plant identification, including brief descriptions and line drawings to help identify some of the more common or characteristic wetland plants of the Northeast and Great Lakes region. For a book of this kind, an exhaustive treatment of wetland plants is not appropriate. Consequently, it includes most of the dominant and common species in the region (i.e., more than 250 species). A set of simple keys is also provided as an aid to identification. Three tables are included to help quicken the identification process. Tables 10.2 and 10.3 provide listings of common flowering wetland herbs and shrubs, respectively, along with their flower color and size and blooming period. Table 10.4 lists berry-producing species. *Some uncommon to rare, yet conspicuous and characteristic plants are illustrated in Chapter 5* (Figures 5.5 and 5.7). If you can't find a plant in this book, consult one of the field guides or taxonomic manuals listed at the end of the chapter.

As noted already at various points, wetland vegetation can be separated into five major life-form groups:

(1) Aquatic herbs: free-floating species, floating-leaved rooted vascular plants, and submergent plants growing beneath the water's surface. *Some common aquatics in tidal and nontidal waters are illustrated in Chapter 5, Figures 5.3 and 5.9; otherwise they are not included in this guide*, since such areas are obviously wetland.

(2) Emergent herbs: nonwoody plants whose stems and leaves normally extend above the water's surface or grow erect in periodically flooded or saturated soils. They can be divided into three subtypes: ferns, grasses and grasslike plants (e.g., sedges and rushes), and broad-leaved herbs.

Table 10.2. Flower color, blooming period, and relative flower size of some common wetland herbs in the Northeast.

Flower Color	Blooming Period	Flower Size	Herb
White	Spring	Small	Violet (53), Spring Beauty (87)
	Summer	Minute	Lizard's Tail (67), Smartweeds (68), Boneset (78), Bugleweeds (79), Water Parsnip (91), Tall Meadow Rue (92)
		Small	Pipeworts (51), Goldthread (54), Arrowheads (56), Canada Mayflower (63), Flat-topped White Aster (69), Starflower (88), Garlic Mustard (Fig. 8.7)
		Large	Rose Mallow (16)
	Fall	Small	Perennial Salt-Marsh Aster (4), Asters (73)
Yellow/orange	Spring	Small	Trout Lily (52), Marsh Marigold (59), Golden Ragwort (93), Winter Cress (93b)
	Summer	Minute	St. John's-worts (82b), Marsh Yellow Cress (93b)
		Small	Jewelweed (70), Swamp Candles (85), Fringed Loosestrife (86), Moneywort (Fig. 8.7)
		Medium	Spatterdock (58), Yellow Flag (61b)
	Fall	Small	Goldenrods (3,75), Bur Marigold (77), Beggar-ticks (90)
Green	Summer	Minute	Wood Nettle (71), False Nettle (76), Japanese Knotweed (Fig. 8.7)
Pink	Spring	Small	Spring Beauty (87)
	Summer	Minute	Annual Salt-Marsh Fleabane (18), Smartweeds (21, 68), Joe-Pye-weeds (89), Water-willow (83)
		Small	Swamp Milkweed (81), Marsh St. John's-wort (82), Purple Loosestrife (84)
		Medium	Seashore Mallow (16b)
		Large	Rose Mallow (16)
Red	Summer	Small	Cardinal Flower (72), Red Milkweed (81b)
Purple/Blue	Spring	Small	Violets (53), Ground Ivy (Fig. 8.7)
	Summer	Minute	Northern Sea Lavender (17), Mints (79), Blue Vervain (80)
		Small	Pickerelweed (66), Asters (64, 73), Purple Loosestrife (84), Bittersweet Nightshade (Fig. 8.7)
		Medium	Blue Flag (61), Asters (73)
	Fall	Minute	Annual Salt-Marsh Aster (4b)
		Small	New York Ironweed (74), Asters (73)
		Medium	Asters (73)

Note: Large flowers are greater than 4 inches wide; medium flowers 1–4 inches wide; small flowers ¼–1 inch wide; minute flowers less than ¼ inch wide. Species number is given in parentheses following the plant name for easy reference to illustrations and descriptions.

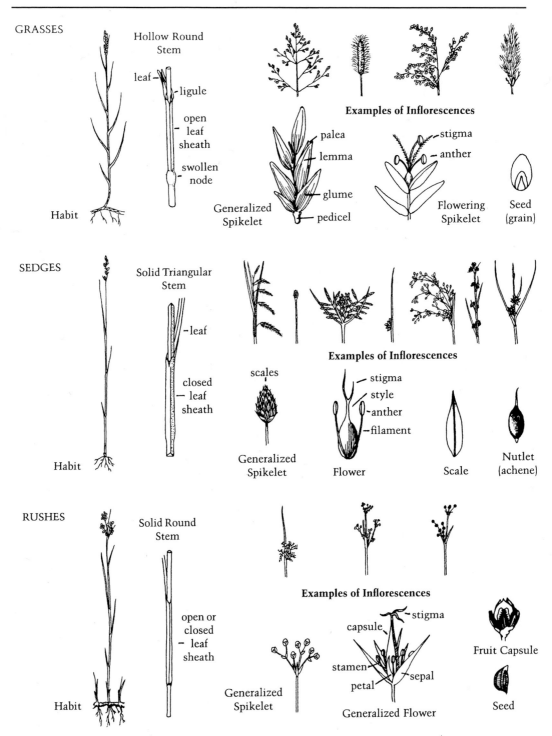

Figure 10.3. Distinguishing characteristics of grasses, sedges, and rushes. (Source: Tiner 1988)

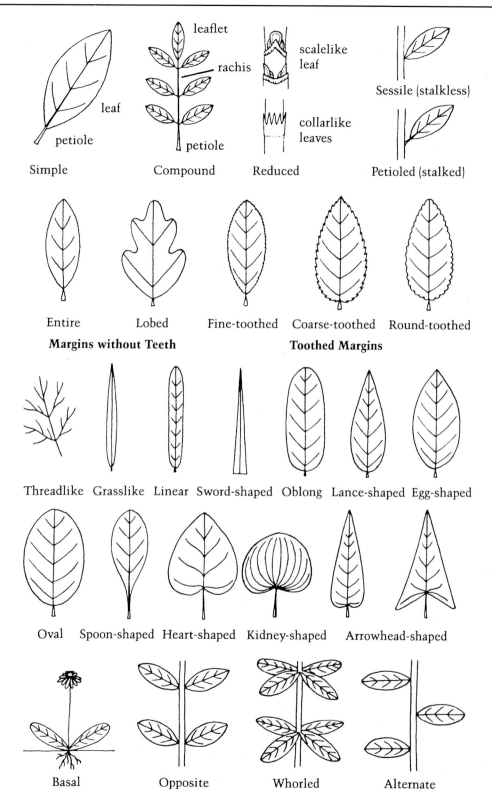

Figure 10.4. Leaf types and arrangements. (Source: Tiner 1987; copyright 1987 Ralph W. Tiner, Jr. Reprinted with permission.)

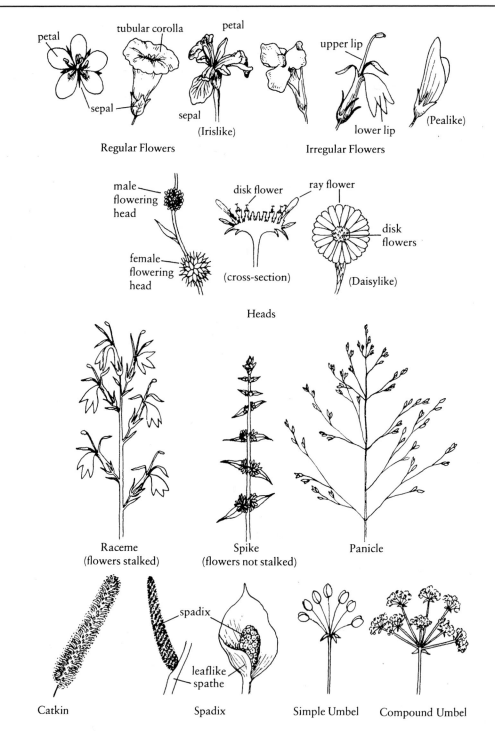

Figure 10.5. Flower types and arrangements. (Source: Tiner 1987; copyright 1987 Ralph W. Tiner, Jr. Reprinted with permission.)

Table 10.3. Flower color, blooming period, and relative flower size of wetland shrubs in the Northeast.

Flower Color	Blooming Period	Flower Size	Shrub
White	Spring–early summer	Small	Oblong-leaf Shadbush (115), Chokeberries (115b), Black Haw (106b), Nannyberry (106b)
	Spring–midsummer	Small	Wild Black Currant (95b), Leatherleaf (100), Highbush Blueberry (124)
	Early summer	Minute	Fetterbush (117), Maleberry (118)
		Small	Staggerbush (122)
	Summer	Minute	Buttonbush (109)
		Small–minute	Elderberry (104), Arrowwoods (106), Wild Raisins (107), Meadowsweets (112), Winterberry (116), Dwarf Huckleberry (121b)
		Small	Multiflora Rose (94b), Mountain Laurel (96b), Inkberry (99), Bog Rosemary (102), Labrador Tea (103), Dogwoods (108)
		Medium	Rosebay Rhododendron (101), Swamp Azalea (126), Multiflora Rose (94b)
	Late summer	Small	Sweet Pepperbush (114), Groundsel Bush (20)
Yellow	Spring	Minute	Spicebush (125)
	Spring–midsummer	Small–minute	Mountain Holly (123)
		Small	Wild Black Currant (95b), Swamp Fly-honeysuckle (107b)
	Summer	Small	Shrubby St. John's-wort*, Shrubby Cinque-foil (Figure 5.4)
	Fall	Small	Witch Hazel*
Pink-purple	Spring	Medium	Pinxter Flower (126b)
	Spring–early summer	Small	Dangleberry (121), Dwarf Huckleberry (121b)
		Medium	Rhodora (120)
	Summer	Small–minute	Dwarf Huckleberry (121b)
		Small	Bristly Black Currant (95), Sheep Laurel (96), Bog Laurel (97), Bog Rosemary (102)
		Medium	Swamp Rose (94)
	Late summer	Small–minute	Steeplebush (111)
Green	Spring–early summer	Small–minute	Dangleberry (121), Common Buckthorn (161b, Fig. 8.7)
	Summer	Small	Bristly Black Currant (95)
Yellow-green	Summer	Minute	Poison Sumac (105), Poison Ivy (128), Glossy Buckthorn (161b, Fig. 8.7)
Reddish	Early summer	Small–minute	Black Huckleberry (121b)

Note: Medium flowers are 1–4 inches wide; small flowers ¼–1 inch wide; minute flowers less than ¼ inch wide; small–minute flowers are around ¼ inch wide. Species number is given in parentheses following the plant name for easy reference to illustrations and descriptions. Species marked by an asterisk (*) are not described in this book.

Table 10.4. Wetland plants producing berries in the Northeast.

Berry Color	Lifeform	Plant
Red	Herb	Wild Calla (55), Jack-in-the-Pulpit (62), Canada Mayflower (63), Bittersweet Nightshade* (Fig. 8.7)
	Low shrub	Bunchberry (108b), Ground Hemlock (137b), Wintergreen*, Partridgeberry*
	Shrub	Currants (95b), Red Chokeberry (115b), Winterberries (116), Mountain Holly (123), Spicebush (125), Roses (94)
	Woody vine	Red-berried Greenbrier (127b), Cranberries (134)
	Tree	American Holly (141)
Green	Herb	Arrow Arum (57), Spring Beauty (87)
White	Woody vine	Creeping Snowberry (134b), Poison Ivy (128)
	Shrub or Tree	Poison Sumac (105), Dogwoods (108b), Poison Ivy (128)
Blue	Herb	Mile-a-minute Weed (Fig. 8.7)
	Shrub	Smooth Gooseberry (95b), Wax Myrtle (98), Arrowwoods (106), Wild Raisins (107), Silky Dogwood (108), Dangleberry (121), Highbush Blueberry (124)
	Woody vine	Common Greenbrier (127), Virginia Creeper (129)
	Tree	Eastern Red Cedar (139b), Black Gum (161)
Dark purple to black	Low shrub	Dwarf Huckleberry (121b), Black Huckleberry (121b), Black Crowberry*
	Shrub	Currants (95), Inkberry (99), Elderberry (104), Arrowwoods (106), Wild Raisins (107), Oblong-leaf Shadbush (115), Chokeberries (115b), Glossy Buckthorn (161b)
	Woody vine	Common Greenbrier (127), Virginia Creeper (129), Swamp Dewberry (131), Laurel-leaved Greenbrier (132), River Grape (135)
	Tree	Black Gum (161), Buckthorns (161b, Fig. 8.7)

Note: Berry color is given to aid in plant identification and is based on ripe berries, as immature berries tend to be green. Species number is shown in parentheses for easy reference to plant illustration or description. Plants not described in this book are marked by an asterisk (*).

(3) Shrubs: woody plants shorter than 20 feet, including young trees (saplings) and true shrubs with multiple woody stems.

(4) Trees: woody plants 20 feet or taller, typically having a single main stem or trunk.

(5) Woody vines: woody plants climbing other plants or trailing on the ground surface.

Three figures provide a general review of important plant characteristics. Figure 10.3 shows diagnostic characteristics of grasses, sedges, and rushes, to allow quick separation of grasses from grasslike plants. Leaf and flower characteristics are illustrated in Figures 10.4 and 10.5, respectively.

How to Use the Plant-Identification Keys

The keys to common wetland plants of the Northeast are provided to facilitate identification of the plants included in this field guide. Four specific keys are provided:

(1) Key A—Salt- and Brackish-Marsh Herbs and Shrubs,

(2) Key B—Marsh and Swamp Herbs,

(3) Key C—Swamp Shrubs and Woody Vines, and

(4) Key D—Swamp Trees.

The keys are based on vegetative characteristics, so that plants can be identified at most times during the growing season and not only when flowers are present. Each key consists of a series of couplets of contrasting statements. Start at couplet 1 of the appropriate key and begin matching couplet statements with the plant at hand. Read both parts of the couplet, and choose the one that best fits your plant. Next to that choice, you will see one or more species reference numbers (e.g., species 1 or species 1–4) or a number that refers to the next numbered couplet to read. You will eventually find a couplet that includes species numbers. At that point, find the appropriate illustrations immediately following the keys. (*Note: Species numbers followed by the letter "b" refer to a plant described as a similar species; it may or may not be illustrated.*)

Review the illustration carefully, paying attention to all features (e.g., leaves, flowers, and fruits), and be sure to examine all the illustrations of plants having the specific features. After locating the illustration that best resembles the plant in question, read the applicable description accompanying the illustrations. Be sure to read about species that resemble or are related to the illustrated species. Sometimes these species occur in different habitats. If the plant is not illustrated or covered under "Similar species," check the illustrations in Chapter 5 (Figures 5.5 and 5.7) for some less-common species characteristic of certain wetland types. Reference to these species is made at the end of each section of the plant descriptions, based on specific plant morphology; these species are not cross-referenced in the keys. If you still have not identified the plant, consult other field guides or ask a botanist for help.

Keys to Common Hydrophytes of the Northeast

1. Plants growing in salt or brackish marshes . . . . Key A
1. Plants growing in freshwater wetlands
   (including slightly brackish marshes) 2
   2. Plants herbaceous (soft-stemmed, nonwoody) . . . . Key B
   2. Plants woody 3
      3. Shrubs, typically less than 20 feet tall at maturity
         and multistemmed, *and* woody vines . . . . Key C
      3. Trees, 20 feet or taller at maturity and
         single-stemmed (main trunk) . . . . Key D

Key A—Salt- and Brackish-Marsh Herbs and Shrubs

1. Plant herbaceous, or if woody, a trailing plant 2
   2. Fleshy leaves or stem . . . . Species 1–4 (Figure 5.1)
   2. Parts not fleshy 3
      3. Grasslike 4
         4. Plant four feet or taller . . . . Species 5–11
         4. Plant shorter . . . . Species 12–15; Species 5,
                                                                                   39, 43, and 47
      3. Not grasslike 5
         5. Plant 4 feet or taller . . . . Species 3 and 16
         5. Plant shorter . . . . Species 3, 17, 18, and 68
1. Plant woody (shrub or tree) 6
   6. Deciduous 7
      7. Leaves opposite . . . . Species 19
      7. Leaves alternate . . . . Species 20, 98b, and 110
   6. Evergreen 8
      8. Leaves scalelike . . . . Species 139b
      8. Leaves broad . . . . Species 98

Key B—Marsh and Swamp Herbs

1. Stem armed with prickles . . . . Species 21 (Figure 8.7)
1. Stem lacking prickles 2
  2. Fern . . . . Species 22–29
  2. Not a fern 3
    3. Grass or grasslike (see Figure 9.2) 4
      4. Grass 5
        5. Plant 6 feet or taller . . . . Species 6, 7, and 30
        5. Plant shorter . . . . Species 6b, 8, and 31–37
                    (Figure 5.5)
      4. Grasslike 6
        6. Sedge 7
          7. Plant 4 feet or taller . . . . Species 15b and 44–45
          7. Plant shorter . . . . Species 10b and 38–43
                    (Figure 5.5)
        6. Not a sedge 8
          8. Rush . . . . Species 14b and 46–47
          8. Other grasslike plant 9
            9. Plant 6 feet or taller . . . . Species 48 and 49b
            9. Plant shorter . . . . Species 49, 50, 51,
                    and 61
    3. Not grasslike, but a broad-leaved herb 10
    10. Leaves all or mostly basal 11
      11. Plant less than 1 foot tall . . . . Species 52–55, 59b,
                    63, 88, and 108b (Figures 5.5 and 5.7)
      11. Plant taller 12
        12. Leaves simple . . . . Species 55–61 (Figure 5.5)
        12. Leaves compound . . . . Species 62
    10. Leaves arranged along a stem 13
      13. Leaves simple 14
      14. Leaves alternately arranged 15
        15. With entire margins 16
          16. Plant less than one foot tall . . . . Species 63–64 (Figures 5.7 and 8.7)
          16. Plant taller . . . . Species 61, 65–69, and 75b (Figures 5.5, 5.7, and 8.7)
        15. With toothed margins . . . . Species 16, 67, 70–75, and 93 (Figure 8.7)
      14. Leaves oppositely arranged or whorled 17
        17. Leaves opposite 18
          18. With toothed margins . . . . Species 70, 71b, 76–80, and 90b (Figure 8.7)
          18. With entire margins 19
            19. Plant 1 foot or taller . . . . Species 81–86
            19. Plant shorter . . . . Species 87
        17. Leaves whorled 20
          20. With entire margins . . . . Species 83–84, and 88
          20. With toothed margins . . . . Species 89
      13. Leaves compound 21
        21. Leaves oppositely arranged . . . . Species 90
        21. Leaves alternately arranged 22

Key C—Swamp Shrubs and Woody Vines

Key D—Swamp Trees

*If thorns are present, the tree is one of the following: if simple, fine-toothed leaves with short thorns on tip of branches = Common Buckthorn (*Rhamnus cathartica;* FACU in Midwest but rather common in wetlands; Figure 8.7); if simple, toothed and lobed leaves with prominent thorns = Hawthorns (*Crataegus* spp.; FACW to FACU−); if opposite, compound leaves with variable-sized coarse-toothed leaflets, thorns on trunk, branches, and leaf stalks, and flowers and black berries borne on much-branched, terminal inflorescence = Devil's Walking Stick or Hercules' Club (*Aralia spinosa;* FAC); and if opposite, compound leaves with eighteen or more leaflets, stout branched or unbranched thorns on trunk and branches, and elongate, somewhat flattened fruit pods = Honey Locust (*Gleditsia triacanthos;* FAC−). The former three species occur as both shrubs and short trees. None of these species are illustrated; the above descriptions should be sufficient for identification.

Illustrations and Descriptions of Common Hydrophytes

Illustrated wetland plants are briefly described in this section. Each species has been assigned a unique number that cross-references the illustration with its corresponding description. The descriptions include the common name of the plant, its scientific name, a brief overview of some major features, flowering period, habitats, regional wetland-indicator status for the Northeast (according to the 1988 list; *please note that status may vary for some species in the Great Lakes region*), and range along the East Coast. Related species or plants that may be confused with the described plant are listed under "Similar species." Some similar species are also illustrated; they can be located in the text by following the numbering code for the principal species; for example white grass (illustration 35b) can be located under the description of rice cut-grass (species 35). More-detailed descriptions can be found in various taxonomic manuals or field guides listed at the end of the chapter. These books should also prove useful for identifying plants not included in this guidebook. Photographs of most, if not all, of the plants referenced in this book are posted on various websites. Two helpful sites are the U.S. Department of Agriculture's Plants Database (http://plants .usda.gov) and the Robert W. Freckmann Herbarium at the University of Wisconsin-Stevens Point (http://wisplants.uwsp.edu). Also note that some species are illustrated in Chapter 5 (Figures 5.4, 5.5, 5.7, and 5.10). They are referred to at the end of the applicable section of plant descriptions (e.g., sundews and pitcher plants listed at end of Herbs with All or Mostly Basal Leaves). Scientific names are consistent with the 1988 national wetland plant list.

Salt- and Brackish-Marsh Herbs and Shrubs

Fleshy Herbs

1. Marsh Orach (*Atriplex patula*): a prostrate herb up to 3½ feet long or more, with grooved stems, light green arrowhead- or triangle-shaped leaves (to 3 inches long) mostly alternately (sometimes oppositely) arranged, and many tiny green flowers borne in ball-shaped clusters; July–November; high salt marshes; FACW; Nova Scotia to Florida. Similar species: Coast-blite Goosefoot (*Chenopodium rubrum*) has somewhat triangle-shaped leaves, but their margins are coarse-toothed or lobed; FACW; New Jersey north. Sea Blites (*Suaeda* spp.) have grooved stems and bear similar flowers, but their leaves are linear (about 2 inches long and somewhat rounded in cross-section) with pointed tips; OBL.

2. Common Glasswort (*Salicornia europaea*): up to 20 inches tall (usually 6–8 inches), no apparent leaves, and fleshy stems (less than ⅕ inch wide) turn red in the fall; August–November; high salt marshes, pannes, salt barrens, and sometimes low marshes; OBL; Newfoundland to Florida. Similar species: Bigelow's Glasswort (*S. bigelovii*) has thicker stems (⅕–¼ inch wide) that turn yellowish orange in the fall; OBL. Woody Glasswort (*S. virginica*) has mostly unbranched woody-cored stems; OBL.

3. Seaside Goldenrod (*Solidago sempervirens*): up to 4 feet tall or more, with thick, fleshy, lance-shaped leaves (to 16 inches long at base, gradually reduced in size toward top of stem) and yellow flowers blooming in late summer and fall; August–October; irregularly flooded salt and brackish marshes and sand dunes; FACW; Newfoundland to Florida.

4. Perennial Salt-Marsh Aster (*Aster tenuifolius*): up to 2¼ feet tall, with few, somewhat fleshy linear leaves (to 6 inches long) and white, pale blue, or purplish daisylike flowers (to 1 inch wide); August–October; high salt marshes and brackish marshes; OBL; New Hampshire to Florida. Similar species: Annual Salt-Marsh Aster (*A. subulatus*) bears small purplish to bluish flowers (less than ½ inch wide) that lack petals; OBL.

Tall Graminoids

5. Smooth Cordgrass (*Spartina alterniflora*): perennial grass up to 8 feet tall (much shorter—to 1½ feet—in high marsh), with overlapping sheaths, soft base of stem, and narrow terminal inflorescence (to 12 inches long) comprising many dense spikes with overlapping spikelets; July–September; low and high salt and brackish marshes; OBL; Newfoundland to Florida.

6. Big Cordgrass (*Spartina cynosuroides*): perennial grass up to 10 feet tall, with elongate leaves (to over 2 feet long), very rough leaf margins, and a terminal inflorescence of twenty to fifty one-sided flowering spikes; August–October; upper edges of salt marshes,

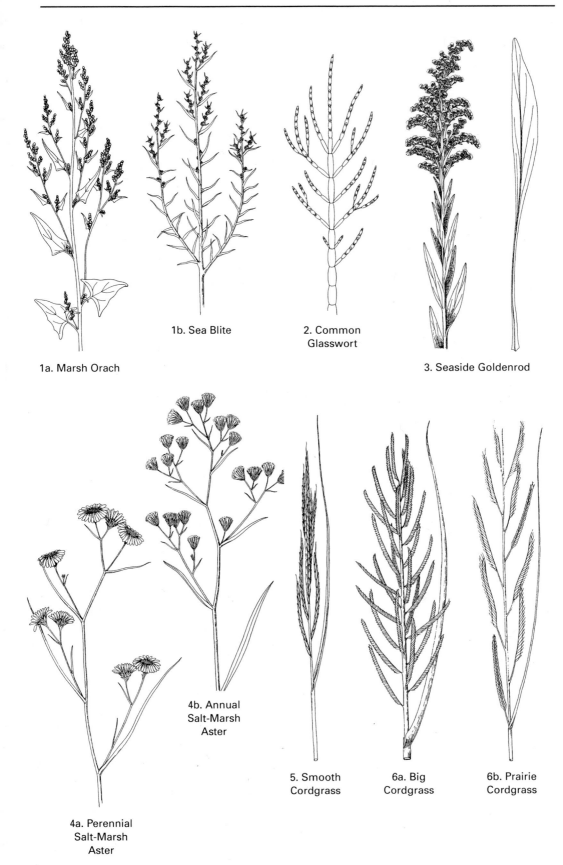

1a. Marsh Orach

1b. Sea Blite

2. Common
Glasswort

3. Seaside Goldenrod

4a. Perennial
Salt-Marsh
Aster

4b. Annual
Salt-Marsh
Aster

5. Smooth
Cordgrass

6a. Big
Cordgrass

6b. Prairie
Cordgrass

and brackish and tidal fresh marshes; OBL; Massachusetts to Florida. Similar species: Prairie Cordgrass (*S. pectinata*), a shorter species (to 6½ feet), occurs along the upper edges of salt marshes and in fresh marshes from Newfoundland to New Jersey; its inflorescence is more open, bearing ten to twenty spikes with bristle-tipped spikelets, and its leaves are long, with elongate threadlike tips; July–September; OBL.

7. Common Reed (*Phragmites australis*): perennial grass up to 14 feet or more tall, often forming dense colonies, with long, tapering, conspicuouly two-ranked leaves and purplish to brownish (sometimes whitish), feathery, plumelike inflorescence; late July–October; brackish and tidal fresh marshes, nontidal marshes, dikes, and disturbed soils; FACW; Nova Scotia to Florida.

8. Switchgrass (*Panicum virgatum*): perennial grass up to 6½ feet tall (usually about 4 feet), in dense clumps, with open branched inflorescence; July–September; brackish marshes and upper edges of salt marshes, nontidal marshes, and moist fields; FAC; Nova Scotia to Florida.

9. Narrow-leaved Cattail (*Typha angustifolia*): perennial grasslike plant up to 6 feet tall, with long narrow leaves (to ½ inch wide) and inconspicuous flowers borne in terminal spike (persists through winter), composed of two parts (male flowers above and female flowers below) separated by a distinct gap; May–July; brackish marshes and fresh tidal and nontidal marshes; OBL; Maine to Florida along coast, also inland along Great Lakes and Lake Champlain. Similar species: Broad-leaved Cattail (*T. latifolia*) grows in tidal and nontidal fresh marshes; it is taller (to 10 feet), has wider leaves (to 1 inch wide), and its flowering spike lacks the gap found in Narrow-leaved Cattail; OBL. Southern cattail (*T. domingensis*) resembles narrowleaved cattail, but is taller (8–13 feet) and occurs from Delaware south; OBL.

10. Olney's Three-square (*Scirpus americanus*): perennial sedge up to 7 feet tall with triangular stems lacking leaves; June–September; upper edges of salt marshes and brackish and tidal fresh marshes; OBL; New Hampshire to Florida. Similar species: Common Three-square (*S. pungens*) grows to 4 feet tall; its triangular stems do not have deeply concave sides, and it grows in similar habitats plus nontidal marshes; FACW+; Newfoundland to Florida.

11. Black Needlerush (*Juncus roemerianus*): perennial rush up to 6½ feet tall, with olive-brown stems; May–October; brackish marshes and upper salt marshes; OBL; southern Delaware to Florida (reportedly in southern New Jersey).

Short Graminoids

12. Salt-hay Grass (*Spartina patens*): perennial grass up to 3 feet tall, often forming dense windswept or "cowlicked" mats in the high marsh, with slender stems and wirelike leaves with margins rolled inward, and an open terminal inflorescence (to 8 inches long) composed of three to six spikes with overlapping spikelets; late June–October; salt and brackish marshes and sand dunes; FACW+; Quebec to Florida.

13. Salt Grass (*Distichlis spicata*): perennial grass up to 1½ feet tall, often forming dense mats in association with salt-hay grass, with conspicuously two-ranked leaves, overlapping leaf sheaths, and a dense terminal spikelike inflorescence (to 2½ inches long); August–October; salt and brackish marshes; FACW+; New Brunswick and Prince Edward Island to Florida.

14. Black Grass or Black Rush (*Juncus gerardii*): perennial rush up to 2 feet tall, often occurring in dense stands in the high marsh and giving the marsh a dark brown or blackish appearance when in flower, with one or two narrow leaves (to 8 inches long and round in cross section), inconspicuous flowers borne in clusters near top of stem, and round, persistent fruit capsules; June–September; salt and brackish marshes; FACW+; Newfoundland to Virginia (also reported in Florida). Similar species: Baltic rush (*J. balticus*) occurs along the upper edges of salt marshes and in nontidal calcareous marshes from Newfoundland to Pennsylvania; it has a single leafless stem (up to 3 feet high) resembling that of Soft Rush (*J. effusus;* species 46); FACW+.

15. Salt-Marsh Bulrush (*Scirpus robustus*): perennial sedge up to 3½ feet tall, with stout triangular stems, several elongate grasslike leaves, and three or more budlike spikes borne near top of plant and surrounded by leaflike bracts; July–October; irregularly flooded salt and brackish marshes, occasionally tidal fresh marshes; OBL; Nova Scotia to Florida. Similar species: River Bulrush (*S. fluviatilis*) occurs in tidal fresh and nontidal marshes; it grows to 5 feet tall, its budlike spikelets are mostly borne on long stalks,

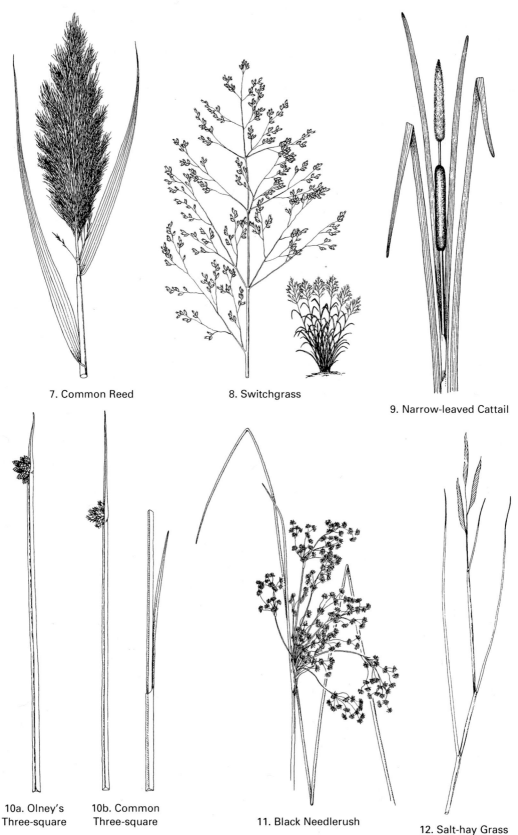

7. Common Reed

8. Switchgrass

9. Narrow-leaved Cattail

10a. Olney's
Three-square

10b. Common
Three-square

11. Black Needlerush

12. Salt-hay Grass

and its uppermost leaves droop; OBL; Quebec to Virginia. Salt-Marsh Sedge (*Carex paleacea*), a sedge resembling Fringed Sedge (*C. crinita*; species 38), grows along the upper edges of salt marshes and in brackish marshes from Massachusett north; its spikelets are borne on long drooping stalks; OBL.

See also Smooth Cordgrass (species 5), Twig-Rush (species 39), and Spikerushes (species 43).

Flowering Herbs

16. Rose Mallow (*Hibiscus moscheutos*): perennial up to 7 feet tall, with egg-shaped to weakly tree-lobed leaves having rounded or heart-shaped bases, large, showy, pink or white five-petaled flowers (4 to 6½ inches wide) having purple or red center, and persistent five-celled fruit capsules (flowers and capsules look like those of its ornamental relative Rose of Sharon); July–September; brackish, tidal fresh, and nontidal marshes; OBL; Massachusetts to Florida. Similar species: Seashore Mallow (*Kosteletzkya virginica*) has smaller pink flowers (to 2½ inches wide) and rough, hairy, triangular leaves; OBL; Long Island to Florida.

17. Northern Sea Lavender (*Limonium nashii*): low-growing perennial, locally called "sea heather," up to 2 feet tall, with cluster of spoon-shaped to lance-shaped basal leaves (to 6 inches long) and many-branched terminal spike (to 2 feet tall) bearing many tiny light purplish (lavender) narrow tubular flowers (¼ inch long) with hairs at base; July–September; irregularly flooded salt marshes, occasionally in low salt marshes; OBL; Labrador to Florida. Similar species: Carolina Sea Lavender (*L. carolinianum*) occurs mostly from Long Island south; its flowers lack hairs at base; OBL.

18. Annual Salt-Marsh Fleabane (*Pluchea purpurascens*): annual up to 3 feet tall, with aromatic (camphor-scented when crushed) leaves (to 5 inches long) alternately arranged, and a showy cluster of pink or purplish flowers (¼ inch long); August–September; high salt and brackish marshes, sometimes tidal fresh and nontidal marshes; OBL; southern Maine to Florida.

See also Seaside Goldenrod (species 3).

Shrubs with Opposite Leaves

19. High-tide Bush or Marsh Elder (*Iva frutescens*): deciduous shrub common on tidal ditch banks, usually less than 6 feet tall, with fleshy leaves (to 3½ inches long) mostly oppo-

site, but smaller upper leaves alternately arranged, and small greenish white flowers in heads borne on erect leafy spikes; August–October; irregularly flooded salt and brackish marshes, especially along mosquito ditches and upper high marshes; FACW+; Nova Scotia and southern Maine to Florida.

Shrubs with Alternate Leaves

20. Groundsel Bush (*Baccharis halimifolia*): deciduous shrub up to 10 feet tall, with coarse-toothed, somewhat leathery leaves, bears cottony "flowers" in the fall; August–November; salt, brackish, and tidal fresh marshes, and nontidal swamps; FAC; Massachusetts to Florida.

See also Sweet Gale (species 109) and Wax Myrtle (species 98).

Marsh and Swamp Herbs

Prickly Herbs

21. Arrow-leaved Tearthumb (*Polygonum sagittatum*): reclining herb with jointed, prickly stems, leaves to 4 inches long, and clusters of pink (sometimes white or green), five-lobed flowers; June–September; tidal fresh and nontidal marshes; OBL; Newfoundland to Florida. Similar species: Halberd-leaved Tearthumb (*P. arifolium*) has more triangle-shaped leaves and pink, four-lobed flowers; OBL. Mile-a-minute Weed (*P. perfoliatum*), an invasive annual plant, bears recurved barbed hooks (prickles) on stems, leaf stalks, and on main veins of triangle-shaped leaves, cup-shaped leaf sheaths on stems, and blue berries (about ⅕ inch wide) in clusters; FAC (Figure 8.7).

Ferns

22. Cinnamon Fern (*Osmunda cinnamomea*): up to 5 feet tall, with dense cluster of sterile fronds surrounding separate, shorter, cinnamon-colored fertile frond (that withers soon after flowering) and woolly lower stem; spring; marshes, wet meadows, and forested swamps; FACW; Labrador to Florida. Similar species: Virginia Chain Fern (*Woodwardia virginica*) lacks woolly lower stem, and its upper leaflets are fertile, bearing dotlike spores on undersides; OBL. Interrupted Fern (*O. claytoniana*) closely resembles cinnamon Fern, but it has spore-bearing leaflets between the sterile leaflets and lacks woolly lower stem; FAC.

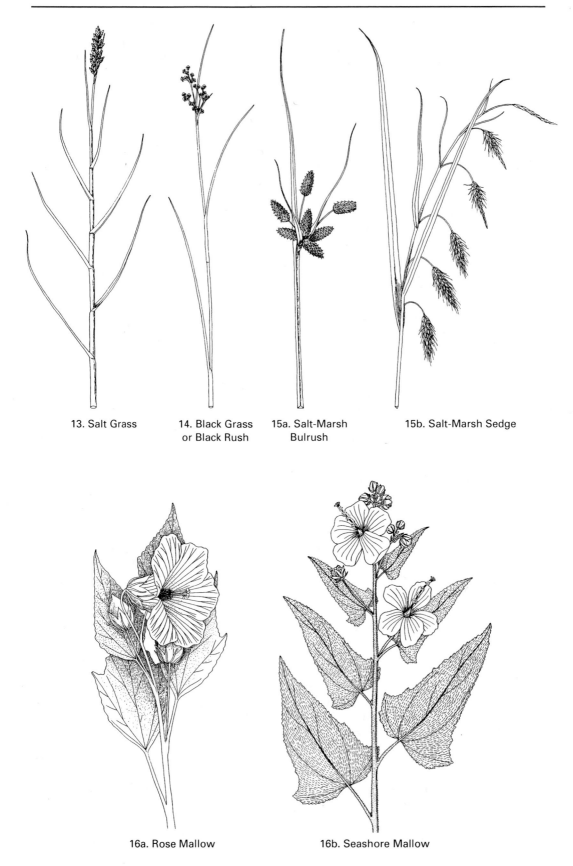

13. Salt Grass

14. Black Grass
or Black Rush

15a. Salt-Marsh
Bulrush

15b. Salt-Marsh Sedge

16a. Rose Mallow

16b. Seashore Mallow

17. Northern Sea Lavender

18. Annual Salt-Marsh
Fleabane

19. High-tide Bush or Marsh Elder

20. Groundsel Bush

23. Virginia Chain Fern (*Woodwardia virginica*): up to 4 feet tall, with purplish brown stalks, resembles Cinnamon Fern, but lacks separate fertile frond, spores being borne in two rows on underside of fertile leaflets; midsummer–early fall; bogs, forested wetlands, and shrub swamps; OBL; Nova Scotia to Florida.

24. Net-veined Chain Fern (*Woodwardia areolata*): up to 2½ feet tall, with deeply lobed sterile frond having fine-toothed leaf margins, and separate frond bearing linear fertile leaflets; midsummer–fall; forested swamps; OBL; Nova Scotia to Florida.

25. Sensitive Fern (*Onoclea sensibilis*): up to 3½ feet tall, with light green fronds with entire margins, and separate fertile stalk bearing beadlike leaflets; summer–fall; wet meadows, marshes, forested wetlands, and moist woods; FACW; Newfoundland to Florida. Similar species: Net-veined Chain Fern (*Woodwardia areolata*) has sterile fronds with finely toothed margins and fertile frond with linear leaflets; OBL.

26. Ostrich Fern (*Matteuccia struthiopteris*): up to 5 feet tall, with plumelike sterile fronds tapering greatly toward base and widest near top, surrounding separate featherlike fertile fronds (to 2 feet tall) composed of dark brown (when mature) podlike sporangia; midsummer–fall; floodplain forested wetlands and moist woodlands; FACW; Newfoundland to southern New England, in eastern mountains to Virginia.

27. Royal Fern (*Osmunda regalis*): up to 6 feet tall, with light brown fertile leaflets at top of frond; spring–early summer; marshes, wet meadows, and swamps; OBL; Newfoundland to Florida. Similar species: Interrupted Fern (*O. claytoniana*) has fertile spore-bearing leaflets located between the sterile leaflets of the frond; FAC.

28. Marsh Fern (*Thelypteris thelypteroides*): up to 2½ feet tall, with black rhizomes, light green to yellow-green leaves, undersides of leaflets with forked veins, and upper fertile leaflets bearing fruit dots near midvein; summer–fall; marshes, wet meadows, and swamps; FACW+; Newfoundland to Florida. Similar species: Massachusetts Fern (*T. simulata*) has unbranched veins on undersides of leaflets; FACW; Nova Scotia to Virginia. Tapering Fern or New York Fern (*T. noveboracensis*) has its lower leaflets tapering to very small leaflets at base of stem; FAC; Newfoundland to New York.

29. Crested Fern (*Dryopteris cristata*): evergreen fern up to 3 feet tall, with sterile evergreen leaves composed of somewhat triangle-shaped, many-lobed leaflets and separate, taller, deciduous fertile frond bearing many dotlike sori (spore clusters) on upper leaflets between midvein and leaf margin; summer–fall; nontidal marshes, wet meadows, shrub swamps, and forested wetlands; FACW+; Newfoundland to North Carolina. Similar species: Lady Fern (*Athyrium filix-femina*) bears crescent-shaped sori that resemble eyebrows; FAC.

Grasses

30. Wild Rice (*Zizania aquatica*): annual up to 10 feet tall, with large, flat soft leaves (to 2 inches wide) having rough margins and inflorescence divided into two parts, appressed upper female portion and more open lower male portion with dangling yellow male flowers; June–September; slightly brackish and tidal fresh marshes, and nontidal marshes; OBL; Nova Scotia to Florida .

31. Bluejoint (*Calamagrostis canadensis*): perennial up to 5 feet tall, with flat leaves (to ⅘ inch wide); June–August; marshes, wet meadows, shrub swamps, bogs, and moist soils; FACW+; Greenland to Delaware.

32. Wood Reed (*Cinna arundinacea*): perennial up to 5 feet tall, with slightly rough-margined flat leaves (to ½ inch wide) with midvein off-center and inflorescence with ascending or spreading branches; late July–October; forested wetlands; FACW+; Maine to Georgia. (Note: Usually has a "wrinkled" zone along the outer leaf margin near the bottom of the leaf.)

33. Walter Millet (*Echinochloa walteri*): annual up to 6½ feet tall, usually shorter, with long tapering leaves (to 1 inch wide), short, coarse-haired leaf sheaths, and spikelets covered with long bristles; August–October; marshes, swamps, and shallow water; FACW+; Massachusetts to Florida. Similar species: Barnyard Grass (*E. crusgalli*) has spikelets that usually lack bristles; FACU.

34. Nerved or Fowl Manna Grass (*Glyceria striata*): perennial up to 4 feet tall, with flattened or folded leaves (rough above), rough leaf sheaths, and many-branched terminal inflorescence with small spikelets (less than ⅒ inch wide); June–September; marshes, shrub swamps, and forested wetlands; OBL; Newfoundland to Florida. Similar species: Rattlesnake Grass or Manna Grass (*G. canadensis*)

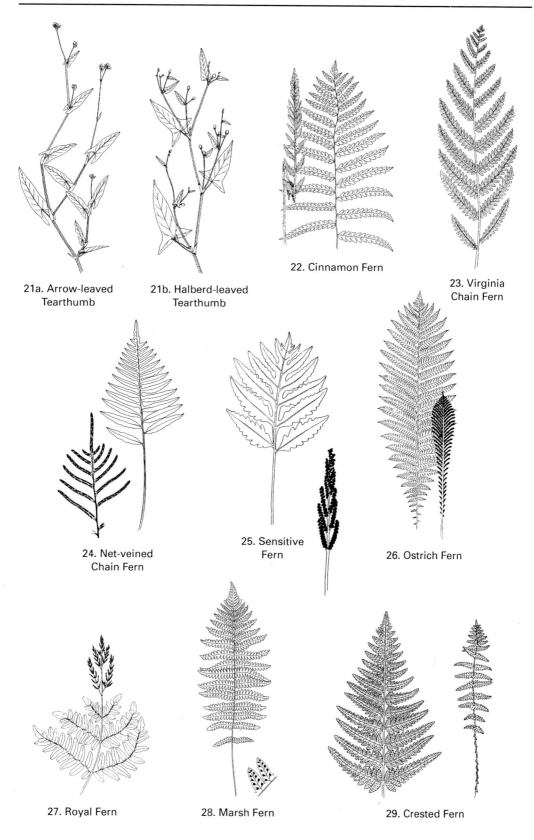

21a. Arrow-leaved
Tearthumb

21b. Halberd-leaved
Tearthumb

22. Cinnamon Fern

23. Virginia
Chain Fern

24. Net-veined
Chain Fern

25. Sensitive
Fern

26. Ostrich Fern

27. Royal Fern

28. Marsh Fern

29. Crested Fern

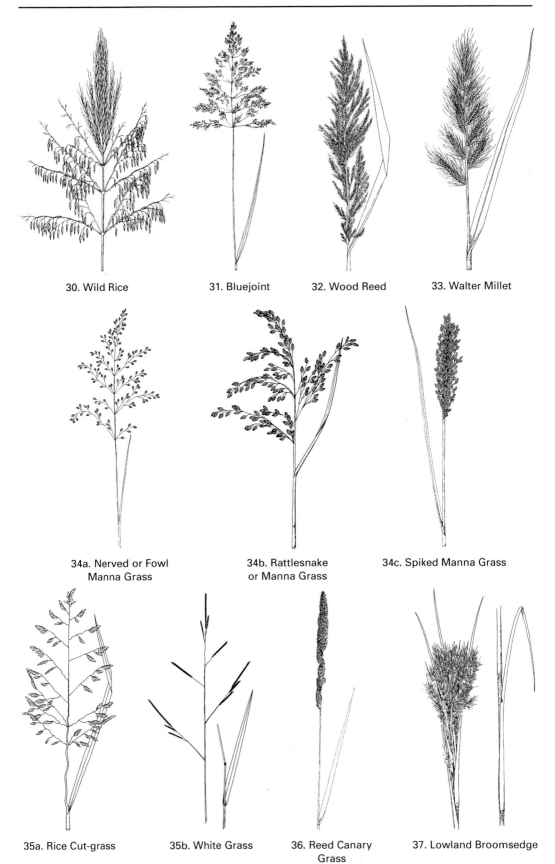

30. Wild Rice

31. Bluejoint

32. Wood Reed

33. Walter Millet

34a. Nerved or Fowl
Manna Grass

34b. Rattlesnake
or Manna Grass

34c. Spiked Manna Grass

35a. Rice Cut-grass

35b. White Grass

36. Reed Canary
Grass

37. Lowland Broomsedge

has larger egg-shaped spikelets (about ⅕ inch wide); New Jersey and western Maryland north; OBL. Spiked Manna Grass (*G. obtusa*) has a dense terminal spike (to 7 inches long); Nova Scotia south to North Carolina along the coast and to eastern Pennsylvania inland; OBL.

35. Rice Cut-grass (*Leersia oryzoides*): perennial up to 5 feet tall, with rough, hairy stems, hairy nodes, rough-margined light green leaves and sheaths, and an open-branched terminal inflorescence, usually with a squiggly lower stem rising from leaf sheath; June–October; marshes and wet meadows; OBL; Nova Scotia to Florida. Similar species: White Grass (*L. virginica*) occurs in temporarily flooded forested wetlands and moist woods; its slightly rough-margined stem has hairy nodes, and its spikelets are appressed along branches of the open terminal inflorescence; FACW. Japanese Stilt Grass (*Eulalia viminea*), an invasive species, flowers in late summer; its leaves are yellowish green and rough when rubbed, the center of the leaf (midvein area) appears as a smooth wide band, its stem lacks hairy nodes, its leaf sheaths bear somewhat stiff hairs, and its seeds are often enclosed within leaf sheaths in late summer and fall; FAC. (Note: In spring and early summer, this grass often appears as a low-growing annual forming a thick carpet on the forest floor; in late summer, it may be 3 feet tall or more.)

36. Reed Canary Grass (*Phalaris arundinacea*): perennial to 5 feet high, with tapered leaves (to ⅘ inch wide) and compressed or open, few-short-branched inflorescence; June–August; marshes, wet meadows, swales, and moist woods; FACW+; Newfoundland to North Carolina. (Note: It is widely cultivated in the Northeast.)

37. Lowland Broomsedge (*Andropogon glomeratus*): perennial up to 5 feet tall, occurring in dense clumps, with feathery inflorescence; August–October; marshes and savannas; Maine to Florida.

See also Big Cordgrass (species 6), Common Reed (species 7), and Switchgrass (species 8).

Sedges

38. Many sedges of the genus *Carex* occur in Northeast wetlands—too many to include in this book. They have triangular stems, and most are found in wet meadows and marshes, while some also occur in forested wetlands. Most sedges are difficult to identify without referring to taxonomic plant manuals. Some common and conspicuous types are illustrated. For others, refer to other field guides or plant manuals listed at the end of this chapter. Tussock Sedge (*Carex stricta*): perhaps the most common marsh, wet-meadow, and swamp sedge in the Northeast, forming dense clumps ("tussocks") up to 3½ feet tall, with fertile stems much longer than leaves; OBL. Lurid Sedge (*C. lurida*): a common wet-meadow sedge to 3½ feet tall, with yellow-green leaves and terminal inflorescence bearing one linear male spike and two to four female spikes (with inflated sacs); OBL. Fringed Sedge (*C. crinita*): marsh, wet-meadow, and swamp species to 4½ feet tall, with distinctive drooping inflorescence; OBL. Bladder Sedge (*C. intumescens*): a forested-wetland species to 3½ feet tall, with dark green leaves and terminal inflorescence bearing one stalked male spike and one to three female spikes with five to fifteen inflated sacs; FACW+. Hop Sedge (*C. lupulina*): a sedge of marshes, meadows, and swamps to 4 feet tall, with many large inflated sacs borne on each female spike; OBL. Long Sedge (*C. lonchocarpa*): distinctive sedge to 4 feet tall, with yellow-green leaves and terminal inflorescence bearing two types of flowering spikes, one terminal male spike and two to five female spikes, usually separated by a considerable distance; OBL. Fox Sedge (*C. vulpinoidea*): a wet-meadow species to 3½ feet tall, with terminal inflorescence bearing many cylindrical spikes, each subtended by a bristle; OBL. Stalk-grained Sedge (*C. stipata*): resembles Fox Sedge, but has winged triangular stems; OBL. Bull Sedge (*Carex bullata*): common sedge of wet pitch-pine lowlands, to 3 feet tall, with one or two well-separated female spikes (to 1½ inches long and less than 1 inch wide) and one to three male spikes on long stalk extending well beyond female spike; OBL. Bearded Sedge (*C. comosa*): a sedge of forested wetlands to 5 feet tall; OBL. Bristlebract Sedge (*C. tribuloides*): a common woodland and marsh species to 3½ feet tall; FACW+. Beaked Sedge (*C. rostrata*): a sedge of shallow water and wet meadows to 4 feet tall; OBL. Woolly-fruited Sedge (*C. lasiocarpa*): a fen and bog species that also occurs in shallow water, with characteristic wirelike or threadlike leaves (to 4 feet long); OBL. Woolly Sedge (*C. lanuginosa*): a marsh and wet meadow species, is similar to Woolly-fruited Sedge, but does not have long, threadlike leaves; OBL. Three-seeded Sedge (*C. trisperma*): sedge of

spruce bogs and evergreen forested swamps, to 2¼ feet tall, forming a grasslike carpet on hummocks; OBL. Few-seeded Sedge (*Carex oligosperma*): a common sedge in northern bogs, fens, and shallow water to 4 feet tall; OBL. Poor Sedge (*C. paupercula*): a bog and wet meadow species to 3 feet tall; OBL (Mud Sedge, *C. limosa*, looks very similar). Little Prickly Sedge (*C. echinata*) of wet meadows and fens grows to 1½ feet tall; OBL. Yellow sedge (*C. flava*): another wet meadow and fen species, grows to 2½ feet tall; see Figure 5.4. Some other bog sedges include Hoary Sedge (*C. canescens;* to 3 feet tall) and Coast Sedge (*C. exilis;* to 2½ feet tall).

39. Twig-rush (*Cladium mariscoides*): perennial up to 3 feet tall, with roundish-triangular stems, long narrow leaves (to 8 inches long) grooved near base and rolled inward at tip, slightly rough leaf margins, and inconspicuous flowers forming three to ten or more, brown to reddish brown spikelets borne on upright branches from upper leaf axils and one terminal; August–October; brackish and tidal fresh marshes, nontidal marshes, fens, savannas, shores, and pond margins; OBL; Newfoundland to Florida.

40. Three-way Sedge (*Dulichium arundinaceum*): perennial up to 3½ feet tall, with hollow, jointed, round stems, leaves distinctly three-ranked, and inconspicuous flowers borne in several stalked spikes; July–October; marshes, bogs, swamps, and pond margins; OBL; Newfoundland to Florida.

41. Cotton-grasses (*Eriophorum* spp.): perennials up to 4 feet tall, with flat linear leaves, inconspicuous flowers borne on spikes in terminal inflorescence subtended by two to five leaflike bracts extending above the inflorescence, and bristly seeds forming distinctive cottony masses; September–October; marshes, fens, wet meadows, and bogs; OBL; Newfoundland to Florida. (See Figure 5.5.)

42. White Beak-rush (*Rhynchospora alba*): perennial to 2¼ feet tall, often forming dense colonies, with very slender stems and clusters of whitish flowers borne on short stalks at end of elongate stalks arising from leaf axils and terminally; June–September; bogs and wet sand; Newfoundland to North Carolina; OBL. Similar species: Other beak-rushes have brownish spikelets.

43. Spikerushes (*Eleocharis* spp.): numerous species with a characteristic leafless flowering stalk bearing a single budlike spike at the top; tidal and nontidal marshes, wet meadows, shallow water, and exposed mud; most are OBL, with a few being FACW. Similar species: When not in flower, the single terminal spikelet of Yellow-eyed Grass (*Xyris* spp.) resembles that of spikerushes, but at the base of the flowering stalk there is a cluster of linear, grasslike leaves; when in bloom, the spikelet bears many small yellow flowers; occurs in sandy and peaty soils from Maine south.

44. Wool-grass (*Scirpus cyperinus*): perennial usually 4 to 6 feet tall in dense clumps, with round to roundish triangular stems, long tapered leaves having very rough margins, and many woolly (at maturity) rusty-colored spikelets forming a terminal inflorescence; August–September; marshes, wet meadows, and swamps; Newfoundland to Florida; FACW+. Similar species: Black-girdled Bulrush (*S. atrocinctus*) closely resembles wool-grass, but fruits earlier and has olive or brownish woolly (not rusty) spikelets, a black band at the top of its stem immediately below its terminal leaves (leafy bracts), and blackish leafy bracts below its spikelets; June–July; FACW+. Most other bulrushes have strongly triangular stems. Green Bulrush (*S. atrovirens*), another earlier-blooming bulrush of wet meadows and forested wetlands, has fewer leaves, a distinctly triangular stem, and dark green or brown (when mature) round clusters of nonwoolly spikelets borne on mostly ascending branches; June–August; OBL. Red-tinged Bulrush (*S. microcarpus*), a wet meadow species (up to 3 feet tall) from New England and New York, has stem marked with red and green horizontal stripes and red-tinged leaf sheaths; OBL. River Bulrush (*S. fluviatilis*), a robust emergent to 5 feet tall or more, has uppermost leaves that droop and its budlike spikelets are mostly borne on long stalks; OBL; Quebec to Virginia, inland to Washington. Common Three-square (*S. pungens*) has a nearly leafless stem with a cluster of spikelets borne near the top of the stem (see species 10); OBL.

45. Soft-stemmed Bulrush (*Scirpus validus*): perennial up to 10 feet tall, with soft, round, usually grayish green stems, no apparent leaves, and many stalked budlike spikes borne on drooping branches near top of stem; June–September; brackish and freshwater marshes and shallow water; OBL; Newfoundland to Florida. Similar species: Hard-stemmed Bulrush (*S. acutus*) has dark green stems, and its spikelets form an almost terminal inflorescence; OBL.

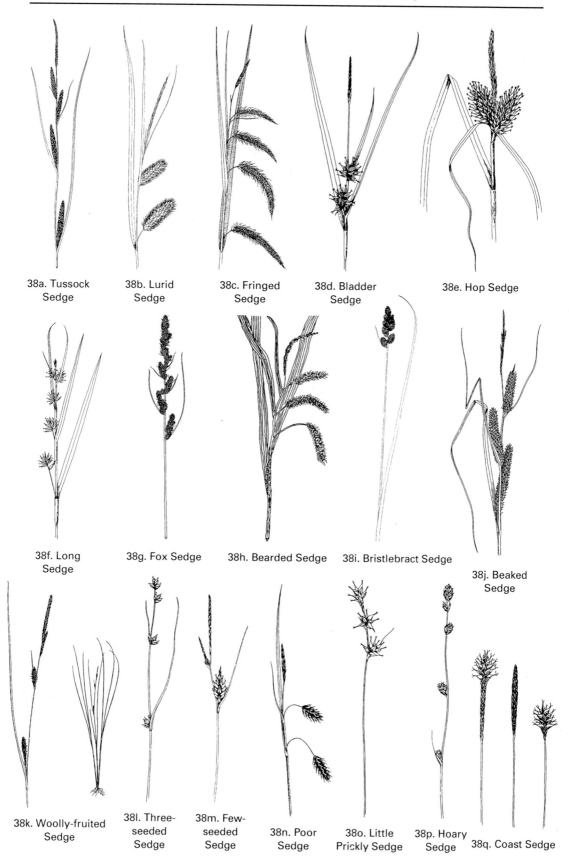

38a. Tussock
Sedge

38b. Lurid
Sedge

38c. Fringed
Sedge

38d. Bladder
Sedge

38e. Hop Sedge

38f. Long
Sedge

38g. Fox Sedge

38h. Bearded Sedge

38i. Bristlebract Sedge

38j. Beaked
Sedge

38k. Woolly-fruited
Sedge

38l. Three-
seeded
Sedge

38m. Few-
seeded
Sedge

38n. Poor
Sedge

38o. Little
Prickly Sedge

38p. Hoary
Sedge

38q. Coast Sedge

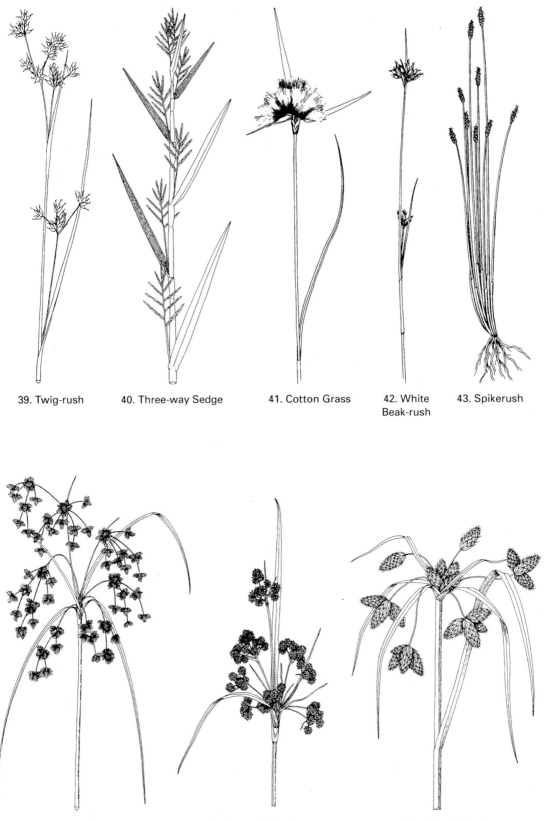

39. Twig-rush

40. Three-way Sedge

41. Cotton Grass

42. White Beak-rush

43. Spikerush

44a. Wool-grass

44b. Green Bulrush

44c. River Bulrush

Water Bulrush (*Scirpus subterminalis*) is a shallow water aquatic species with roundish stems (usually floating but sometimes emerging slightly out of the water), a single spikelet near top of stem, and long, threadlike leaves arising from base of plant; OBL; throughout Northeast and Great Lakes region.

See also Common Three-square, listed under Olney's Three-square (species 10), and River Bulrush, listed under Salt-Marsh Bulrush (species 15).

Rushes and Other Graminoids

46. Soft Rush (*Juncus effusus*): perennial up to 3½ feet tall, lacking leaves, with soft, round, green stems filled with whitish tissue, forms dense clumps, with clusters of flowers borne on upper half of stems and later bearing small fruit capsules; July–September; marshes, wet meadows and pastures, and shrub swamps; FACW+; Newfoundland to Florida. Similar species: Bayonet Rush (*J. militaris*), a shallow-water emergent of acidic ponds and lakes like those of Pine Barrens, Adirondacks, and northern New England, has a single hollow leaf (with distinctive cell walls) arising from the middle of the stem and has reddish stem bases; OBL; Newfoundland to eastern Maryland. Baltic Rush (*J. balticus*) occurs along freshwater shores, especially calcareous marshes; it does not form dense clumps, and its solitary leafless stems may be bluish green and bear flower clusters near the top; FACW+.

47. Canada Rush (*Juncus canadensis*): perennial up to 3¼ feet tall growing in clumps, with few elongate leaves (round in cross section, with distinctive cell walls), small flowers borne in dense brown heads, and fruit capsules; July–October; marshes, swamps, wet shores, and upper edges of salt marshes; OBL; Nova Scotia to Georgia. Similar species: Bog Rush (*J. pelocarpus*), a bog species, has an open branched inflorescence; OBL.

48. Broad-leaved Cattail (*Typha latifolia*): perennial up to 10 feet tall, with elongate, flattened basal-leaved and fertile stalk bearing distinctive terminal inflorescence (male part above female part); May–July; marshes and shallow water; OBL; Newfoundland to Florida. Similar species: See Narrow-leaved Cattail (species 9). Blue Cattail (*T. × glauca*) is a hybrid of Broad-leaved Cattail and Narrow-leaved Cattail; its leaves are intermediate in width (²⁄₅–³⁄₅ inch), the male and female parts of the inflorescence are usually separated up to 1½ inches, and the female spike is longer than 6 inches; OBL.

49. Eastern Bur-reed (*Sparganium americanum*): perennial up to 3½ feet tall, with soft, linear, light green to yellow-green leaves (triangular in cross section), clasping leaves forming sheaths at stem, and several greenish to whitish flower balls and later ball-like fruit clusters; May–August; marshes, muddy shores, and shallow water; OBL; Newfoundland to Florida. Similar species: Great or Giant Bur-reed (*S. eurycarpum*) grows to 7 feet tall, has stiffer leaves that tend to be medium to dark green in color, and can be mistaken for cattail when not in bloom; OBL.

50. Sweet Flag (*Acorus calamus*): perennial up to 4 (rarely to 7) feet tall, with aromatic irislike leaves having midrib slightly off center, and fingerlike fleshy appendage (spadix) bearing many small yellowish flowers; May–August; marshes, wet meadows, and shallow water; OBL; Nova Scotia to Florida.

51. Pipeworts (*Eriocaulon* spp.): perennials with cluster of grasslike basal leaves and separate flower stalk with roundish head of small white flowers at top; July–October; marshes, savannas, mud flats, and shallow water; OBL; Nova Scotia to Florida.

See also Blue Flag (species 61), which has grasslike leaves.

Herbs with All or Mostly Basal Leaves

52. Trout Lily or Dog-toothed Violet (*Erythronium umbilicatum*): early spring–blooming perennial up to 10 inches tall, common on floodplains, with green-brown-purple mottled leaves and bearing a single, nodding yellow flower (with recurved petals) on a long stalk; March–June; temporarily inundated floodplains and rich, moist upland woods; FAC; Nova Scotia to Florida.

53. Violets (*Viola* spp.): spring-blooming perennials up to about 6 inches tall, with kidney- or heart-shaped leaves and bearing either white flowers (White Violet, *V. blanda*, FACW; or the fragrant Sweet White Violet, *V. pallens*, OBL) or violet-blue flowers (Marsh Blue Violet, *V. cucullata*, with club-shaped white hairs on inner parts of petals, FACW+; or Common Blue Violet, *V. papilionacea*, with long slender white hairs on petals, FAC); April–July; forested wetlands, seeps, wet meadows, and streamsides; throughout region.

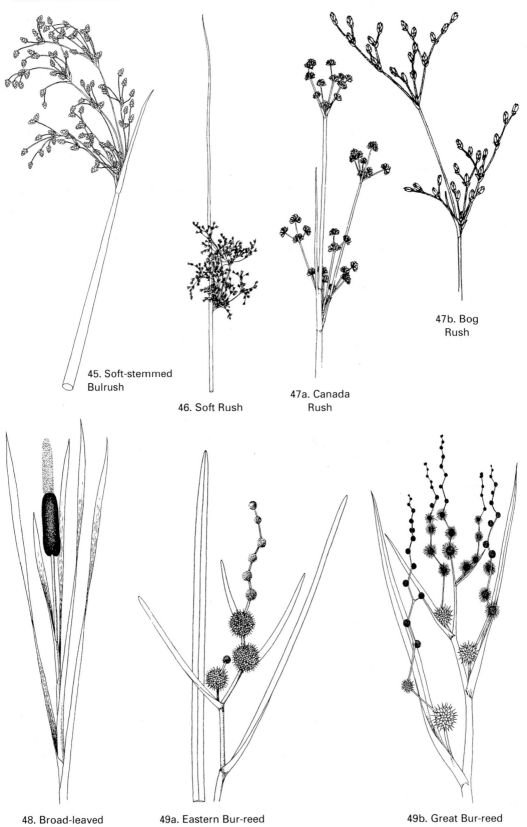

45. Soft-stemmed
Bulrush

46. Soft Rush

47a. Canada
Rush

47b. Bog
Rush

48. Broad-leaved
Cattail

49a. Eastern Bur-reed

49b. Great Bur-reed

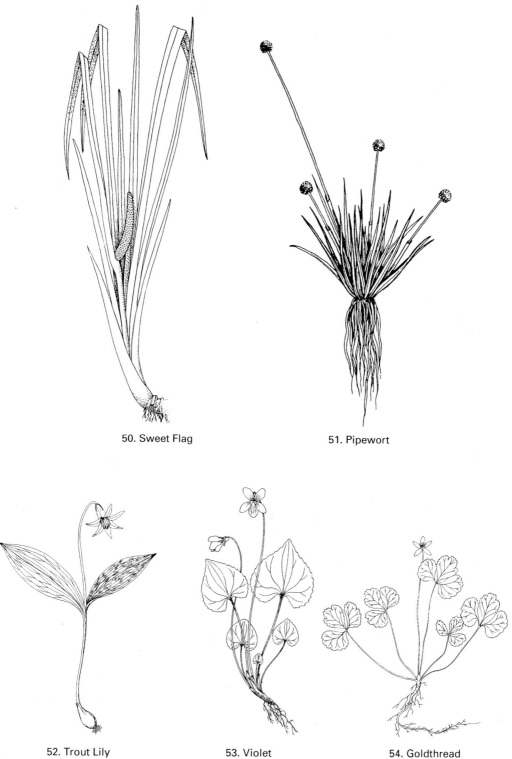

50. Sweet Flag

51. Pipewort

52. Trout Lily

53. Violet

54. Goldthread

54. Goldthread (*Coptis trifolia*): perennial up to 6 inches tall, with yellow-orange ("gold") roots, compound leaves composed of three shiny evergreen toothed cloverlike leaflets, and small white five- to seven-petaled starlike flowers (about ½ inch wide) borne on long separate stalks; May–July; evergreen forested wetlands and cool evergreen upland woods; FACW; Labrador to northern New Jersey, in eastern mountains to North Carolina. Similar species: Wild Strawberry (*Fragaria virginiana*), an upland species, has similar-shaped deciduous leaves and wider flowers (½ to 1 inch wide), and bears a strawberry fruit; FACU.

55. Wild Calla (*Calla palustris*): perennial to 2 feet tall, usually appearing much shorter, with thick heart-shaped leaves (to 8 inches long), small flowers borne on thick cylinder-shaped fleshy stalk (spadix) surrounded by a whitish leaf (spathe), and red berries; April–August; margins of bogs and shallow water; OBL; Newfoundland to northern New Jersey, in eastern mountains to western Maryland.

56. Arrowheads (*Sagittaria* spp.): perennials with flowering stalk up to 4 feet tall and leaves variably shaped (but most species have three-lobed arrowhead-shaped leaves), bearing many white three-petaled flowers (about 1½ inches wide) on separate flowering stem; July–September; marshes and shallow water; OBL; Nova Scotia to Florida. Similar species: Water Plantains (*Alisma* spp.) have oval-shaped leaves with somewhat heart-shaped bases and prominent tips, and a greater number of tiny three-petaled flowers (less than ⅗ inch wide) on a many-branched separate flowering stem; OBL.

57. Arrow Arum (*Peltandra virginica*): somewhat fleshy leaved perennial up to 2 feet tall, with large arrowhead-shaped leaves (to 12 inches long at flowering, growing larger thereafter), inconspicuous flowers borne on fleshy spike (spadix) enclosed within an elongate fleshy green structure (spathe), and slimy greenish pealike berries; May–July; slightly brackish to tidal fresh marshes, nontidal marshes, shallow water, and swamps; OBL; southern Maine to Florida. Similar species: May be confused with Arrowheads (*Sagittaria* spp., species 56) and Pickerelweed (*Pontederia cordata*, species 66).

58. Spatterdock or Yellow Pond Lily (*Nuphar luteum*): floating-leaved or emergent perennial up to 16 inches tall, with fleshy heart-shaped leaves (up to 16 inches long), bearing a single five- or six-petaled yellow flower (to 3 inches wide); May–October; shallow water, tidal fresh and nontidal marshes, and swamps; OBL; southern Maine to Florida.

59. Marsh Marigold (*Caltha palustris*): perennial up to 2 feet tall, with kidney- to heart-shaped toothed leaves borne on long stems (mostly basal, but some appearing alternately arranged), and clusters of showy, bright yellow, shiny flowers (up to 1 inch wide) with five or six (sometimes as many as nine) petals; April–June; seasonally flooded forested wetlands, shrub swamps, seeps, stream banks, and wet meadows; OBL; Newfoundland to Virginia. Similar species: Robin-run-away or Dewdrop (*Dalibarda repens*) has similar-shaped leaves, but is much shorter (to 4 inches tall) and bears five-petaled white flowers (½ inch wide) in summer; it grows in evergreen forested wetlands from Nova Scotia to western Connecticut and in eastern mountains to North Carolina; FAC.

60. Skunk Cabbage (*Symplocarpus foetidus*): foul-smelling perennial herb up to 2 feet tall that covers the ground in many forested wetlands from spring into summer, with large oval to heart-shaped leaves (skunklike odor when crushed) and inconspicuous flowers borne on an oval-shaped to roundish fleshy spike (spadix) enclosed within a thick, fleshy, purple-and-green-striped hood (spathe); February into May (beginning before leaves emerge); seasonally flooded forested wetlands, shrub swamps, tidal fresh and nontidal marshes, and seeps; OBL; Quebec and Nova Scotia to southern Manitoba, south to North Carolina and Iowa, in eastern mountains to Georgia and Tennessee. Similar species: False Hellebore (*Veratrum viride*, species 65) also has large leaves, but they are alternately arranged and deeply ridged (folded), and no parts are foul-smelling; FACW+.

61. Blue Flag (*Iris versicolor*): perennial up to 4 feet tall, growing in dense clump, with somewhat fleshy, pale bluish green, mostly basal sword-shaped leaves (few alternately arranged leaves on flowering stem), bluish to purplish iris flowers (to 4 inches wide), and three-angled elongate fruit pods; May–July; slightly brackish to tidal fresh marshes, nontidal marshes, wet meadows, and forested wetlands; OBL; Newfoundland to Virginia. Similar species: Slender Blue Flag (*I. prismatica*) has narrow grasslike leaves (less than ¼ inch wide) and bears bluish flowers; OBL. Yellow Flag (*I. pseudacorus*), an

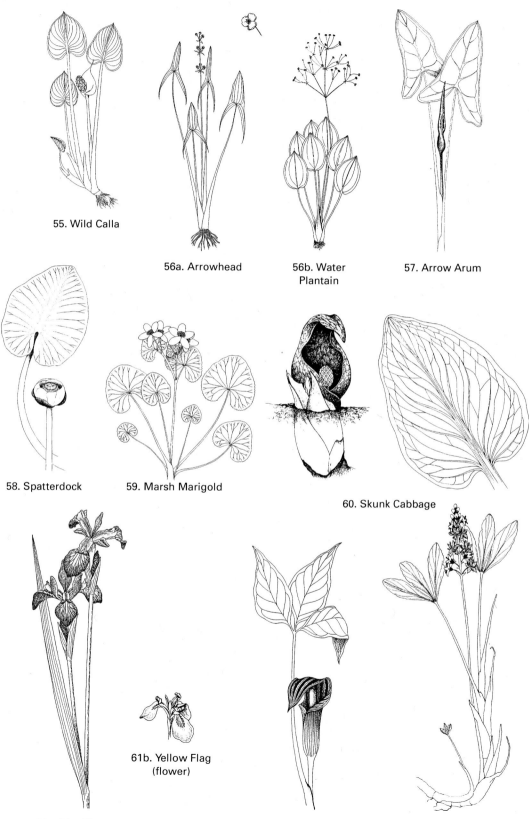

55. Wild Calla

56a. Arrowhead

56b. Water Plantain

57. Arrow Arum

58. Spatterdock

59. Marsh Marigold

60. Skunk Cabbage

61b. Yellow Flag (flower)

61a. Blue Flag

62a. Jack-in-the-pulpit

62b. Buckbean

escaped garden plant introduced from Eurasia, has greener leaves and bears yellow iris flowers; OBL.

62. Jack-in-the-pulpit (*Arisaema triphyllum*): perennial up to 3 feet tall, with compound leaf divided into three (sometimes five) leaflets, inconspicuous flowers borne on fleshy finger-like spike enclosed within a leafy tubular hood (spathe), and cluster of red berries; March–June; forested wetlands and rich, moist upland woods; FACW−; Nova Scotia to Florida. Similar species: Wake-robin or Purple Trillium (*Trillium erectum*), a moist-woods species, has a single whorl of three broad leaves (widest near middle) and bears a distinctive three-petaled purplish flower; FACU−. Buckbean (*Menyanthes trifoliata*), a northern species of bogs, fens, cedar swamps, and shallow water, has compound leaves composed of three somewhat oval-shaped leaflets, leaves alternately arranged but arising from close to the thick base of plant (appearing basal); many five-petaled, white (pink- to purple-tinged) flowers covered with hairs, borne on a leafless stalk; May–July; OBL; south to New Jersey, Ohio, and Indiana.

See also Mayflower (species 63), plus Sundews and Pitcher Plant in Figure 5.7.

Herbs with Simple, Entire, Alternate Leaves

63. Canada Mayflower (*Maianthemum canadense*): low perennial up to 8 inches tall (usually 5 inches or less), with one to three short-stalked to sessile leaves (up to 4 inches long) having heart-shaped bases, small four-petaled white flowers borne in terminal spike, and clusters of red berries; May–June; forested wetlands, hemlock swamps, and moist upland woods; FAC−; Labrador and Newfoundland to Delaware, in eastern mountains to Georgia. (Note: Often observed with only one leaf appearing as basal leaf.) Similar species: Three-leaved False Solomon's-seal (*Smilacina trifolia*), a northern bog species, usually has three leaves clasping the stem and small six-petaled white flowers borne on long stalks in a terminal inflorescence; OBL. Sessile-leaved Bellflower (*Uvularia sessilifolia*) of upland woods has narrower leaves (less than ½ inch wide), one or two yellowish six-petaled drooping flowers, and a triangular fruit pod; FACU−.

64. Bog Aster (*Aster nemoralis*): perennial up to 2 feet tall, often less than 1 foot, with rough hairy stems, many (forty or more) leaves (to 2½ inches long) with somewhat wrinkled surfaces and rough above, and one or more pink to light purple daisylike flowers; bogs and pond shores; FACW+; Newfoundland to New York and southern New England.

65. False Hellebore (*Veratrum viride*): perennial up to 6½ feet tall, with large, deeply grooved or creased, pale green leaves (to 12 inches long and 6 inches wide) and many small, six-petaled, yellow-green flowers (½ inch wide) borne in branched terminal inflorescence (to 20 inches long); May–July; forested wetlands, seeps, and wet meadows; FACW+; New Brunswick to North Carolina, in mountains to Georgia. Similar species: Might be confused with Skunk Cabbage (species 60), but the foul smell of that plant is distinctive.

66. Pickerelweed (*Pontederia cordata*): somewhat fleshy-leaved perennial up to 3½ feet tall, with heart-shaped (sometimes lance-shaped) leaves (to 7¼ inches long) and many small, showy, purple tubular and lobed flowers borne on a single terminal spike (to 4 inches long); June–November; marshes and shallow water; OBL; Nova Scotia to northern Florida.

67. Lizard's Tail (*Saururus cernuus*): perennial up to 4 feet tall, with heart-shaped leaves (to 6 inches long and 3 inches wide) and many small, white, fragrant flowers borne on one or two long terminal spikes (to 8 inches long); June–September; marshes, swamps, and shallow water; OBL; southern New England to Florida. (Note: Leaf margins sometimes wrinkled, appearing toothlike.)

68. Smartweeds (*Polygonum* spp.): annual herbs of variable height (mostly to 3½ feet tall), with jointed stems, leafy or hairy sheaths at nodes, mostly lance-shaped leaves, and small pink, white, or green flowers borne in dense or loose spikes; July–October; marshes and wet or damp soils; OBL to UPL. Some common wetland smartweeds in the Northeast are Pinkweed (*P. pensylvanicum*), up to 6½ feet tall, with dense pink-flowered spikes (FACW); Nodding Smartweed (*P. lapathifolium*), up to 5 feet tall with nodding white-flowered spikes (FACW+); Mild Water Pepper (*P. hydropiperoides*), with more open, purple- to pink-flowered spikes, and seeds with mild peppery taste (OBL); Water Pepper (*P. hydropiper*), with greenish or red-tipped flowers in loose spikes and peppery-tasting seeds (OBL); and Dotted Smartweed (*P. punctatum*), with green to greenish white flowers in loose,

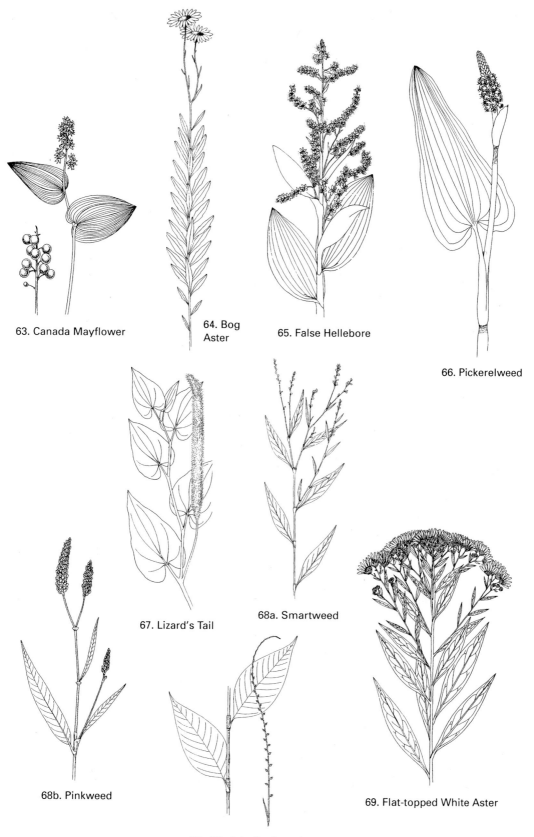

63. Canada Mayflower

64. Bog Aster

65. False Hellebore

66. Pickerelweed

67. Lizard's Tail

68a. Smartweed

68b. Pinkweed

68c. Virginia Knotweed

69. Flat-topped White Aster

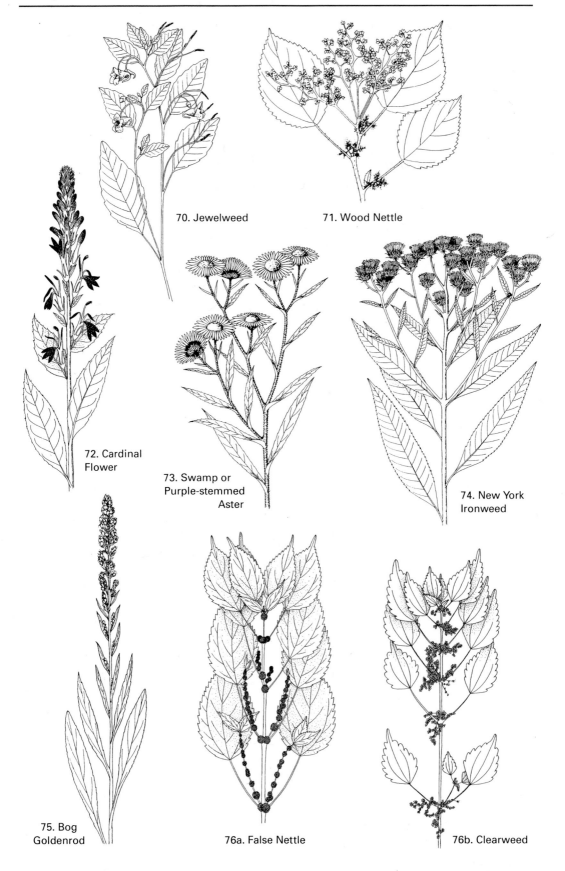

70. Jewelweed

71. Wood Nettle

72. Cardinal Flower

73. Swamp or Purple-stemmed Aster

74. New York Ironweed

75. Bog Goldenrod

76a. False Nettle

76b. Clearweed

erect spikes, and seeds without peppery taste (OBL). The last species is common in slightly brackish marshes as well as nontidal marshes and shores. Virginia Knotweed or Jumpseed (*P. virginicum*), a floodplain species up to 4 feet tall, bears many tiny greenish white flowers on a terminal inflorescence (to 20 inches long) at the end of a tall flowering stem (dry fruit pop off the stem when touched); FAC; New Hampshire to Florida. Water Smartweed (*P. amphibium*) is an aquatic species with floating leaves (sometimes with rounded or heart-shaped bases) and many small, bright pink flowers borne in dense, short terminal spikes (to 1½ inches long and ¼ inch wide); OBL (see Figure 5.9 for illustration).

69. Flat-topped White Aster (*Aster umbellatus*): perennial up to 6½ feet tall, with rough upper leaf surfaces and many white daisylike flowers (up to ¼ inch wide) borne in leafy branches, often forming a somewhat flat-topped terminal inflorescence; July–September; forested wetlands, shrub swamps, dry thickets, and borders of woods and fields; FACW; Nova Scotia to North Carolina, in eastern mountains to Georgia.

See also Swamp Aster (species 73) and illustrations of wetland orchids in Figure 5.7.

Herbs with Simple, Toothed, Alternate Leaves

70. Jewelweed or Spotted Touch-me-not (*Impatiens capensis*): annual usually 2 to 4 feet tall, with smooth, somewhat succulent stems, leaves sometimes opposite, orange or yellow-orange uniquely shaped flowers (with reddish brown to reddish spots) borne on long drooping stalks, and seed pods that pop open when ripe (hence the name touch-me-not); June–September; swamps, wet meadows, seeps, stream banks, marshes, and moist upland woods; FACW; Newfoundland to Florida. (Note: Ruby-throated hummingbirds may be seen visiting these flowers.) Similar species: Pale Touch-me-not (*I. pallida*) has yellow flowers; FACW.

71. Wood Nettle (*Laportea canadensis*): perennial up to 5½ feet tall, with stinging hairs and small greenish flowers borne on open branched clusters from upper leaf axils; July–September; temporarily inundated floodplain forests and moist alluvial woods; FACW; Nova Scotia to Georgia. Similar species: Stinging Nettle (*Urtica dioica*) has opposite leaves; FACU.

72. Cardinal Flower (*Lobelia cardinalis*): perennial up to 5 feet tall, with many showy bright red, two-lipped tubular flowers borne on terminal spikes; July–October; tidal fresh and nontidal marshes, forested wetlands, riverbanks, and seeps; FACW+; New Brunswick to Florida. (Note: Ruby-throated hummingbirds may be seen feeding on these flowers.)

73. Swamp or Purple-stemmed Aster (*Aster puniceus*): perennial to over 6 feet tall, usually less, with often hairy, purplish stems, clasping leaves (usually toothed, with rough upper surfaces), and light purplish to bluish daisylike flowers (to 1½ inch wide); August–November; wet meadows, marshes, shrub swamps, and open forested wetlands; OBL; Newfoundland to Georgia. Similar species: New York Aster (*A. novi-belgii*) has only slightly clasping leaves; FACW+. New England Aster (*A. novae-angliae*) has clasping entire leaves with heart-shaped bases and often deep purplish flowers (to 2 inches wide); FACW–. Small White Aster (*A. vimineus*) has smaller white flowers (to ½ inch wide) and weakly toothed leaves; FAC. Panicled or Lowland White Aster (*A. simplex*) has white flowers (to 1 inch wide) and toothed or entire leaves; FACW.

74. New York Ironweed (*Vernonia noveboracensis*): perennial up to 7 feet tall, with rough-surfaced leaves and many purple flowers borne in loose, open, somewhat flat-topped inflorescence; August–October; wet meadows, marshes, swamps, and stream banks; FACW+; Massachusetts to Georgia west to Ohio. Similar species: Smooth Ironweed (*V. fasciculata*), a Midwest marsh and wet prairie species (Ohio to Minnesota), has smooth leaves; FAC+ (FACW in Midwest).

75. Bog Goldenrod (*Solidago uliginosa*): perennial up to 5 feet tall, with very long basal leaves (to 16 inches), upper leaves reduced in size, and numerous small yellow flowers in heads borne terminally on one-sided branches from upper leaf axils; July–October; bogs, forested wetlands, and wet meadows; OBL; Nova Scotia to Maryland, in eastern mountains to North Carolina. Similar species: Rough-leaved Goldenrod (*S. patula*) also has different-sized upper and lower leaves, but they have very rough (sandpapery) upper surfaces; OBL. Rough-stemmed Goldenrod (*S. rugosa*) has rough hairy stems and leaves that are of the same general size; FAC. A few other goldenrods occur in wet meadows; consult one of the field guides listed at the end of this chapter.

See also Golden Ragwort (species 93), which has simple toothed basal leaves.

Herbs with Simple, Toothed, Opposite Leaves

76. False Nettle (*Boehmeria cylindrica*): perennial up to 3 feet tall, with unbranched stems and minute greenish flowers borne on axillary spikes; July–September; marshes, forested wetlands, wet meadows, floodplain forests, and moist, shaded soils; FACW+; Quebec to Florida. Similar species: Clearweed (*Pilea pumila*) is also a floodplain species, but it is shorter and has somewhat translucent stems; FACW.

77. Bur Marigold (*Bidens laevis*): annual or perennial up to 3½ feet tall, with showy yellow daisylike flowers (to 2½ inches wide) having seven or eight petals, and barbed seeds; tidal fresh marshes, nontidal marshes, and pond shores; OBL; New Hampshire to Florida, mostly along coast. Similar species: Nodding Bur Marigold or Beggar-ticks (*B. cernua*) has nodding yellow flowers (to 2¼ inches wide) and shorter petals (to ⅔ inch long); OBL. Swamp Beggar-ticks (*B. connata*) has yellow or orange disk flowers (lacking petals) in heads (to ½ inch wide); OBL.

78. Boneset (*Eupatorium perfoliatum*): perennial up to 5 feet tall, with triangle-shaped leaves that join at base (perfoliate), and dense clusters of white flowers in heads borne on a flat-topped inflorescence; late July–October; wet meadows, marshes, shrub swamps, and shores; FACW+; Nova Scotia to Florida. (Note: This is a good plant for butterfly watching.)

79. Wild Mint (*Mentha arvensis*): perennial up to 1½ feet tall or more, with square stems, aromatic (minty when crushed) stems and leaves, and many small, lavender to white, two-lipped tubular flowers borne in dense ball-like clusters in axils of upper leaves; July–September; wet meadows and marshes; FACW; Labrador to Virginia. Similar species: Spearmint (*M. spicata*) is also aromatic when crushed, but has sessile leaves (without stalks) and four-lobed pinkish to purplish flowers borne on slender, loose spikes; FACW+. Peppermint (*M. piperita*) has stalked leaves (with petioles) and pinkish to purplish flowers borne in dense spikes; FACW+. Water Horehounds or Bugleweeds (*Lycopus* spp.) have square stems and many small four-petaled tubular white flowers borne in ball-like clusters in leaf axils, but these plants do not produce a minty odor when crushed; OBL.

80. Blue Vervain (*Verbena hastata*): perennial to 6 feet tall, with rough hairy stems and leaves (to 7¼ inches long) and dense terminal spikes of bluish to violet five-lobed tubular flowers; June–October; wet meadows, marshes, open shrub swamps, and moist fields; FACW+; Nova Scotia to Florida.

See also Stinging Nettle listed under Wood Nettle (species 71).

Herbs with Simple, Entire, Opposite or Whorled Leaves

81. Swamp Milkweed (*Asclepias incarnata*): perennial to 6 feet tall, with milky sap, many uniquely shaped pink to purplish red flowers borne in several terminal clusters, and elongate, erect fruit pods; June–August; marshes, wet meadows, shrub swamps, forested wetlands, and shores; OBL; Nova Scotia to Florida. Similar species: Lance-leaved Red Milkweed (*A. lanceolata*) occurs in brackish and fresh marshes from New Jersey south; it has long, narrow, lance-shaped leaves (less than 1 inch wide) and orange or red flowers; OBL. Red Milkweed (*A. rubra*) grows in wet pine swamps and bogs from Long Island south; it has broad lance-shaped leaves (1 to 2½ inches wide) with round or heart-shaped bases and reddish to purplish flowers; OBL. (Note: Milkweeds attract many butterflies and other insects.)

82. Marsh St. John's-wort (*Triadenum virginicum*): perennial to 2½ feet tall, with reddish stems, egg-shaped leaves (to 2½ inches long), numerous five-petaled pinkish (fleshy) to light purplish flowers borne on stalks from upper leaf axils or terminally, and dark red, tapered fruit capsules; July–August; marshes, bogs, shrub swamps, and shores; OBL; Nova Scotia to Florida inland to Michigan. Similar species: Other St. John's-worts have yellow flowers. The common Dwarf St. John's-wort (*Hypericum mutilum*) grows to 3 feet tall, with somewhat clasping leaves (to 1 inch long); FACW. Marsh St. John's-wort (*T. fraseri*) is almost identical, except that its sepals are shorter (less than ⅕ inch) than those of *T. virginicum* (⅕–⅓ inch long); its inland range is wider, to Minnesota and Nebraska; OBL.

83. Water-willow or Swamp Loosestrife (*Decodon verticillatus*): perennial usually to 4 feet high, with leaves in whorls of threes or fours, angled stems rooting at tips giving the

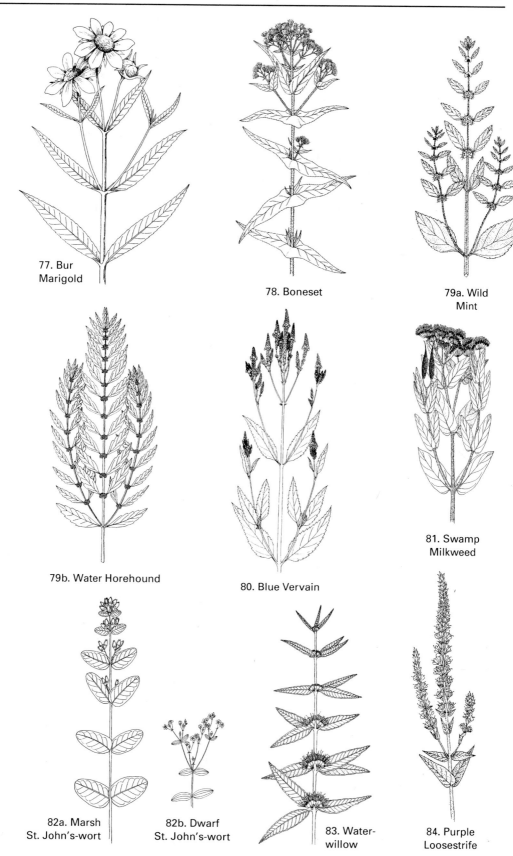

77. Bur Marigold

78. Boneset

79a. Wild Mint

79b. Water Horehound

80. Blue Vervain

81. Swamp Milkweed

82a. Marsh St. John's-wort

82b. Dwarf St. John's-wort

83. Water-willow

84. Purple Loosestrife

plant a whiplike, arching appearance (especially in winter), bases of stems woody, many showy pinkish to purplish bell-shaped flowers (to ¾ inch wide) borne in clusters from axils of upper leaves, and roundish three- to five-celled fruit capsules (⅕ inch wide); July–September; tidal fresh and nontidal marshes, shrub swamps, bogs, pond and lake margins, and forested wetlands; OBL; central Maine to Florida.

84. Purple Loosestrife (*Lythrum salicaria*): extremely showy perennial usually to 5 feet tall, with leaves in whorls of threes or opposite, bases of leaves often heart-shaped and somewhat clasping, many pinkish to purplish five- or six-petaled flowers (to ¾ inch wide) borne in dense, leafy spikes (up to 16 inches long); June–September; tidal fresh and nontidal marshes, wet meadows, and borders of rivers and lakes; FACW+; Quebec to Maryland. (Note: This is an invasive species that has proliferated in many areas of the Northeast.)

85. Swamp Candles or Yellow Loosestrife (*Lysimachia terrestris*): perennial up to 3 feet tall, with many red-centered, five-petaled yellow flowers (½ inch wide) borne singly on long stalks in a terminal spike; June–August; marshes, wet meadows, open shrub swamps, and forested wetlands; OBL; Newfoundland to Georgia.

86. Fringed Loosestrife (*Lysimachia ciliata*): perennial up to 3½ feet tall, with leaf stalks fringed with hairs, and nodding, five-petaled yellow flowers (¾ inch wide), petals with abrupt tips, borne on long stalks; June–August; marshes, shrub swamps, forested wetlands, and stream banks; FACW; Nova Scotia to Florida.

87. Spring Beauty (*Claytonia virginica*): low-growing, spring-blooming perennial less than 12 inches tall, with underground bulb, grasslike leaves (to 8 inches long), showy five-petaled pinkish to white (with pink stripes) flowers borne on long stalks in loose clusters, and green berrylike fruits; March–May; temporarily inundated floodplain forests, forested wetlands, and damp woods; FACU; Nova Scotia to Georgia.

88. Starflower (*Trientalis borealis*): low-growing perennial less than 1 foot tall, with a single whorl of several lance-shaped, unequal-sized leaves, bearing one or two white star-shaped flowers (½ inch wide); May–August; evergreen forested wetlands and northern evergreen woods; FAC; Labrador to Connecticut and New York, in eastern mountains to Virginia and West Virginia.

Herbs with Simple, Toothed, Whorled Leaves

89. Hollow-stemmed Joe-Pye-weed (*Eupatoriadelphus fistulosus*): perennial up to 7 feet tall, with somewhat waxy, purplish, hollow stem, leaves in whorls of mostly fours to sevens, and many small purplish to pinkish flowers in heads forming a round-topped terminal inflorescence; July–September; wet meadows, shrub swamps, forested wetlands, and moist fields and woods; FACW; southern Maine to Florida. Similar species: Spotted Joe-Pye-weed (*E. maculatus*) has purple or purple-spotted stems, leaves in whorls of fours and fives, and a flat-topped inflorescence; FACW. Purple Joe-Pye-weed (*E. purpureus*) has a solid stem with purplish nodes, and leaves in threes and fours; FACU. Eastern Joe-Pye-weed (*E. dubius*) has a purple-spotted stem, and leaves with three main veins (one midrib and two prominent lateral veins); FACW. (Note: These plants are good for watching butterflies.)

Herbs with Compound, Opposite Leaves

90. Devil's Beggar-ticks (*Bidens frondosa*): annual up to 4 feet tall, with yellow to orangish disk flowers borne in dense heads (to ½ inch wide), and later bearing two-barbed seeds ("stickers"); June–October; marshes, wet meadows, floodplain forests, ditches, pastures, waste places, and wet soils; FACW; Newfoundland to Georgia.

Herbs with Compound, Alternate Leaves

91. Water Parsnip (*Sium suave*): perennial up to 7 feet tall, with grooved stems and many minute white flowers borne in umbels (to 5 inches wide), forming a terminal inflorescence; July–September; slightly brackish, fresh tidal and nontidal marshes, swamps, and muddy shores; OBL; Newfoundland to Florida; listed as poisonous by some sources. Similar species: Water Hemlock (*Cicuta maculata*) has purple-streaked stems and coarser-toothed leaflets (sometimes twice-divided); OBL; poisonous. Poison Hemlock (*Conium maculatum*) has spotted stems and parsleylike leaves (divided three or four times), and its umbel (to 2½ inches wide)

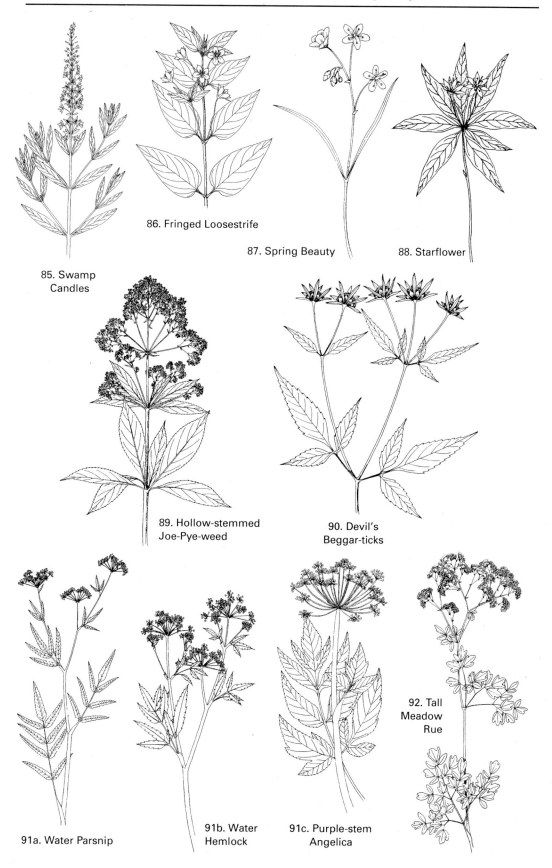

85. Swamp
Candles

86. Fringed Loosestrife

87. Spring Beauty

88. Starflower

89. Hollow-stemmed
Joe-Pye-weed

90. Devil's
Beggar-ticks

91a. Water Parsnip

91b. Water
Hemlock

91c. Purple-stem
Angelica

92. Tall
Meadow
Rue

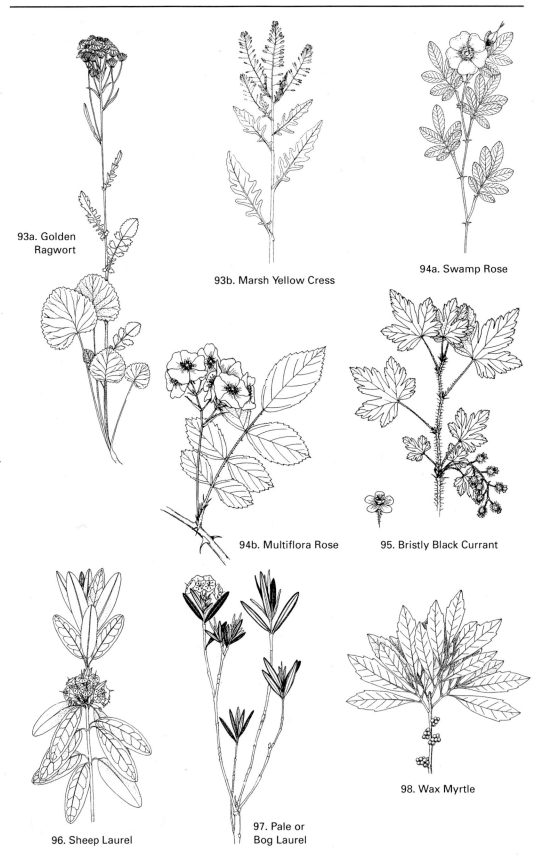

93a. Golden Ragwort

93b. Marsh Yellow Cress

94a. Swamp Rose

94b. Multiflora Rose

95. Bristly Black Currant

96. Sheep Laurel

97. Pale or Bog Laurel

98. Wax Myrtle

is subtended by small leaflike bracts; FACW; poisonous. Hemlock Parsley (*Conioselinum chinensis*) has parsleylike leaves and rough-margined leaf sheaths, and its umbel lacks leaflike bracts at the base; FACW. Purple-stem Angelica (*Angelica atropurpurea*) occurs in wet meadows, shrub swamps, and forested wetlands from Maryland and Delaware north; it has smooth purplish stems (to 1⅔ inches wide and to 10 feet tall), large inflated leaf sheaths, and very large umbels (to 10 inches wide); OBL.

92. Tall Meadow Rue (*Thalictrum pubescens*): perennial up to 6 feet tall or more, with uniquely shaped leaves composed of round-toothed leaflets, and many small white flowers borne in clusters, forming a terminal inflorescence; June–July; marshes, wet meadows, forested wetlands, and stream banks; FACW+; Labrador to North Carolina.

93. Golden Ragwort (*Senecio aureus*): perennial up to 3 feet tall, with both compound and simple leaves (lower leaves simple round-toothed), and tall flowering stem bearing several to many small yellow daisylike flowers (about ¾ inch wide); April–June; deciduous forested wetlands, wet meadows, and bogs; FACW; Nova Scotia to Florida. Similar species: Two other herbs with compound leaves produce yellow flowers, but they are in the Mustard family and lack the roundish simple, basal leaves of Golden Ragwort and their flowers are not daisylike. Marsh Yellow Cress (*Rorripa palustris*) has compound leaves with an irregularly coarse-toothed terminal leaflet, four-petaled flowers (less than ⅕ inch wide) borne terminally and in leaf axils on elongate inflorescences, and short fruit pods (less than ½ inch long) borne on stalks; June–September; marshes, wet meadows, and streambanks; OBL. Yellow Rocket or Common Winter Cress (*Barbarea vulgaris*), a Eurasian introduction, occurs in wet meadows, ditches, and disturbed lands; the terminal leaflet of its lower leaves is heart-shaped to roundish arrowhead-shaped (much larger than lower leaflets on same leaf), its flowers are larger (⅕–½ inch wide), and its fruit pods are long and narrow (over ¾ inch long); April–June; FACU.

See also Marsh Cinquefoil in Figure 5.5.

Swamp Shrubs

Thorny or Prickly Shrubs

94. Swamp Rose (*Rosa palustris*): thorny deciduous shrub up to 7 feet tall, with compound leaves divided into seven fine-toothed leaflets with narrow stipules along base of leaf stalk, scattered thorns on branches, large, showy, pink five-petaled flowers (about 2 inches wide), and hairy, reddish, fleshy rose hips; June–October; marshes, forested wetlands, and shrub swamps; OBL; Nova Scotia to Florida. Similar species: Virginia Rose (*R. virginiana*) looks almost identical to Swamp Rose, but has much wider stipules (appearing leaflike) along the leaf stalk; FAC. Shining Rose (*R. nitida*), more common in northern New England, has shiny, dark green leaflets and very bristly stems and branches; FACW+. Multiflora Rose (*R. multiflora*), an Asian exotic escaped from cultivation and widespread throughout the Northeast, has coarse-toothed leaves, smaller, white flowers (to 1½ inches wide) and smaller, berrylike rose-hip fruits (less than ⅓ inch wide); FACU. The familiar raspberries and blackberries (*Rubus* spp.) may also occur in wetlands, especially wet meadows and edges of swamps; they have three or five leaflets and characteristic berries. Other prickly shrubs found in wetlands are currants or gooseberries (*Ribes* spp.); they have three- to seven-lobed simple leaves and thorny or bristly stems (see species 95).

95. Bristly Black Currant (*Ribes lacustre*): northern shrub up to 5 feet tall, with bristly and prickly stems and branches (longer spines at nodes), branches foul-smelling (skunklike odor) when broken, deeply three- to five-lobed leaves (to 4 inches long), small green, pinkish, or purple five-petaled flowers borne in drooping clusters, and purplish black bristly berries (also foul-smelling); May–August; forested wetlands, moist woods, and thickets; FACW; Newfoundland to New York and western Massachusetts, in eastern mountains to Tennessee. Similar species: Skunk Currant (*R. glandulosum*) also is foul-smelling and has bristly flowers and berries (red), but lacks prickly stems; FACW. Other gooseberries or currants have flowers and berries lacking bristles. Smooth Gooseberry (*R. hirtellum*) has smooth or prickly stems (young stems prickly) with peeling bark, greenish yellow to dull purplish flowers borne on smooth stalks, and bluish black berries; FAC. The following species lack bristles on branches: Swamp Red Currant (*R. triste*), an introduced shrub, has several pink or purplish flowers borne on branched clusters, and red berries; OBL. Wild Black Currant (*R. americanum*) grows along stream banks and in moist woods to about 5 feet

tall; it has hairy young branches and bears whitish or dull yellowish flowers and black berries; FACW.

Evergreen Shrubs with Simple, Entire, Opposite Leaves

96. Sheep Laurel (*Kalmia angustifolia*): usually less than 3 feet tall, but growing up to 6 feet tall, leaves green above and pale green below, round branchlets, cluster of pink to reddish purple five-lobed flowers (½ inch wide) borne below new leaf growth, and five-celled fruit capsules; May–June; forested wetlands, bogs, and acidic upland woods; FAC; Newfoundland and Labrador to Virginia. Similar species: Mountain Laurel (*K. latifolia*), typically an upland plant but found in Pine Barrens swamps, is much taller (up to 20 feet), with thick, dark green leaves, mostly alternately arranged, and clusters of mostly white flowers (sometimes pinkish); FACU.

97. Pale or Bog Laurel (*Kalmia polifolia*): ericaceous shrub up to 2 feet tall, with two-sided branchlets, distinct nodes on branches, narrow leaves shiny dark green above and whitish hairy below, small five-lobed pale pink flowers (½–⅗ inch wide) borne at top of branches, and small five-valved round fruit capsules; June–July; shrub bogs; Newfoundland to Pennsylvania and northern New Jersey; OBL. Similar species: Bog Rosemary (*Andromeda glaucophylla*) looks similar when flowers are absent, but has alternately arranged leaves.

Evergreen Shrubs with Simple, Toothed, Alternate Leaves

98. Wax Myrtle (*Myrica cerifera*): shrub or tree to 36 feet tall, with smooth, waxy young twigs, leathery yellow-green leaves (few-toothed at end or entire) that are fragrant when crushed, and small, waxy, hard berries; March–June; tidal and nontidal shrub swamps and forested wetlands along the coast; FAC; southern New Jersey to Florida. Similar species: Its relatives also have fragrant parts when crushed. Northern Bayberry (*M. pensylvanica*) has few-toothed or entire deciduous leaves, hairy young twigs, and hairy waxy fruits (when young); FAC. Sweet Gale (*M. gale*) is a northern plant of bogs and lakeshores (also upper edges of New England salt marshes) ranging into northern New Jersey; it is a low, compact deciduous shrub less than 6 feet tall, bearing small pine-conelike female catkins; OBL.

99. Inkberry or Gallberry (*Ilex glabra*): shrub usually less than 6 feet tall (rarely to 15 feet), with smooth thick, somewhat leathery leaves (up to 2 inches long) having one to three pairs of coarse teeth near the tip, small six- to eight-lobed white flowers borne singly or in clusters from leaf axils, and black berries; June–July; shrub swamps, forested wetlands, and sandy woods; FACW−; Nova Scotia to Florida.

100. Leatherleaf (*Chamaedaphne calyculata*): low, compact shrub up to 5 feet tall, with small leaves (to 2 inches long) often pointing upward, small five-lobed bell-shaped flowers hanging from short stalks at the ends of branches, and persistent five-celled round fruit capsules; April–June; shrub bogs and acidic pond margins; OBL; Newfoundland to North Carolina. (Note: Older leaves are actually shed after new leaves have matured; leaves are maroon-colored through winter.)

Evergreen Shrubs with Simple, Entire, Alternate Leaves

101. Rosebay Rhododendron (*Rhododendron maximum*): ericaceous shrub or small tree up to 35 feet tall, forming dense streamside thickets, with leathery leaves (to 10 inches long) shiny dark green above and whitish hairy below, large, showy, white or pinkish five-lobed bell-shaped waxy flowers (1½ inches wide) borne in dense clusters, and elongate fruit capsules; June–July; stream banks (in mountains), forested wetlands, and moist woods; FAC; Nova Scotia to northern New Jersey, in eastern mountains to Georgia.

102. Bog Rosemary (*Andromeda glaucophylla*): ericaceous shrub up to 2½ feet tall, with round branchlets, thick, leathery, linear leaves (up to 2¼ inches long) bluish green above and whitish hairy below with margins rolled inward, small white to pinkish urn-, turban-, or vase-shaped flowers (about ¼ inch long) borne in drooping clusters, and small five-valved round fruit capsules; May–July; shrub bogs; Newfoundland to Pennsylvania and northern New Jersey, in eastern mountains to West Virginia; OBL.

103. Labrador Tea (*Ledum groenlandicum*): ericaceous shrub up to 3 feet tall, with thick, leathery, oblong leaves (up to 2 inches long) dark green above, rusty-hairy below with margins rolled inward, aromatic (when crushed), five-petaled white flowers (about

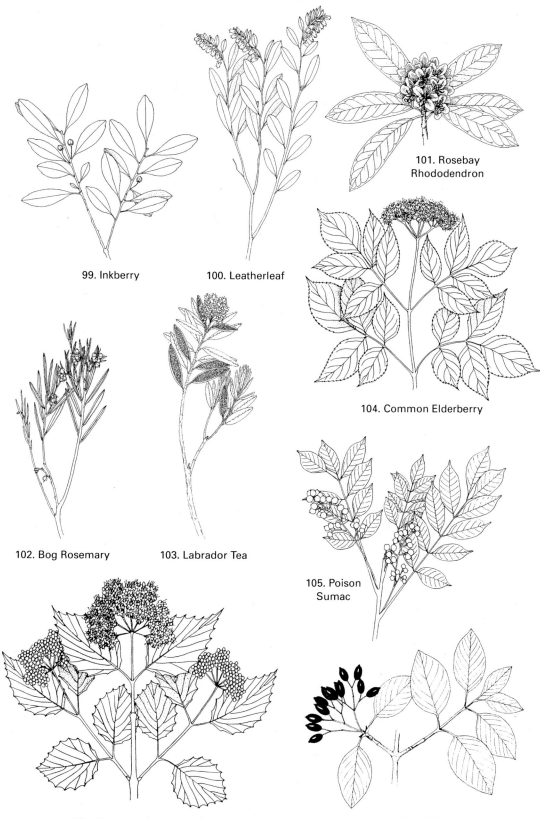

99. Inkberry

100. Leatherleaf

101. Rosebay Rhododendron

104. Common Elderberry

102. Bog Rosemary

103. Labrador Tea

105. Poison Sumac

106a. Northern Arrowwood

106b. Black Haw

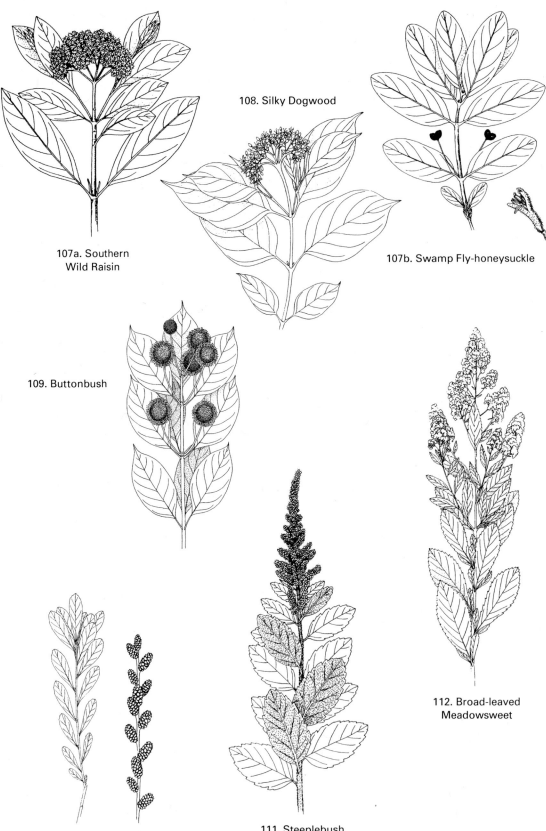

108. Silky Dogwood

107a. Southern
Wild Raisin

107b. Swamp Fly-honeysuckle

109. Buttonbush

112. Broad-leaved
Meadowsweet

110. Sweet Gale

111. Steeplebush

⅓ inch wide) borne in terminal clusters, and cylinder-shaped five-part fruit capsules; May–July; shrub bogs and adjacent forested wetlands; Newfoundland to Pennsylvania and northern New Jersey; OBL.

See also Sheep Laurel (species 96), Wax Myrtle (98), Leatherleaf (species 100), Sand Myrtle under Large Cranberry (species 134), and Ground Hemlock under Eastern Hemlock (species 137).

Deciduous Shrubs with Compound Leaves

104. Common Elderberry (Sambucus canadensis): up to 12 feet tall, with light brown bark covered with raised bumps, leaves with five to eleven (usually seven) leaflets, many small white flowers borne in somewhat flat-topped inflorescences, and dark purplish berries; June–July; marshes, wet meadows, swamps, and moist areas; FACW−; Nova Scotia to Florida.

105. Poison Sumac (Toxicodendron vernix): poisonous tall shrub or low tree up to 20 feet high, milky sap, leaves composed of seven to thirteen leaflets, reddish leaf stalks, small greenish yellow flowers in dense clusters borne on long stalks from leaf axils, and small whitish berries; May–July; seasonally flooded forested wetlands and borders of marshes; OBL; Maine to Florida.

Deciduous Shrubs with Simple, Toothed, Opposite Leaves

106. Northern Arrowwood (Viburnum recognitum): up to 15 feet tall, with twigs somewhat angled, many small white five-lobed flowers borne on long-stalked clusters at ends of branches, and bluish black berries; May–July: shrub swamps, forested wetlands, and moist thickets; FACW−; New Brunswick to New Jersey, in mountains to Georgia. Similar species: Southern Arrowwood (V. dentatum) is common along the Coastal Plain; its twigs are velvety-hairy, and its leaves are sometimes velvety-hairy beneath; FAC. Black Haw (V. prunifolium) has twigs that are often short and stubby with many leaf scars (look like apple twigs), fine-toothed leaves (1–2 inches long near flowers) with red- to purple-tinged stalks, and small fragrant three- to five-lobed white flowers that bloom in April and May; FACU. Nannyberry (V. lentago) has fine-toothed, somewhat egg-shaped leaves

(2–4 inches long) with long-pointed tips and winged stalks; FAC.

See also Southern Wild Raisin (species 107).

Deciduous Shrubs with Simple, Entire, Opposite Leaves

107. Southern Wild Raisin or Possum Haw (Viburnum nudum): up to 20 feet tall, with somewhat leathery leaves (sometimes toothed) having shiny upper surfaces, many small white flowers borne in flat-topped axillary inflorescences, and bluish black berries; May–July; marshes, shrub swamps, and forested wetlands; OBL; Connecticut to Florida. Similar species: Northern Wild Raisin or Withe Rod (V. cassinoides) grows up to 13 feet, its upper leaf surfaces are dull green, and its leaves are entire, toothed, or wavy-margined; Newfoundland to Maryland, in eastern mountains to Alabama; FACW. Swamp Fly-honeysuckle (Lonicera oblongifolia), a northern shrub of bogs, fens, and cedar and larch swamps, grows to 6 feet tall, its leaves are oblong (to 3½ inches long) with soft hairy undersides, its yellowish two-lipped flowers are borne on long stalks from leaf axils, and its berries are orange or red; May and June; OBL; Pennsylania to Minnesota and north. Mountain Fly-honeysuckle or Velvet Honeysuckle (L. caerulea var. villosa) is similar in form and in range, but is shorter to 3½ feet tall with hairy twigs and leaves, yellowish five-lobed tubular flowers borne on short stalks (less than ½ inch), and blue berries; FACW+.

108. Silky Dogwood (Cornus amomum): up to 20 feet tall, with smooth reddish to gray twigs (often streaked with purple), a brownish or buff pith, many small white flowers in somewhat flat-topped terminal inflorescences, blue and bluish white berries, and leaves having main veins that remain attached when leaf is gently pulled apart; May–July; wet meadows, shrub swamps, forested wetland, riverbanks, and moist woods; FACW; Quebec to South Carolina. Similar species: Red Osier Dogwood (C. stolonifera) has bright red twigs, a white pith, and white berries; FACW+. Red-panicled or Gray Dogwood (C. foemina ssp. racemosa) has gray twigs, red-stalked paniceled flowers and berries, and white berries; FAC. Bunchberry (C. canadensis), a low-growing dogwood (to 8 inches tall), occurs in bogs from New Jersey north; it has a whorl of four to eight leaves at the top of the stem and bears reddish berries in

a dense cluster; FAC−. (Note: Dogwoods can usually be distinguished from other shrubs with similar leaves by gently pulling apart a leaf. Dogwoods have somewhat elastic veins that remain intact even when the leaf is divided into two pieces; the pieces are held together by these veins.)

109. Buttonbush (*Cephalanthus occidentalis*): usually 4–6 feet tall, but growing up to 20 feet with flaky, grayish brown older bark, somewhat purplish young twigs marked with elongated dots (lenticels), leaves usually in whorls of threes and fours, small whitish tubular flowers in ball-shaped clusters (about 1 inch wide), and nutlet-bearing fruit balls; May–August; marshes, shrub swamps, forested wetlands, and shores of rivers, lakes, and ponds; OBL; New Brunswick and Quebec to Florida.

Deciduous Shrubs with Simple, Toothed, Alternate Leaves

110. Sweet Gale (*Myrica gale*): low, compact shrub usually less than 3 feet tall, but growing up to 6 feet, with entire or few-coarse-toothed (at tips), aromatic (bayberry-scented) leaves (to 2½ inches long) having hairy undersides and wedge-shaped bases, somewhat oval-shaped fruiting catkins bearing many nutlets, and small, closed, pineconelike catkins (to ¾ inch long) persisting through winter; April–June; shrub bogs, marshes, fens, margins of lakes and ponds, and upper edges of salt marshes; Newfoundland to northern New Jersey, in eastern mountains to North Carolina; OBL.

111. Steeplebush (*Spiraea tomentosa*): up to 3½ feet tall, with woolly twigs, leaves (to 2 inches long) having white or rusty woolly undersides, and many small, five-petaled pink flowers borne in dense terminal steeplelike inflorescence; July–September; wet meadows, marshes, shrub swamps, and moist open fields; FACW; Nova Scotia to North Carolina. Similar species: Other *Spiraea* lack woolly twigs and undersides of leaves.

112. Broad-leaved Meadowsweet (*Spiraea latifolia*): up to 6 feet tall, with purplish brown to reddish brown twigs, leaves up to 1½ inches wide, many small five-petaled white flowers borne in clusters in open (mostly smooth) inflorescences at end of upper branches and top of stem, and persistent five-parted fruit capsules; June–September; wet meadows, shrub swamps, moist open fields, and roadsides; FAC+. Simi-

lar species: Narrow-leaved Meadowsweet (*S. alba*) has yellow-brown to dull brown twigs, narrower leaves to ¾ inch wide, and fine-haired inflorescences; FACW+.

113. Willows (*Salix* spp.): variable shrubs up to 20 feet tall, with toothed or entire leaves and flowers borne on catkins that develop before leaves; late February–June; shrub swamps, wet meadows, marshes, and shores; mostly OBL to FACW. Many willows can be found in Northeast wetlands, especially in more northern areas (from New Jersey north). Large Pussy Willow (*S. discolor*) is a northern species that extends into Delaware; it has smooth twigs and toothed leaves that are smooth on both sides, with large leaflike stipules (in leaf axils), and is perhaps the earliest-blooming shrub; FACW. Silky Willow (*S. sericea*) has fine-toothed leaves that are white and silky-hairy below; OBL. Sandbar willow (*S. interior*) occurs on mud flats, sandbars, and moist floodplains in western New England and eastern mountains; it has long, narrow, somewhat spiny-toothed leaves (to 5½ inches long); OBL. Autumn willow (*S. serissima*) grows in calcareous wetlands; its toothed leaves are whitish below and its fruit capsules open in late summer or fall; OBL (see Figure 5.4). Bog willow (*S. pedicellaris*) is one of the willows with small entire leaves (to 2½ inches long) and is less than 3 feet tall; OBL. Hoary willow (*S. candida*), a calcareous-loving species (to 6 feet tall), also has entire leaves, but they are elongate (to almost 5 inches long) and are white woolly beneath; OBL (see Figure 5.4). Many willows have prominent grayish conelike galls on their branches.

114. Sweet Pepperbush (*Clethra alnifolia*): up to 13 feet tall, with leaves to 4 inches long (entire near base), many fragrant small white five-petaled flowers borne in terminal spikes, and persistent five-celled fruit capsules; July–September; forested wetlands, shrub swamps, and sandy woods along the coast; FAC+; southern Maine to Florida.

115. Oblong-leaf Shadbush (*Amelanchier canadensis*): shrub to small tree up to 25 feet tall, usually growing in clumps, with leaves up to 3¼ inches long having rounded tips and rounded to heart-shaped bases, showy five-petaled white flowers (about ¾ inch wide) on stalks borne in clusters, petals much longer than wide, and dark purple to black berries; March–June; shrub swamps, forested wetlands, and upland woods; FAC; Newfoundland to Florida. Similar species: Chokeberries (*Aronia* spp.)

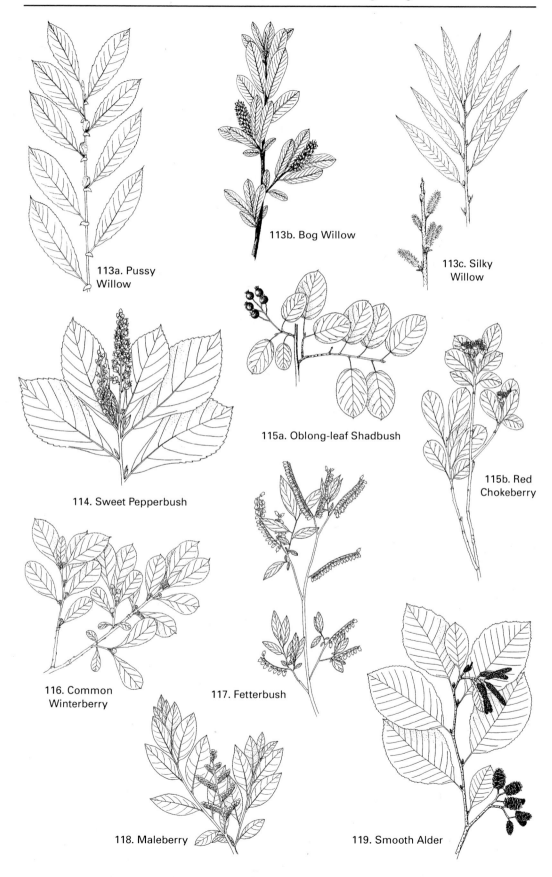

113a. Pussy Willow

113b. Bog Willow

113c. Silky Willow

114. Sweet Pepperbush

115a. Oblong-leaf Shadbush

115b. Red Chokeberry

116. Common Winterberry

117. Fetterbush

118. Maleberry

119. Smooth Alder

also have five-petaled white flowers that bloom April–July (petals about as long as wide; to ½ inch wide), but their leaves usually have sharp-pointed tips and have small dark glands that can be seen on the midvein of the upper leaf surface. Red Chokeberry (*A. arbutifolia*) has woolly young branches, flower stalks, and lower leaf surfaces and and produces red berries; FACW. Black Chokeberry (*A. melanocarpa*) has virtually hairless branches, flower stalks, and lower leaf surfaces, and black berries; FAC. Purple Chokeberry (*A. prunifolia*) has woolly young branches, flower stalks, and lower leaf surfaces, and purplish to blackish berries; FACW.

116. Common Winterberry (*Ilex verticillata*): up to 16 feet tall, with leaves (to 4 inches long) with dull upper surfaces and distinctly-pointed tips, small four- to six-lobed white flowers borne singly or in clusters from leaf axils, and bright red (rarely yellow) berries; May–August; shrub swamps and forested wetlands; FACW; Newfoundland to Georgia. Similar species: Smooth Winterberry (*I. laevigata*) has somewhat thick leaves that are shiny above and orangish red berries; OBL.

117. Fetterbush (*Leucothoe racemosa*): up to 13 feet tall, with smooth gray bark, very fine toothed leaves to 3¼ inches long, many small white, urn-shaped flowers borne in dense, one-sided clusters from leaf axils (flowers typically pointing downward), five-parted fruit capsules with persistent spiny dry sepals (giving a star-like appearance, with the sepals representing the points of the star), and overwintering reddish flower buds; May–June; shrub swamps, forested wetlands, and moist acidic woods; FACW; eastern Massachusetts to Florida.

118. Maleberry (*Lyonia ligustrina*): up to 13 feet tall, with zigzag light brown twigs, many small, globe-shaped white flowers borne in dense, branched clusters (flowers generally arranged on the upper side of the branches), and persistent, round, five-parted fruit capsules with starlike openings; May–July; shrub swamps, forested wetlands, wet sandy soil, and moist thickets; FACW; Maine to Florida.

119. Smooth Alder (*Alnus serrulata*): up to 20 feet tall, with fine-toothed leaves (to 5¼ inches long), early-flowering catkins, persistent pinelike cones (about ½ inch long), over-wintering flower-bud spikes, and sticky buds; March–May; marshes, shrub swamps, forested wetlands, and riverbanks; OBL; Maine to Florida. Similar species: Speckled Alder (*A. rugosa*),

the common alder of New England, has double-toothed leaves (coarse and fine teeth intermixed) and stems marked with conspicuous light-colored spots (lenticels); FACW+.

See also currants and gooseberries under Bristly Black Currant (species 95), Common Buckthorn under Black Gum (species 161), and Northern Bayberry under Wax Myrtle (species 98).

Deciduous Shrubs with Simple, Entire, Alternate Leaves

120. Rhodora (*Rhododendron canadense*): ericaceous shrub up to 3 feet tall, with stiff ascending branches, somewhat egg-shaped leaves (to 2½ inches long) gray-green above and paler somewhat hairy below, with short-pointed tips and wedge-shaped bases, showy pinkish to purplish two-lipped flowers (about 1¼ inches wide), and persistent elongate five-parted fruit capsules; April–June (before leaves emerge); shrub bogs and rocky slopes; FACW; Newfoundland to northeastern Pennsylvania and northern New Jersey.

121. Dangleberry or Tall Huckleberry (*Gaylussacia frondosa*): up to 6 feet tall, with smooth young twigs, dull blue-green to dull green leaves covered with yellow resin dots on undersides only, green to purplish bell-shaped flowers borne on long stalks (to 1 inch long) in drooping leafy clusters, and light blue edible berries dangling from long stalks; April–June; forested wetlands and dry woods; FAC; southern New Hampshire to Florida. Similar species: Dwarf Huckleberry (*G. dumosa*) is an occasional bog species that is shorter (less than 20 inches) and has hairy young twigs, thick, dark green, shiny leaves with resin dots mostly on undersides, white or pink bell-shaped flowers, and black berries; FAC. Black Huckleberry (*G. baccata*), more typical of uplands but occurring in some wetlands (e.g., Pine Barrens), is usually less than 3½ feet tall, with sticky flowers and leaves (covered with resin dots on both sides), pinkish or reddish flowers (sometimes partly greenish), and black to blue fruits; FACU. Blueberries (*Vaccinium* spp.) lack resinous glands on leaves. (Note: An easy way to distinguish huckleberries from blueberries is to gently rub the bottom of a leaf on a white sheet of paper. Huckleberries leave a yellow stain; blueberries do not.)

122. Staggerbush (*Lyonia mariana*): up to 7 feet tall, with zigzag twigs, small white or

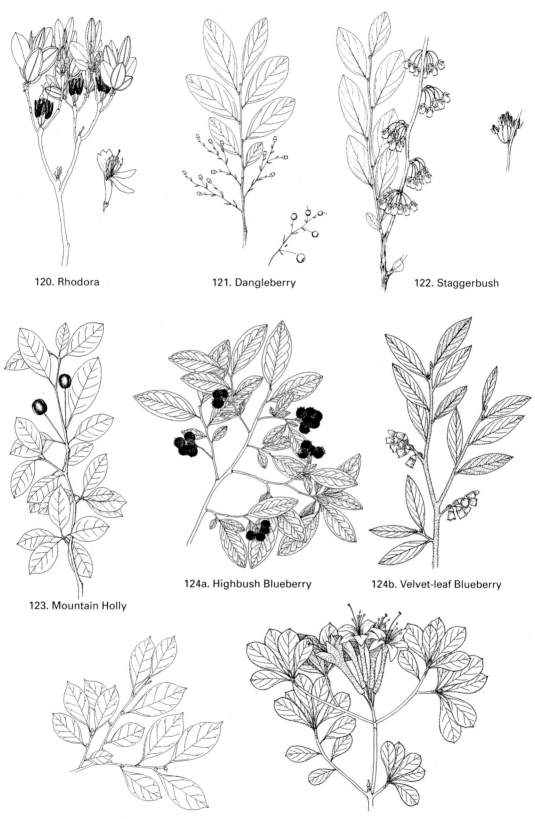

120. Rhodora

121. Dangleberry

122. Staggerbush

123. Mountain Holly

124a. Highbush Blueberry

124b. Velvet-leaf Blueberry

125. Spicebush

126. Swamp Azalea

pinkish flowers in drooping clusters on long stalks from leaf axils of tips of branches, and persistent five-celled oval to cone-shaped fruit capsules on long stalks; May–June; savannas, shrub swamps, forested wetlands, and sandy soils; FAC−; Rhode Island to Florida.

123. Mountain Holly (*Nemopanthus mucronata*): up to 13 feet tall, with often purple-tinged leaf stalks, leaves to 1 inch long with bristle tip, small four- or five-petaled white or yellowish flowers borne on long stalks, and dull reddish berries; late April–June; bogs, shrub swamps, and forested wetlands; OBL; Newfoundland to West Virginia.

124. Highbush Blueberry (*Vaccinium corymbosum*): usually 6–10 feet tall, but growing up to 20 feet, with older bark shredding, twigs gray on top and greenish to reddish below, leaves green above and paler below, small white urn-shaped flowers (to ¼ inch long) borne in dense clusters (first appearing before leaf-out), and dark blue edible berries; April–July; forested wetlands, shrub swamps, bogs, and upland woods; FACW−; Nova Scotia to Florida. Similar species: Velvet-leaf Blueberry (*V. myrtilloides*) is a low-growing blueberry of northern bogs (less than 2 feet tall) with hairy-margined velvety leaves and velvety, warty stems; FAC. Lowbush Blueberry (*V. angustifolium*) is also low growing, but has fine-toothed leaves and lacks velvety stems and leaves; FACU−. See also Dangleberry (species 121) above.

125. Spicebush (*Lindera benzoin*): up to 16 feet tall, with older bark dark gray and covered with pimplelike bumps, and leaves, twigs, and berries aromatic (lemon-scented) when crushed, small six-petaled yellow flowers borne in clusters along last year's twigs, and red berries; March–May; forested wetlands, moist upland woods, and moist rocky woodlands; FACW−; southern Maine to Florida.

126. Swamp Azalea (*Rhododendron viscosum*): up to 12 feet tall, with shiny green leaves (to 2½ inches long) having fine hairy margins, fragrant, showy, white, sticky five-lobed flowers (about 1 inch long) borne in clusters at end of branches, and persistent elongate five-part fruit capsules; May–September; forested wetlands and edges of marshes; FACW+; Maine to Florida. Similar species: Pinxter Flower or Pink Azalea (*R. periclymenoides*) also occurs in swamps from Massachusetts to South Carolina; it grows to 10 feet and bears odorless pinkish to purplish (rarely white) flowers that bloom be-

fore leaves emerge; March–May; FAC. Dwarf Rhododendron (*R. atlanticum*), a low-growing species (to 3 feet tall), occurs in pine lowlands and flatwoods from southern New Jersey to South Carolina; it also produces early-blooming (April–June) white, pink, or purplish flowers, but they are fragrant; FAC.

See also Northern Bayberry under Wax Myrtle (species 98), Sweet Gale (species 110), and Willows (species 113).

Swamp Woody Vines

Thorny or Prickly Vines

127. Common Greenbrier (*Smilax rotundifolia*): woody vine with many stout thorns (green bases with dark tips), rounded to heart-shaped leaves, and bluish black to black berries; forested wetlands and open woods and thickets; FAC; Nova Scotia to Florida. Similar species: Red-berried Greenbrier (*S. walteri*) has very prickly lower stems, somewhat narrower lance-shaped leaves, and bright red berries; OBL; New Jersey to Florida. Cat Greenbrier or Sawbrier (*S. glauca*) is found in pine lowlands and flatwoods from southern New England to Florida; its stems are also prickly, but its somewhat egg-shaped leaves are distinctly whitish below, and its berries are mostly blue; FACU. See also Laurel-leaved Greenbrier (species 132).

Vines with Compound Leaves

128. Poison Ivy (*Toxicodendron radicans*): woody vine or erect shrub up to 10 feet tall, with older vines densely covered by dark fibers, leaves divided into three leaflets, milky sap, small yellowish flowers, and small grayish white berries; May–July; tidal and nontidal marshes, swamps, upland woods and thickets, and probably the woods behind your house; FAC; Nova Scotia to Florida.

129. Virginia Creeper (*Parthenocissus quinquefolia*): vine climbing by adhesive disks at end of tendrils, leaves divided into three to seven (usually five) leaflets, small greenish to whitish flowers borne in terminal umbellike clusters, and dark blue to black berries; June–August; shrub swamps, forested wetlands, upland woods, moist thickets, and fence rows; FACU; Maine to Florida.

130. Trumpet Creeper (*Campsis radicans*): vine with leaves divided into five to thirteen leaflets, large, showy, orange-red tubular flowers

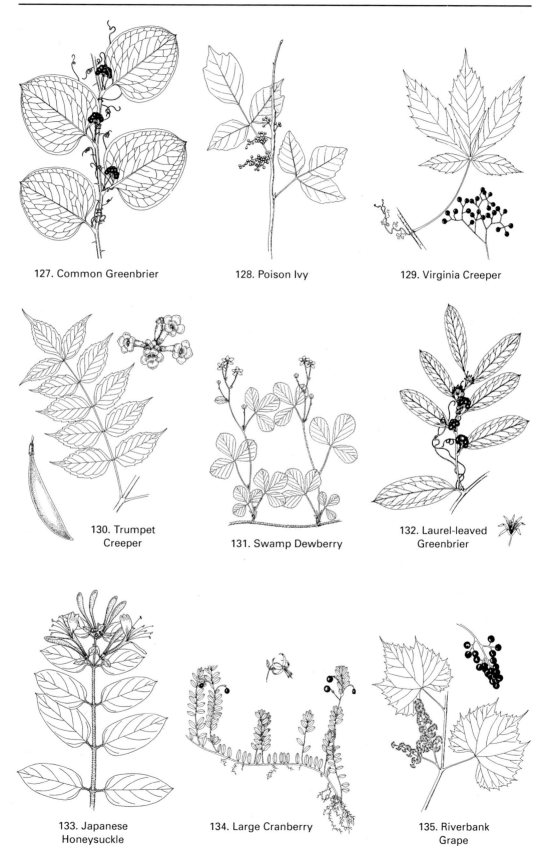

127. Common Greenbrier

128. Poison Ivy

129. Virginia Creeper

130. Trumpet Creeper

131. Swamp Dewberry

132. Laurel-leaved Greenbrier

133. Japanese Honeysuckle

134. Large Cranberry

135. Riverbank Grape

borne in terminal clusters, and podlike fruit capsules (to 6 inches long); June–August; tidal and nontidal swamps, floodplains, thickets, fence rows, and moist upland woods; FAC; New England to Florida.

131. Swamp Dewberry (*Rubus hispidus*): trailing vinelike shrub, with bristly stems, leaves divided into three shiny dark green leaflets (look like strawberry leaves) that turn reddish in winter, five-petaled white flowers (to ¾ inch wide) borne in clusters, and blackberry-like fruits; June–September; forested wetlands, bogs, shrub swamps, marshes, and moist uplands; FACW; Nova Scotia to South Carolina.

Vines with Simple Leaves

132. Laurel-leaved Greenbrier (*Smilax laurifolia*): high-climbing vine with stems unarmed except for spiny prickles near base, evergreen leaves (to 8 inches long) with pointed tips, and black berries; July–September; tidal and nontidal swamps; OBL; New Jersey to Florida.

133. Japanese Honeysuckle (*Lonicera japonica*): trailing or climbing vine, with hairy twigs, evergreen leaves, fragrant showy white to yellowish two-lipped flowers (to 1⅔ inches long) borne in pairs, and black berries; forested wetlands (mostly temporarily flooded), upland woods, fields, and fence rows; FAC−.

134. Large or American Cranberry (*Vaccinium macrocarpon*): trailing evergreen with small leaves (⅜–⅝ inch long) that are shiny dark green above and reddish in winter, small pinkish white flowers with four recurved petals, and red berries (½–⅞ inch wide); summer; bogs, acidic marshes, interdunal swales, and cultivated bogs; OBL; Newfoundland to Virginia, in mountains to North Carolina. Similar species: Small Cranberry (*V. oxycoccus*), a northern bog plant that ranges into northern New Jersey, has smaller leaves (less than ⅜ inch long) and smaller berries (¼ inch wide); OBL. In the New Jersey Pine Barrens, Sand Myrtle (*Leiophyllum buxifolium*) may resemble cranberry; however, it is a low-growing shrub (to 3 feet tall) of pitch-pine lowlands, with clusters of small five-petaled white flowers (¼ inch wide; May blooming) that are distinctly different from those of cranberries; FACU−. Creeping Snowberry (*Gaultheria hispidula*), a creeping vine with small evergreen leaves, occurs in bogs and evergreen forested wetlands from New Jersey north; its leaves are aromatic (wintergreen-scented when crushed), and its berries are white; FACW. Mountain

Cranberry or Lingberry (*Vaccinium vitis-idaea*), a northern species of bogs and evergreen forests, has leathery oval-shaped leaves with rounded tips bearing dark glands on undersides, bell-shaped white to pink four-lobed flowers borne in clusters at end of branches, and red berries; June and July; FAC; Maine to Minnesota and north.

135. Riverbank Grape (*Vitis riparia*): high-climbing vine, with reddish brown to purplish older bark shredding into loose strips, coarse-toothed leaves (to 7 inches long) usually three- to five-lobed, and purplish black to black grapes covered by whitish waxy coat; May–July; forested wetlands on floodplains, alluvial woods, and moist upland woods and thickets; FACW; New Brunswick to Virginia. Similar species: Fox Grape (*V. vulpina*) has broader-toothed leaves that are usually not lobed, and glossy black grapes lacking waxy coat; FAC.

Swamp Trees

Needle-leaved Evergreens

136. Pitch Pine (*Pinus rigida*): up to 85 feet tall, with needles (to 5 inches long) in bundles of threes, and cones armed with stout, sharp prickles; varied habitats including lowlands, flatwoods, cedar swamps, bogs, and dry sandy soils; FACU; southern Maine to Virginia. Similar species: Loblolly Pine (*P. taeda*), a southern species ranging into Cape May County, New Jersey, has yellowish green, often twisted needles (usually less than 10 inches long) and longer cones (to 5 inches long); FAC−. Eastern White Pine (*P. strobus*), a northern species, has needles in bundles of five and more elongate, open cones; FACU.

137. Eastern Hemlock (*Tsuga canadensis*): up to 100 feet tall, with blunt-tipped flattened leaves (to ⅝ inch long) that are shiny dark green above, green with two whitish lines below, and attached to branches by short stalks, and small cones (to 1 inch long); forested wetlands, stream banks, river gorges, mountain slopes on shallow rocky soils, and moist upland woods; FACU; New Brunswick to Delaware, in mountains to Georgia. Similar species: Balsam Fir (*Abies balsamea*), the common Christmas tree in many homes, is a northern species that dominates some forested wetlands; it has aromatic short flat needles (to 1⅓ inches long) that are whitish below, and elongate cylinder-shaped cones

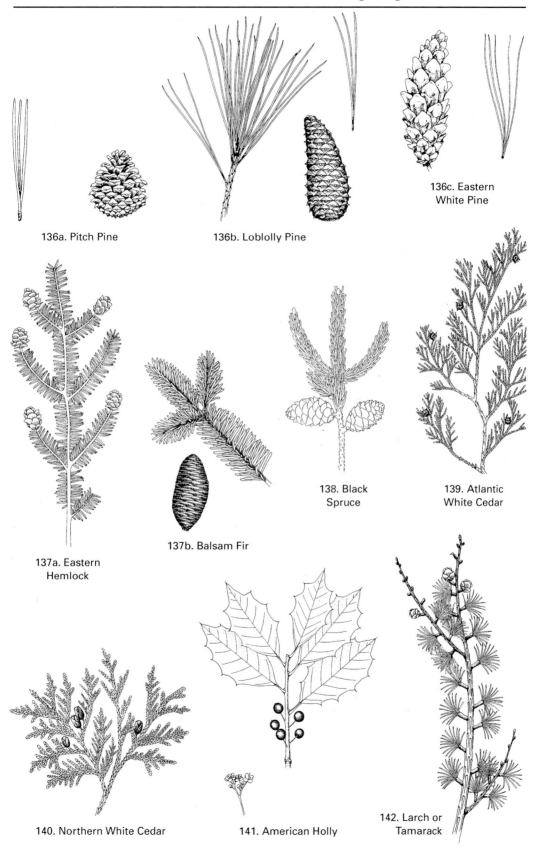

136a. Pitch Pine

136b. Loblolly Pine

136c. Eastern
White Pine

137a. Eastern
Hemlock

137b. Balsam Fir

138. Black
Spruce

139. Atlantic
White Cedar

140. Northern White Cedar

141. American Holly

142. Larch or
Tamarack

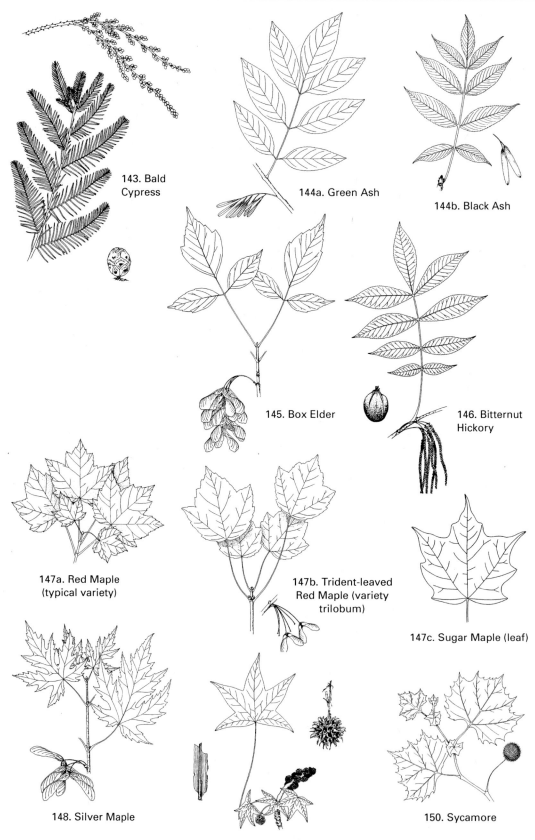

143. Bald Cypress

144a. Green Ash

144b. Black Ash

145. Box Elder

146. Bitternut Hickory

147a. Red Maple (typical variety)

147b. Trident-leaved Red Maple (variety trilobum)

147c. Sugar Maple (leaf)

148. Silver Maple

149. Sweet Gum

150. Sycamore

(up to 3½ inches long); FAC. Ground Hemlock or American Yew (*Taxus canadensis*), a low-growing shrub (up to 6 feet tall), occurs in bogs and coniferous woods from Maine to Pennsylvania; its leaves have sharp tips and are green above and below; FAC.

138. Black Spruce (*Picea mariana*): up to 100 feet tall but usually much shorter in bogs, with hairy branches, pale bluish green short needles (to ⁷/₁₆ inch long), and small grayish brown, somewhat roundish cones (to 1¼ inches long) that remain on tree for some time; bogs, evergreen forested wetlands. and mountain slopes; FACW−; Newfoundland to northern New Jersey, in eastern mountains to Virginia and West Virginia. Similar species: Red Spruce (*P. rubens*), a typical upland species that occurs in bogs in the Poconos, western Maryland, and elsewhere; its needles are dark green or yellow-green and are longer (to ⅝ inch long), and its reddish brown, somewhat cylinder-shaped cones (to 1½ inches long) hang from short stalks and fall off when mature; FACU. White Spruce (*P. glauca*), another typical upland species, has smooth branches, very aromatic (almost pungent when crushed) needles, and pale brown cones; FACU.

Scale-leaved Evergreens

139. Atlantic White Cedar (*Chamecyparis thyoides*): up to 90 feet tall, with flattened or four-angled twigs, bluish green to pale green leaves, and small globe-shaped cones (¼ to ⅜ inch wide) with short points; tidal and nontidal swamps, shrub bogs, and stream banks; OBL; central Maine to Florida. Similar species: Eastern Red Cedar (*Juniperus virginiana*) bears soft, dark blue berries and may have scalelike leaves or needlelike leaves; FACU.

140. Northern White Cedar (*Thuja occidentalis*): up to 65 feet tall, with yellowish green leaves (to about ¾ inch long) and small, reddish brown oblong cones (about ½ inch long); forested wetlands, wooded fens, and old fields and rocky areas in limestone regions; FACW; Nova Scotia to western New England and New York, in eastern mountains to North Carolina.

Broad-leaved Evergreens

141. American Holly (*Ilex opaca*): to 70 feet tall or more, with leathery, stiff leaves, shiny

above and pale below and having prickly teeth along margins, small four-petaled white flowers (¼ inch wide) borne in clusters, and red berries; May–June; forested wetlands, moist, sandy upland woods, and alluvial woods; FACU; Maine to Florida.

Needle-leaved Deciduous Trees

142. Larch or Tamarack (*Larix laricina*): a northern bog species up to 90 feet tall, with bluish green needles (to 1 inch long) that turn yellow in the fall, and small cones (to ¾ inch long); bogs and forested wetlands, and uplands (northern part of range); FACW; Labrador to northern New Jersey, in eastern mountains to West Virginia. Similar species: European Larch (*L. decidua*), an introduced ornamental, has larger hairy cones (1 to 1¼ inch long); UPL.

143. Bald Cypress (*Taxodium distichum*): the characteristic southern swamp tree up to 140 feet tall, with "knees" (erect upright growths of the roots) when growing in seasonally flooded or wetter areas, buttressed trunks (swollen bases), featherlike leaf sprays, and round cones (1 inch wide); tidal swamps and nontidal river swamps; OBL; southern Delaware to Florida (also reported from southern New Jersey), in Mississippi Valley north to southern Illinois and Indiana.

Broad-leaved Deciduous Trees with Compound, Opposite Leaves

144. Green Ash (*Fraxinus pennsylvanica* var. *subintegerrima*): up to 80 feet tall, with smooth grayish twigs having shallow notched leaf scars, leaves divided into five to nine (usually seven) leaflets borne on short stalks, and winged fruits; tidal and nontidal swamps; FACW; Maine to Florida. Similar species: Red Ash (*F. pennsylvanica*) has velvety, hairy twigs and hairy leaf stalks and lower leaf surfaces; FACW. White Ash (*F. americana*) has deeply notched leaf scars on twigs; FACU. Black Ash (*F. nigra*), the common ash in northern swamps (New England and New York to Minnesota), has stalkless leaflets; FACW.

145. Box Elder (*Acer negundo*): to 75 feet tall or more, with smooth green twigs, leaves divided into three to five (sometimes seven) leaflets, and winged maplelike fruits (samaras); April–May; temporarily flooded forested

wetlands, floodplains, and moist soils; FAC+; New Hampshire to Florida.

Broad-leaved Deciduous Trees with Compound, Alternate Leaves

146. Bitternut Hickory (*Carya cordiformis*): up to 135 feet tall, with yellow-green leaves (to 14 inches long) divided into seven to eleven (usually nine) leaflets (to 6 inches long), small flowers borne on long slender spikes, and four-valved, husk-covered, bitter-tasting nuts; floodplains, forested wetlands, and dry and moist woods; FACU+; New Hampshire and southern Quebec to Florida. Similar species: A few other trees with compound leaves occur on floodplains. Shellbark Hickory (*C. laciniosa*) usually has seven leaflets and edible nuts; FAC. Shagbark Hickory (*C. ovata*) usually has five leaflets and shaggy bark shredding in vertical strips; FACU−. Black Walnut (*Juglans nigra*) has seven to twenty or more leaflets that smell spicy when crushed; FACU. Two thorny trees with compound leaves occur in Northeast wetlands. Honey Locust (*Gleditsia triacanthos*) bears stout, often branched thorns on trunk and branches and has compound leaves divided into many leaflets, each bearing many pairs of small leaflets (about ¾ inches long); FAC−. Devil's Walking Stick (*Aralia spinosa*), a shrub or short tree (to 40 feet tall), has many stout thorns and prickles on trunk, branches, and leaf stalks; FAC.

Deciduous Trees with Simple, Toothed, Opposite Leaves

147. Red Maple (*Acer rubrum*): up to 120 feet tall, with smooth, gray young bark, leaves typically with five-lobes, small red flowers in clusters, and winged fruits; March–May; tidal and nontidal swamps, marshes, alluvial soils, and moist uplands; FAC. Similar species: Variety *drummondii* has leaves with hairy undersides; FACW+. Variety *trilobum* has leaves with three, not five, lobes; FACW+. Sugar Maple (*A. saccharum*), the typical upland maple, has five-lobed leaves lacking many coarse teeth; FACU−.

148. Silver Maple (*Acer saccharinum*): up to 120 feet tall, with deeply five-lobed leaves bright green above and silvery below, greenish or reddish flowers, and winged flattened fruits; February–May; temporarily flooded forested wetlands, alluvial woods, and riverbanks; FACW; New Brunswick to Florida.

Deciduous Trees with Simple, Lobed, Toothed, Alternate Leaves

149. Sweet Gum (*Liquidambar styraciflua*): up to 140 feet tall, with twigs having corky wings, star-shaped leaves with five to seven lobes (aromatic when crushed), and spiny fruit balls; April–May; forested wetlands, bottomlands, moist upland woods, old fields, and clearings; FAC; southern Connecticut to Florida.

150. Sycamore (*Platanus occidentalis*): up to 175 feet tall, with characteristic flaky bark mottled with gray, green, brown, white, or yellow, large somewhat grapelike leaves (to 10 inches long), and round fruit balls (to 1½ inch wide); forested wetlands and floodplains; FACW−; southern Maine to Florida.

151. Pin Oak (*Quercus palustris*): up to 110 feet tall, with lower branches usually drooping, deeply lobed, shiny green leaves having bristle tips, flowering catkins, and small acorns (½ inch long); forested wetlands, especially temporarily flooded, and alluvial woods; FACW; Vermont to North Carolina. Similar species: Red oaks (*Q. rubra* and *Q. falcata*) have somewhat similar leaves, but lack drooping branches; FACU−.

Deciduous Trees with Simple, Lobed, Entire, Alternate Leaves

152. Swamp White Oak (*Quercus bicolor*): up to 100 feet tall, with flaky light gray bark, round-lobed or round-toothed leaves having five to ten pairs of lobes or teeth along the margins, flowering catkins, and large acorns (to 1¼ inches long) borne on long stalks (longer than leaf stalks); swamps, bottomlands, and moist woods; FACW+; Maine to Virginia. Similar species: Basket or Swamp Chestnut Oak (*Q. michauxii*), a southern species that ranges into southern New Jersey, has leaves with seven to seventeen pairs of round or sharp teeth or lobes along margins; FACW. Bur or Mossy-cup Oak (*Q. macrocarpa*), a Midwestern species, has similar leaves, but they have a distinctly deeper sinus near the middle of the leaf separating the upper part from the lower part and a unique acorn with a hairlike fringed cap; floodplains and wooded swamps; FAC−; western New York and Ohio to the Dakotas, also in central Maine, Hudson River valley, New York, south-central Pennsylvania to eastern West Virginia. White Oak (*Q. alba*), a common upland

151a. Pin Oak

151b. Northern
Red Oak

151c. Southern
Red Oak
(leaves)

152a. Swamp White Oak

152b. Basket Oak

152c. Bur Oak

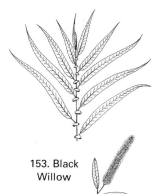

153. Black
Willow

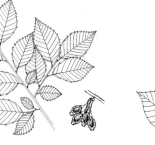

154. American Elm

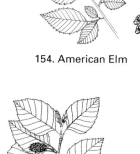

155a. Yellow Birch

155b. Gray Birch (leaf)

156. River Birch

157. Ironwood

oak that may occur in coastal-plain wetlands, has deeply lobed leaves; FACU−.

Deciduous Trees with Simple, Toothed, Alternate Leaves

153. Black Willow (*Salix nigra*): to 70 feet tall or more, with deeply grooved, brownish to blackish bark, orangish twigs in winter, leaves green above and light green below and having stipules at their bases, flowering catkins, and somewhat pear-shaped fruit capsules; April–June; tidal and nontidal swamps, bottomlands, marshes, wet meadows, and riverbanks; OBL; southern Canada to Florida. Similar species: Weeping Willow (*S. babylonica*), a Eurasian introduction, has yellowish twigs in fall and winter, drooping branches, and leaves whitish to pale below; FACW−. Crack Willow (*S. fragilis*) lacks leafy stipules and has very brittle twigs that easily break at their bases and leaves green on both sides; FAC+.

154. American Elm (*Ulmus americana*): up to 125 feet tall, with scaly gray bark, leaves from 4 to 6 inches long having rough upper surfaces and unequal bases, clusters of small greenish flowers borne on drooping stalks, and flattened winged fruits with hairy margins; March–early May; swamps, bottomlands, and rich, moist upland woods; FACW−; Newfoundland to Florida. Similar species: Slippery Elm (*U. rubra*) has fragrant mucilaginous (slimy) inner bark, leaves with very rough upper surfaces, and roundish winged fruits with smooth wavy margins; FAC.

155. Yellow Birch (*Betula alleghenensis*): up to 100 feet tall, with silvery to golden peeling bark, fragrant (root beer scented) twigs when crushed, and flowering catkins; April–May; forested wetlands, hemlock swamps, black-spruce swamps, rocky mountain soils, and rich, moist upland woods; FAC; Newfoundland to Delaware. Similar species: Black Birch (*B. lenta*) may occur in Atlantic white-cedar swamps; its aromatic bark is black, usually with a few cracks and callouses when old, but not peeling; FACU. Gray Birch (*B. populifolia*) has dull whitish bark that does not peel, and triangle-shaped leaves; FAC. Paper Birch (*B. papyrifera*) has white peeling bark; FACU.

156. River Birch (*Betula nigra*): up to 100 feet tall, with reddish brown, peeling young bark, black platelike older bark, and flowering catkins; April–May; swamps, floodplains, and bottomlands; FACW; New Hampshire to Flor-

ida. Similar species: Gray Birch (*B. populifolia*) occasionally occurs in swamps, especially in the Pine Barrens; it has nonpeeling whitish to grayish bark with black chevronlike markings, and toothed triangle-shaped leaves; FAC.

157. Ironwood or Blue Beech (*Carpinus caroliniana*): up to 40 feet tall, with conspicuous musclelike, smooth, dark bluish gray bark, flowering catkins, and uniquely shaped flattened fruits; temporarily inundated forested wetlands, alluvial woods, bottomlands, and moist woods; FAC; Nova Scotia to Florida.

158. Cottonwood (*Populus deltoides*): up to 100 feet tall or more, with gray deeply grooved bark, distinctive triangle-shaped leaves with pointed tips, and cottony seed masses in spring; floodplains and rich woods; FAC; Quebec and western New England to Florida. Similar species: Swamp Cottonwood (*P. heterophyllum*) has leaves with heart-shaped bases and blunt tips; FACW+. Quaking or Trembling Aspen (*P. tremula*, formerly *P. tremuloides*) has more-rounded leaves with pointed tips, and smooth light grayish, greenish, or whitish bark; FACU. Balsam Poplar (*P. balsamifera*), a northern swamp and floodplain species (to Connecticut and northern Pennsylvania), has somewhat triangle-shaped leaves with rounded bases (fragrant when crushed) and very sticky buds (also fragrant when crushed); FACW.

See also Swamp White Oak (species 152).

Deciduous Trees with Simple, Entire, Alternate Leaves

159. Willow Oak (*Quercus phellos*): up to 100 feet tall, with thick, bristle-tipped, narrow willowlike leaves (to 5 inches long and to 1 inch wide) green above and paler below, flowering catkins, and small roundish acorns (about ½ inch long); forested wetlands, floodplains, and bottomlands; FAC+; southern New York to Florida, mostly on the coastal plain. Similar species: Laurel Oak (*Q. laurifolia*), a coastal-plain swamp species from Cape May, New Jersey, south, has thin leaves lacking bristle tips; FACW−. Water Oak (*Q. nigra*), another coastal-plain species from Delaware south, has bristle-tipped leaves that are widest just below the tip (sometimes with two or three shallow lobes) and have long wedge-shaped bases; FAC. Shingle Oak (*Q. imbricata*), a Midwestern species ranging onto floodplains in New Jersey and northern Delaware, has thin leaves, shiny

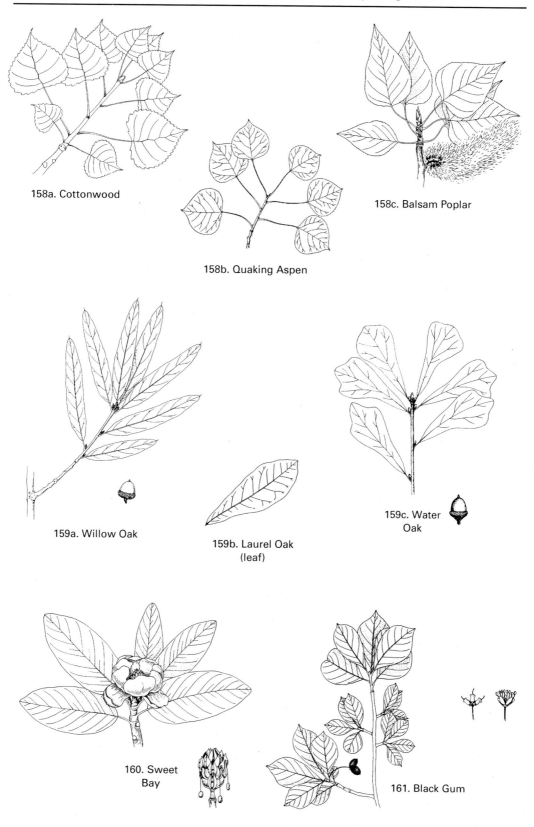

158a. Cottonwood

158b. Quaking Aspen

158c. Balsam Poplar

159a. Willow Oak

159b. Laurel Oak
(leaf)

159c. Water
Oak

160. Sweet
Bay

161. Black Gum

green above and hairy below, with a yellow midrib; FAC.

160. Sweet Bay (*Magnolia virginiana*): up to 70 feet tall, with smooth dark green twigs (aromatic when crushed), leaves shiny green above and whitish below, large, fragrant, nine- to twelve-petaled white flowers, and pink or red conelike fruit clusters; May–July; forested wetlands and moist, sandy woods; FACW+; Long Island (New York) to Florida.

161. Black Gum (*Nyssa sylvatica*): up to 125 feet tall, with deeply furrowed dark brown to grayish bark, branches usually extending outward in planes (perpendicular to the trunk and parallel to the ground), twigs with chambered pith, leaves sometimes with a few irregular teeth above middle; small greenish flowers, and bluish black oval berries (about ½ inch wide) borne in clusters at end of long stalks; April–May; swamps, bottomlands, cedar swamps, moist upland woods, and dry woods; FAC; Maine to Florida. Similar species: Persimmon (*Diospyros virginiana*) occurs in Carolina bay or pothole wetlands on the Delmarva peninsula; its bark is divided into squarish blocks, its pith is essentially solid, and its orangish fleshy fruits are larger (to 1½ inches wide); FAC−. Glossy or European Buckthorn (*Frangula alnus*, formerly *Rhamnus frangula*), an invasive shrub or small tree (to 20 feet), has somewhat similar entire leaves, but its black berries are round (red when immature) and its grayish stems and twigs are usually densely covered with white lenticels (appearing as dots, with many located together forming a short line and sometimes positioned above other such lines resembling a stack of pancakes); its yellow-green flowers bloom throughout summer (May–September) and both flowers and berries can often be observed on the same plant; FAC. Common Buckthorn (*Rhamnus cathartica*), another invasive shrub or small tree that may be a dominant species in Midwestern wetlands, has fine-toothed leaves, thorns at the ends of its twigs, small greenish four-petaled flowers that bloom in spring while leaves are unfolding, and black round berries; FAC. (See Figure 8.7 for illustrations of the two buckthorns.)

Additional Field Guides

Wetland Guides

Chadde, S. W. 1998. *A Great Lakes Wetland Flora.* Pocketflora Press, Calumet, MI.

Fassett, N. C. 1975. *A Manual of Aquatic Plants.* University of Wisconsin Press, Madison, WI.

Hyland, F., and B. Hoisington. 1977. *The Woody Plants of Sphagnous Bogs of Northern New England and Adjacent Canada.* University of Maine, Life Sciences and Agriculture Experiment Station, Orono, ME. Bulletin 744.

Magee, D. W. 1981. *Freshwater Wetlands: A Guide to Common Indicator Plants of the Northeast.* University of Massachusetts Press, Amherst, MA.

Muenscher, W. C. 1972. *Aquatic Plants of the United States.* Cornell University Press, Ithaca, NY.

Tiner, R. W., Jr. 1987. *A Field Guide to Coastal Wetland Plants of the Northeastern United States.* University of Massachusetts Press, Amherst, MA.

Tiner, R. W. Jr. 1988. *Field Guide to Nontidal Wetland Identification.* U.S. Fish and Wildlife Service, Newton Corner, MA, and Maryland Department of Natural Resources, Water Resources Administration, Annapolis, MD. Out of print, but copies available for purchase from Institute for Wetland & Environmental Education & Research, Inc., P.O. Box 288, Leverett, MA 01054.

Tiner, R. W. 1993. *Field Guide to Coastal Wetland Plants of the Southeastern United States.* University of Massachusetts Press, Amherst, MA.

Tiner, R. W. 1994. *Maine Wetlands and Their Boundaries.* State of Maine, Department of Economic and Community Development, Office of Community Development, Augusta, ME.

Ferns

Cobb, B. 1963. *A Field Guide to the Ferns and Their Related Families.* Houghton Mifflin Co., Boston.

Montgomery, J. D., and D. E. Fairbrothers. 1992. *New Jersey Ferns and Fern-Allies.* Rutgers University Press, New Brunswick, NJ.

Grasses

Brown, L. 1979. *Grasses: An Identification Guide.* Houghton Mifflin Company, Boston.

Hitchcock, A. S. 1971. *Manual of the Grasses of the United States.* Dover Publications, New York.

Michener, M. C. 2004. *Graminoids: A Guide to Some Common Grasses, Sedges, and Rushes of the Northeastern USA.* MIST Software Associates, Inc., Hollis, NH.

Wildflowers

Newcomb, L. 1977. *Newcomb's Wildflower Guide.* Little, Brown & Company, Boston.

Niering, W. A., and W. C. Olmstead. 1979. *The Audubon Society Field Guide to North American Wildflowers: Eastern Region.* Alfred A. Knopf, Inc., New York.

Trees and Shrubs

Little, E. L. 1980. *The Audubon Society Field Guide to North American Trees: Eastern Region.* Alfred A. Knopf, Inc., New York.

Petrides, G. A. 1958. *A Field Guide to the Trees and Shrubs.* Houghton Mifflin Company, Boston.

Sargent, C. S. 1965. *Manual of the Trees of North America.* Dover Publications, New York.

Tiner, R. W. 1997. *Winter Guide to Woody Plants of Wetlands and Their Borders: Northeastern U.S.* Institute for Wetland & Environmental Education & Research, Leverett, MA.

CHAPTER 11

Interpreting Hydric Soils

Your first experience with hydric soils probably came upon walking into a marsh or a swamp, when you either felt the spongy or soggy ground underfoot or, in more extreme cases, sank knee-deep or even waist-deep into the muck. It did not take long to realize that these soils were wet. In these situations, you probably would not have to examine the soils to determine that the area was wetland, because good wetland-indicator plants or other signs of wetland hydrology would likely be present. Given the variety of wetlands that exist along the natural soil-wetness gradient, identification of hydric soils becomes more important in drier wetlands (e.g., those saturated for only the early part of the growing season) and for determining wetland boundaries in areas of gentle topographic relief.

Lists of Hydric Soils

The USDA Natural Resources Conservation Service (NRCS) has published national lists of hydric soils and state and county lists of soil-mapping units that contain hydric soils. These lists are helpful for interpreting published county-soil-survey reports, which include maps that show soil-mapping units (see Chapter 13 on uses of these data). As new soils are described, the national list is updated; contact your local NRCS office to learn how to access this information from the Internet.

Describing Soils

In describing soils, soil scientists identify and characterize different layers or horizons (Figure 11.1). Hydric organic soils are typified by one layer, the O-horizon. This horizon is dominated by the decomposed remains of plants that take the form of peat, muck, or a combination of the two. Below the O-horizon is unrelated parent material (the C-horizon) or bedrock (the R-horizon).

Mineral soils usually have more complicated patterns. When examining these soils, different layers or horizons are usually observed. Each layer has a characteristic color, the dominant or matrix color. Some soils, especially hydric soils and soils adjacent to wetlands, often have other additional colors (one or more non-dominant colors, called "mottles"). Describing the soil involves characterizing the soil colors of the matrix and mottles. For hydric soils, the upper 20 inches of soil are usually examined (see below).

The wetter mineral soils often have an organic layer of variable thickness on top of sand, loam, silt, or clay. As in organic soils, this organic layer is called the O-horizon. It is subdivided into three layers based on degree of plant decompositon—Oa (for mucky material, or sapric), Oi (for peaty material, or fibric), and Oe (for peaty muck or mucky peat material, or hemic)—following the same rules used to separate the three types of wet organic soils (see Organic Soils below). Beneath the O-horizon is the uppermost mineral layer: the A-horizon, which is commonly called topsoil. This layer is usually darker than the underlying mineral layer due to enrichment by organic matter. The subsoil may be represented by a B-horizon (with evidence of soil weathering), or an E-horizon (a layer of eluviation where organic matter, aluminum, clay, or other materials are leached out) with a B-horizon beneath it. In evergreen forests, the E-horizon may be subtended by a reddish brown or dark brown layer of variable thickness called the spodic horizon (designated as a Bh-horizon when the layer has a lot of organic matter, or a Bs-horizon when iron and aluminum oxides dominate). Such soils are called spodosols (see discussion later in this chapter). The C-horizon or parent material typically underlies the B-horizon. Floodplains often have buried A-horizons due to periodic sedimentation from

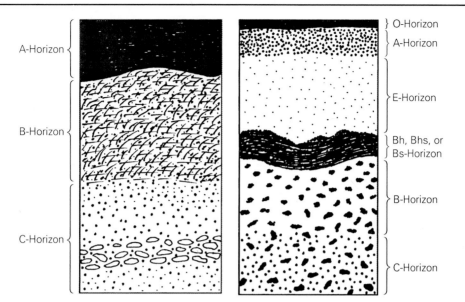

Figure 11.1. Two examples of mineral soil profiles with diagnostic layers, or horizons: nonsandy mineral (left), and mineral spodosol (right). The O-horizon is dominated by organic matter; in organic soils this horizon is typically 16 inches or more thick. The A-horizon is the surface mineral layer, where organic matter mixes with mineral material. The B-horizon is the mineral subsoil, where soil-forming processes are at work. The E-horizon is the eluvial layer, where materials such as iron, aluminum, and organic matter have been leached by organic acids from overlying vegetation. The Bh-, Bs-, or Bhs-horizon is the spodic horizon where these leached materials accumulate. The C-horizon is the parent material layer below the depth at which soil-forming processes occur.

floods, producing a rather interesting combination of soil layers.

Munsell Soil-Color Book

As shown in Figure 11.2, soil colors can tell us a great deal about the wetness regime, especially in mineral soils. Scientists and others examining the soil determine the approximate soil color of different horizons by consulting a Munsell color chart. The so-called Munsell soil-color book (Macbeth Division 1994) contains charts with paint chips of soil colors (Plate 23). Soil colors are identified on the charts by three characteristics: 1) hue, one of the main spectral colors—red, yellow, green, blue, or purple—or various mixtures of them; 2) value, lightness or darkness of the hue; and 3) chroma, purity or saturation of the color.

In the Munsell soil-color book, each hue has its own page, which is subdivided into units for value (on the vertical axis) and chroma (horizontal axis). Although theoretically each soil color represents a unique combination of hue, value, and chroma, the number of combinations common in the soil environment usually is limited. Because of this situation and the fact that accurate reproduction of each soil color is

expensive, the book contains a limited number of combinations. The typical Munsell book has about a dozen pages of different hues (e.g., 10YR, 2.5Y, 5Y, 7.5R), which are mostly combinations of yellows (Y) and reds (R), plus gley charts. A new soil-color book (Color Communications, Inc. 1997) has sixty-two pages of hue-chroma combinations.

Low-chroma colors (2 or less) include black, various shades of gray, and darker shades of brown. These colors are associated with hydric soils and organic-enriched nonhydric soils (usually the A-horizon of the latter soils). The gley charts represent colors associated only with hydric soils (mainly grayish, bluish, and greenish colors), while the other hue charts show both hydric and nonhydric colors.

Using the Munsell Charts

Different colors found in soils are classified by comparing a soil sample with the color chips on the Munsell charts. First try to find a hue page with the correct soil color. In most of the Northeast, begin with the 10YR page. If the soil is yellower or more olive colored, try the 2.5Y or 5Y pages. For redder colors, check 7.5YR, 5YR, and 2.5YR. If the soil is black, gray, blue, or

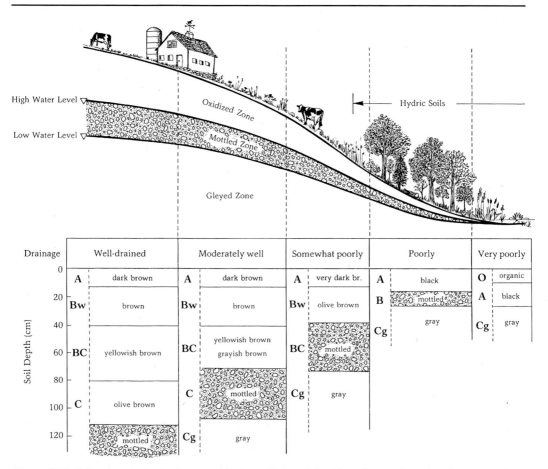

Figure 11.2. Soil colors typically change with slope. Although hydric soils may include a few somewhat poorly drained soils, most are poorly or very poorly drained. The color pattern shown here for somewhat poorly drained soil is that of a nonhydric soil, as there is no evidence of prolonged waterlogging within 6 inches (15 cm) of the surface. (Source: Tiner and Veneman 1995)

green, consult the gley pages. Place the soil sample behind the hue page, look through the holes, and move the sample from hole to hole until you match the dominant soil color to a color chip. The matched color is the matrix color of the soil. The other colors are nondominant mottles. Both matrix and mottle colors are important for recognizing hydric soils.

Be sure you have adequate light to view the soil sample, and do not attempt to read the color charts with sunglasses on. Cloudy days or late fall and winter afternoons may pose problems. In some cases, you must simply return the next day or bring home a sample to evaluate under more favorable conditions.

When recording data for a wetland determination, specialists employ the Munsell notation, an alpha-numeric code for hue, value, and chroma: for example, 5Y 2/1 (hue is 5 yellow,

value is 2, and chroma is 1). Low-chroma colors associated with hydric soils are represented by the left two columns of the hue pages and by the entire gley pages. The rest of the columns are essentially nonhydric-soil colors. The appropriate Munsell color name can be read from the page facing the color chart. (Note: For hydric-soil determinations, be sure to use a 1990 or later edition of the Munsell soil-color book.)

How to Recognize Hydric Soils

Extracting a Soil Sample

To examine the soil, dig a hole about 2 feet deep by 1½ feet wide. If there are more than 8 inches of organic material on the surface (excluding leaf litter), the soil is hydric. If the soil does not

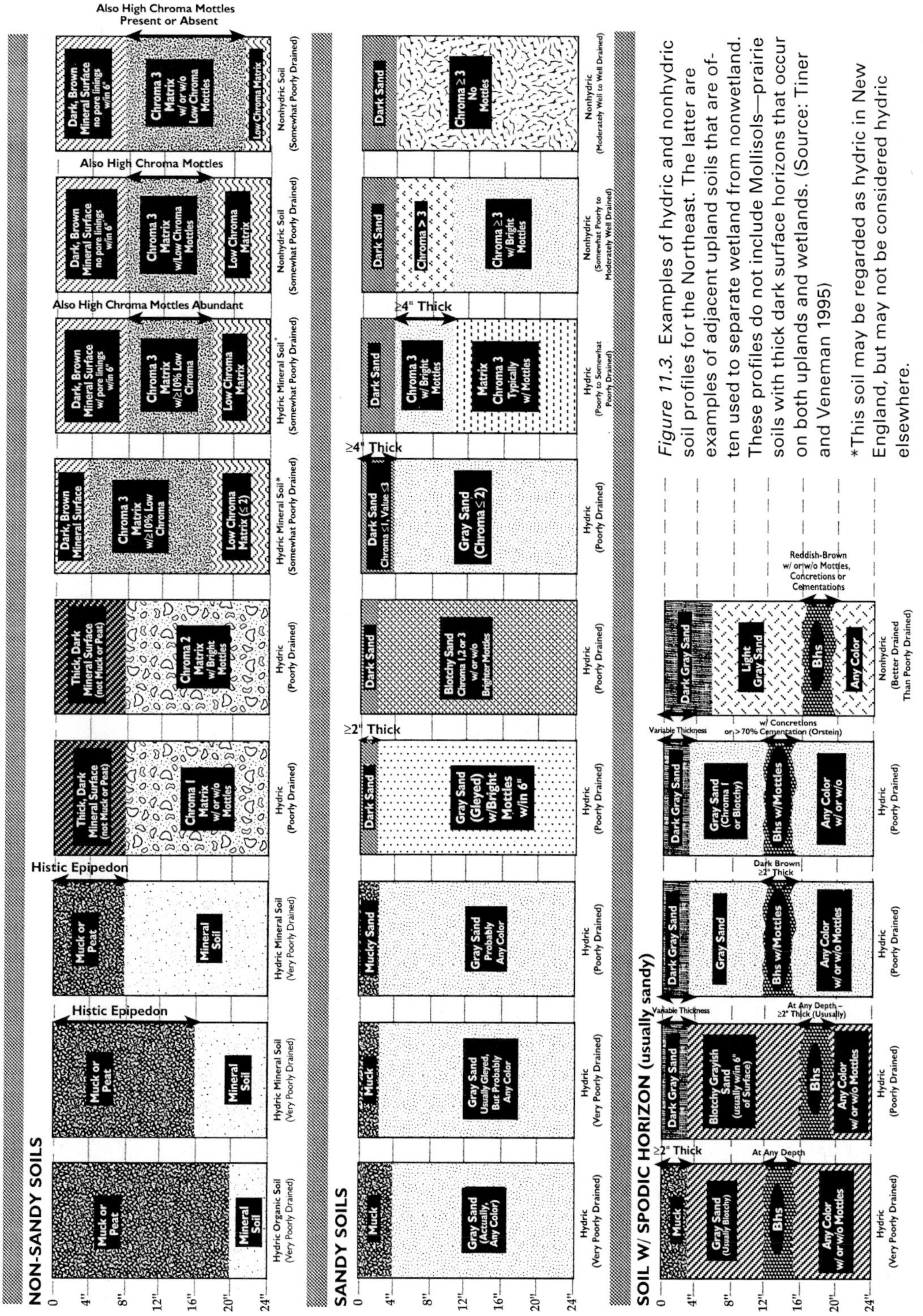

Figure 11.3. Examples of hydric and nonhydric soil profiles for the Northeast. The latter are examples of adjacent upland soils that are often used to separate wetland from nonwetland. These profiles do not include Mollisols—prairie soils with thick dark surface horizons that occur on both uplands and wetlands. (Source: Tiner and Veneman 1995)

*This soil may be regarded as hydric in New England, but may not be considered hydric elsewhere.

have this amount of organic matter, look at the subsoil horizon (often the B-horizon) immediately below the surface layer (A-horizon). The colors below 6 inches must be examined, because many cultivated soils (past and present) have been disturbed above this point by plowing and have been enriched with organic matter, thereby affecting the original soil color. This disturbed zone is called the plow layer. If the subsoil layer is predominantly gray (and you're not looking at a spodosol), the soil has a good chance of being hydric. This layer is easily recognized by an abrupt boundary (straight line) between the A-horizon and B-horizon (see Plate 27).

Typical Hydric-Soil Properties

The following properties usually indicate a hydric soil in the Northeast: 1) a peaty or mucky surface layer 8 inches or thicker, and 2) dominant colors in the mineral-soil matrix with chroma of 2 or less if there are mottles (usually orangish, yellowish, or reddish brown) or 3) dominant colors in the mineral-soil matrix with chroma of 1 or less, if there are no mottles present. Sandy soils have slightly different requirements (see Problematic Hydric Soils below). Figure 11.3 shows typical hydric and adjacent nonhydric soils found in the Northeast. Hydric-soil indicators are still in the development phase. Some indicators developed by New England soil scientists are being used for wetland determination in New England but are not applied elsewhere in the Northeast. The indicators presented in the following discussion include both nationally and regionally recognized ones and are duly noted as such. Photographs of some representative hydric soils are found on color plates 24 through 37 in this book. Other images of hydric soil properties and redoximorphic features are posted on the USDA Natural Resources Conservation Service's Wetland Science Institute webpage (http://www.pwrc.usgs.gov/wli/Training/materials.htm).

Organic Soils

Organic soils typically form in 1) waterlogged depressions (e.g., glacially formed kettle holes, river oxbows, and lake margins) where peat or muck deposits range today from about 1.5 feet to 50 feet or more deep; 2) cold, wet climates like northern New England; and 3) low-lying areas along coastal waters where tidal flooding is frequent and saturation is nearly continuous. In the humid subarctic, the cold climate lowers evaporation and plant transpiration, allowing the development of organic soils on broad lowlands such as the Hudson Bay lowlands (once the Tyrrell Sea) in Canada and former Lake Agassiz from Minnesota north. In these regions, organic soils may cover many miles of the landscape, even rolling terrain with blanket bogs.

Peats and mucks are not the same, although nonscientists commonly use the terms interchangeably. Muck (saprists) consists of organic matter that breaks down into a greasy mass upon rubbing, of which less than one-third of the material can be identified. In peats (fibrists), more than two-thirds of the organic material is identifiable (leaves, stems, roots). There are intergrades between the two—mucky peats and peaty mucks (hemists)—depending on the amount of identifiable material. A fourth group of organic soils (folists) is nonhydric, forming in high mountains in the tropics and in boreal and arctic regions. In the Northeast, folists are limited to mountainous areas in northern New England, where they are generally thin organic soils on bedrock in landscape positions that are obviously not associated with wetlands.

Organic soils can easily be recognized by their characteristic black muck or black to orange-brown peat, which is usually thicker than 16 inches (Plates 24 and 25). Shallow organic soils of variable thickness exist over bedrock. If the organic layer is less than 16 inches thick and overlies mineral material, the soil is classified as mineral. Perhaps the easiest way to identify the presence of organic soil, besides sinking in the muck, is to take a shovel or auger and try to push it into the soil. If the shovel is easily pushed 16 inches or deeper, the soil is organic (Figure 11.4). If the depth of penetration is less, the soil may still be an organic soil if you hit bedrock, but it is more likely that the soil is mineral with a shallow organic layer on top.

While it is not really necessary to separate peats from mucks for wetland identification since both are obviously hydric soils, you might be interested in trying this simple test. When rubbed between your fingers, mucks have almost all of the plant remains decomposed beyond recognition and feel somewhat greasy. Peats are slightly decomposed, and when rubbed between the fingers, most of the plant materials can be recognized as parts of grasses, sedges, and mosses, or types of wood.

Figure 11.4. Professor Peter Veneman demonstrating the simplest technique to detect an organic soil: pushing an auger into the soil. In this case, the entire auger, more than 4 feet long, has been pushed into the soil with little difficulty.

Mineral Soils

The wettest mineral soils often have a layer of organic material (frequently muck) on the surface. When this layer is 8–16 inches thick, it is called a histic epipedon (Plate 26). This is an automatic indicator of hydric soil and wetland (unless the area is effectively drained). Some of these soils produce a strong odor of rotten eggs (hydrogen-sulfide gas), which is another excellent hydric-soil indicator; some mucky organic soils also do this.

For other mineral soils, it is important to recognize different textures, mainly to separate sandy from nonsandy soils. Sandy soils are more permeable and have different hydric properties from those of nonsandy soils. These soils can be distinguished by applying a simple test. Take a small sample of soil, moisten it, and make a ball about one inch in diameter. Gently press the ball between your thumb and index finger. If the ball crumbles, the soil is a sand. If not, it is non-sandy. That's all you need to know for routine hydric-soil assessments, but if you'd like to separate clayey from loamy soils, do the additional steps shown in Figure 11.5.

Gleying and low-chroma mottling (redox depletions) are typical hydric mineral-soil properties. The abundance, size, and color of the mottles usually reflect the duration of the saturation period and indicate whether or not the soil is hydric. In general, the more gray present and the closer the gray layer is to the surface, the wetter the soil and the more likely the soil is hydric (with spodosols a major exception; see below; Plates 27–30). Soils with only gray mottles near the surface may not be wet enough to be considered hydric. Mineral soils lacking gray mottles are usually nonhydric, except for certain sandy soils (Plate 31). Nonhydric mineral soils that are never saturated are usually bright-colored near the surface and are not gray-mottled (Plate 37).

Hydric mineral soils are usually identified by a gray subsoil (redox depletions or depleted matrix) with bright-colored mottles (redox concentrations) within 12 inches of the surface. Depleted matrices are represented by chroma of 2 or less with mottles, or chroma of 1 or less without mottles. Many hydric soils are characterized by a thick, dark surface layer (black or dark brown), a predominantly gray subsurface layer (the low-chroma matrix) with yellow, orange, brown, or reddish mottles, and sometimes iron-oxide concretions near the surface (Plates 26–28, 30). Some hydric mineral soils have reddish brown to orange mottles lining the root channels (oxidized rhizospheres or pore linings; Plate 1e). Black concretions of manganese oxide may also be present near the surface. The gray matrix color of the subsoil (usually within 12 inches of the surface) and thick, dark surface layer are the best indicators of current wetness, since the iron-oxide mottles and concretions are very insoluble, and once formed will often remain indefinitely in the soil as relic mottles.

The wettest mineral soils are typically neutral gray in color (gleyed soil; Plate 27), although occasionally the color may be greenish gray or bluish gray (Plate 28). Sometimes the color may fade when the soil is exposed to air; this usually means that reduced iron is present. Gleyed colors are found on the gley pages of the Munsell soil-color book. A gley matrix within 6 inches of the soil surface is a hydric-soil indicator. These soils are saturated for significant periods to be considered hydric. Soils with a depleted

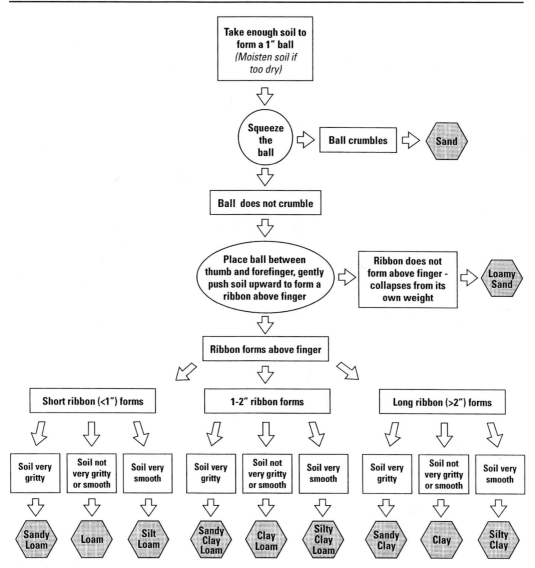

Figure 11.5. Easy steps for texturing soil. (Source: Adapted from Thien 1979)

matrix (chroma 2 or less) at least 6 inches thick starting within 10 inches of the surface are also hydric (Plate 29) (see Dark Prairie Soils section for exceptions). Soils with thick black or dark gray surface horizons (chroma 0—neutral or chroma 1) and depleted matrix subsoils are also hydric (Plate 30). Most nonsandy soils in the Northeast with gray colors dominating the subsoil immediately below the topsoil are generally recognized as hydric. (Note: Beware of gray-colored E-horizons; see spodosols discussion under Problematic Hydric Soils.)

Other hydric-soil indicators for nonsandy soils include 1) hydrogen-sulfide odor detected within 12 inches of the soil surface, 2) stratified layers within 6 inches of the soil surface (one layer must be organic soil or mineral of chroma 1 or less), 3) the presence of 2 percent or more organic bodies of muck or mucky mineral texture (typically attached to the roots) within 6 inches of soil surface (body diameters 0.5 to 1.0 inch; Figure 11.6), 4) a mucky modified mineral surface layer 2 inches or thicker within 6 inches of the soil surface, 5) a muck layer 0.5 inch or thicker within 6 inches of the soil surface (a thickness of 0.75 inch is being tested for the rest of the Northeast), and 6) a loamy mucky mineral layer 4 inches or thicker beginning within 6 inches of the soil surface. Indicators 3, 4, and 5 above are currently permitted for use in the Northeast only on the Delmarva Peninsula, although they are probably also use-

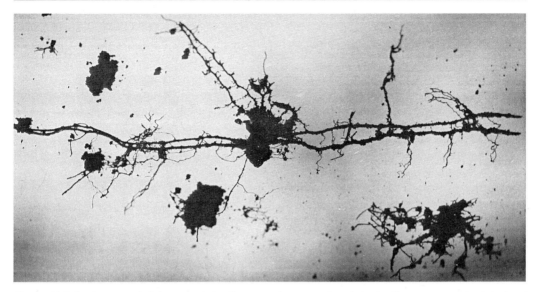

Figure 11.6. In the coastal plain, organic bodies—mucky to mucky-mineral soil masses attached to plant roots—develop in some soils where seasonal waterlogging kills a significant portion of the roots, which then form these ½ inch or larger masses.

ful in the coastal plain of New Jersey. (Note: Mucky modified mineral is an organic-rich mineral layer that feels greasy, making it difficult for some to separate organic soil from mineral soil.) For a list of official indicators for regulatory purposes, consult your state soil scientist or wetland regulator.

Problematic Hydric Soils

Hydric-soil determinations are not always straightforward. There are exceptions to most rules. Certain reddish and other brighter-colored soils are hydric, and certain black- or gray-colored soils are nonhydric. These soils may require considerable expertise for positive identification, unless the vegetation is obviously hydrophytic in the first case and nonhydrophytic in the latter. Consequently, in these and other problematic situations, evidence of wetland hydrology and the presence of wetland vegetation should be used to make a wetland determination rather than worrying about obscure or nonexistent soil properties. Consult a state or local office of the Natural Resources Conservation Service for more information on these soils. Listed below are examples for your general information.

Sandy Soils

Some sandy soils pose the greatest problem in identifying hydric properties, since all or many of the diagnostic characteristics listed above may not be present (Plates 32 and 33). Given their rapid permeability, evidence of prolonged saturation must be found within 6 inches of the surface. Like their nonsandy counterparts, some hydric sandy soils have thick, dark (often black) surface layers with high organic-matter content or a gleyed matrix within 6 inches of the soil surface (Figure 11.3). The former may be a histic epipedon or a sandy mucky mineral layer (at least 2 inches thick) beginning within 6 inches of the soil surface (Plate 32). These soils are readily identified as hydric. The presence of a thin layer of muck (2 inches or thicker) on top of a sandy soil is also evidence of prolonged wetness and a hydric sand. Even an inch of muck is probably enough, although this has not been recognized nationally. Other sandy hydric soils may show evidence of vertical streaking by organic matter below the surface layer. The black or dark gray streaks on a medium or light gray matrix are easily detected (see subsoil in Plate 32).

More difficult to recognize as hydric are blotchy sands (see subsoil in Plate 32). These soils have a blotchy subsoil (mixtures of chroma 1, 2, or 3 colors) due to variable organic coatings around some of the sandy grains. This condition is called either "polychromatic matrix" or "stripped matrix." The presence of organic materials in blotchy sands is determined by a simple test. When rubbed gently on the palm of your hand, organic-coated sand grains from the darker blotches leave a dark-colored, often blackish, stain on your skin. Rubbing the uncoated grains from light blotches leaves almost no stain. Sands with a matrix of chroma 3 with both low-chroma and high-chroma mottles (redoximorphic features: iron depletions and concentrations, respectively) within 6 inches of the soil surface are also hydric. Recently deposited sandy soils, such as sand bars along rivers, may not possess any typical hydric-soil properties, but they can be recognized as hydric soils by their landscape position, associated vegetation, and signs of flooding.

Floodplain Soils

Floodplains are depositional environments, where soils are constantly being buried by new materials brought in by floodwaters. Such soils often have little or no evidence of weathering. These alluvial soils typically have buried surface layers (A-horizons) at various depths. Not all floodplain soils are hydric, since many are only infrequently flooded. Of those that are hydric, some possess typical hydric-soil properties, while others do not. An additional hydric-soil indicator is used along the coastal plain from the Delmarva Peninsula south: loamy floodplain soils with a layer having 40 percent or more chroma 2, with 2 percent or more black (manganese) or reddish brown–orange (iron) mottles occurring as soft masses. Some floodplain soils are predominantly red-colored due to the deposition of soil derived from red parent materials (see below). In general, landscape position, vegetation, and evidence of flooding are useful for distinguishing hydric floodplain soils from nonhydric ones.

Red Parent-Material Soils

Soils derived from red parent materials (e.g., strongly weathered clays and exposed Triassic and Jurassic sandstones and shales) present particular problems for hydric-soil recognition. In the Northeast, examples are found in the Connecticut River valley, central Massachusetts, and central New Jersey. In these areas, glaciation exposed ancient formations that are now eroding. The red colors are attributed to the dominance of the iron mineral hematite. The colors are redder than 10YR and obscure the low-chroma colors that normally develop under anaerobic,

reducing wetland conditions. Some of these hydric soils may have low-chroma colors within 1½ feet of the surface, but many do not. In most instances, the landscape position, resulting vegetation, and certain signs of hydrology are the best indicators for making a wetland determination. Wet-season field checking may be desirable. For a hydric-soil indicator, the federal government is testing the following: within 12 inches of the surface, a layer at least 4 inches thick with a matrix of 7.5YR or redder and a chroma of 3 or less with 2 percent or more redox depletions (chroma 2 or less) or redox concentrations.

Evergreen-Forest Soils

Spodosols are associated with evergreen forests, typically on sandy soils. They are common in northern temperate and boreal regions and along the coastal plain from New Jersey south. Evergreen forests of hemlock, spruces, and pines dominate these regions. Larch, oaks, and beech are also associated with spodosol formation.

Most spodosols (hydric and nonhydric) have a characteristic gray E-horizon (E for eluvial: a leached layer) overlying a diagnostic spodic horizon of accumulated organic matter, iron, and aluminum (see Plates 34–36). The gray layer is not necessarily due to wetness, but is formed by a process called "podzolization" that commonly occurs under pines on sandy soils and under hemlocks and spruces on any soil. Organic acids from the breakdown of leaves of these and other species move down through the soil with rainfall, cleaning the sand grains and depositing organic matter, iron, and aluminum in the next layer (the spodic horizon). This process occurs in both wetlands and uplands. In the Northeast, it takes place mainly in sandy soils. When formed under wet conditions, the spodic horizon is dark brown—the color of coffee grounds—and it is usually thick (greater than 2 inches; Plate 34), whereas in dry situations (nonwetlands), the spodic horizon is more reddish brown and quite thin (Plate 36).

Wet, sandy spodosols may be recognized by typical hydric sandy-soil properties (including a muck layer 2 inches or more thick; Plate 35) or by either: 1) a cemented spodic horizon within 12 inches of the soil surface (Plate 34), 2) high-chroma mottles in the upper part of the spodic horizon, or 3) mottling in the spodic horizon (see Figure 11.3; Plates 34 and 35). Loamy hydric spodosols may also be recognized by these three features.

Dark Prairie Soils

Dark-colored soils called "mollisols" are characteristic prairie soils. They are found in both wetlands and uplands. The dark colors of the topsoil (mollic epipedon) are attributed to high organic matter derived from the decomposition of deep-penetrating roots of grasses in semiarid to subhumid climates. Mollisols are dominant soil types from central Texas north to western Minnesota and the Dakotas, west to eastern Oregon and Washington. In the Great Lakes region, they are the one of the major soil types in northern Illinois, northwestern Indiana, and southern Wisconsin. Mollisols are also common in humid climates along the Atlantic Coastal Plain and on floodplains of major midwestern and southeastern rivers.

The dark colors mask more typical hydric soil properties. Often a combination of landscape position (e.g., depressions and toes of slopes) and vegetation (hydrophytes) are useful for separating the hydric mollisols from the nonhydric ones. This is especially true in the Northeast. In many cases, especially in the Midwest, an examination of a series of soil profiles is required for distinguishing subtle changes in soil properties with increasing wetness. These properties include an increase in darkness of the soil (lower chroma and value; more black) and a thickening of the mollic epipedon. The wetter mollisols may have a layer of peat or muck on the surface. Redoximorphic features like redox concentrations (chroma greater than 2), redox depletions (chroma 2 or less), and oxidized rhizospheres, may also be observed near the surface. The latter features may be only present in the spring. Soil properties below the mollic epipedon often reveal much about a site's wetness—look for a low-chroma matrix (2 or less). Some wet mollisols are among the most difficult hydric soils to identify.

Newly Formed Soils

New hydric soils may form with the help of beavers or humans. Whenever a nonhydric soil becomes flooded for more than one week during the growing season in most years, it is considered hydric by definition (hydric-soil criteria 3 and 4). Of course, the permanence of the beaver dam or human construction (e.g., road impoundments) must be considered before calling any flooded area a hydric soil or wetland. If a beaver dams a road culvert, blocking flow and flooding low-lying nonhydric soils, and someone removes the dam and attempts to keep the beaver out of the area, the situation is temporary, and the area should not be considered to have newly created hydric soil or wetland. If the situation has, however, lasted for some time and wetland vegetation (e.g., cattails and water lilies) is established and upland plants are dying or dead, then the area should be considered to have newly created hydric soil and to be wetland. Given the recent flooding in these cases, the soil properties will not be typical of hydric soils. It usually takes decades and perhaps a century or more for a soil to develop these properties. Again, vegetation and signs of hydrology are the best clues for identifying these wetlands.

Drained Hydric Soils

Where a network of drainage ditches or similar structures (tile drains and dikes, with pump houses) is observed, the effectiveness of the drainage needs to be determined to identify the presence or absence of wetland—drained hydric soils must be distinguished from undrained hydric soils. This is by no means a simple task. In general terms, if the soils are drained to the point where they are no longer capable of supporting the growth and regeneration of wetland vegetation, then they are effectively drained hydric soils and not considered wetland. This exercise is clearly beyond the skills of the average person, yet it is usually easy to recognize that an area has been subjected to drainage (exceptions: tile drains and groundwater withdrawals). The difficulty is determining the magnitude of the hydrologic alteration. To accurately identify the extent of drainage usually requires the services of a wetland scientist or soil-drainage engineer. The detailed soil-wetness studies that may be needed to evaluate such a condition are clearly beyond the scope of this book. One thing to remember, however, is that one shallow ditch is usually not sufficient to effectively drain a large wetland; a network of ditches is typically required to do this.

Additional Readings

Color Communications, Inc. 1997. *Earth Colors Soil Color Book,* Poughkeepsie, NY.

Hurt, G. W., P. M. Whited, and R. F. Pringle (editors). 2003. *Field Indicators of Hydric Soils in the United States. A Guide for Identifying and Delineating Hydric Soils.* Prepared in cooperation with the National Technical Committee for Hydric Soils. USDA Natural Resources Conservation Service, Fort Worth, TX.

Macbeth Division. 1994. *Munsell Soil Color Charts.* Kollmorgen Instruments Corporation, New Windsor, NY.

NEIWPCC Wetlands Work Group. 1995. *Field Indicators for Identifying Hydric Soils in New England.* New England Interstate Water Pollution Control Commission, Wilmington, MA.

Tiner, R. W., and P. L. M. Veneman. 1995. *Hydric Soils of New England.* University of Massachusetts Extension, Amherst, MA. Bulletin C-183R.

Vasilas, L. M., and B. L. Vasilas (editors). 2004. *A Guide to Hydric Soils in the Mid-Atlantic Region.* U.S.D.A. Natural Resources Conservation Service, Morgantown, WV. http://www.epa.gov/reg3esd1/hydricsoils/book.htm.

Vepraskas, M. J. 1996. *Redoximorphic Features for Identifying Aquic Conditions.* North Carolina Agricultural Research Service, North Carolina State University, Raleigh, NC. Technical Bulletin 301.

CHAPTER 12

Identifying Wetland Wildlife

Selected animals found in Northeast and Midwest wetlands are illustrated and briefly described in this chapter. An exhaustive list would require several books, but the animals shown include many of the common or characteristic species. The emphasis is on the more conspicuous, larger organisms: vertebrates (those with backbones), specifically amphibians (salamanders, toads, and frogs), reptiles (turtles, lizards, and snakes), birds, and mammals. Fish are not included in this chapter since they are aquatic species and less easily observed than other vertebrates unless you're a good fisherman (see Figure 6.3).

The book would be incomplete without some mention and illustration of typical lower animals—invertebrates (those without backbones)—since the majority of wetland animals are insects, spiders, crustaceans, and molluscs. Thousands of invertebrates occur in wetlands. They range from tiny water fleas to large crustaceans like the horseshoe crab. Some species are easily recognized or abundant. A few common invertebrates that frequent salt marshes and freshwater wetlands (including pond margins) are illustrated (Figures 12.1–12.2). A list of butterflies found in wetlands is provided in Table 12.1. If you want to know more about these animals and other invertebrates, consult the field guides listed at the end of the chapter.

The following section deals with animal identification. It includes brief descriptions and line drawings to help identify many of the Northeast's more common or characteristic species. For a book of this kind, a thorough treatment of wetland animals is not appropriate, yet over a hundred species are covered. A set of simple keys is provided to aid in recognition.

How to Identify Wetland Animals

Some wetland animals, such as amphibians, reptiles, and invertebrates, can be observed close up, while viewing of more-motile species like birds and mammals usually requires the use of binoculars, which are standard field equipment. A set of keys is provided to help you identify wetland animals. The first key directs you to four other keys for specific groups of animals. Since only a few snakes are common in wetlands, they can be identified using the initial key. No lizards are common in the region's wetlands, so they are not represented. The four specific keys are

(1) Key A—Common Amphibians,
(2) Key B—Common Turtles,
(3) Key C—Common Birds, and
(4) Key D—Common Mammals.

Using these keys, you should be able to identify most animals throughout the year, recognizing that birds may pose problems due to changing plumage from summer (breeding season) to winter (nonbreeding season). The process of identification involves following the keys, reviewing referenced illustrations, and then consulting the brief descriptions of the particular species. Note that the illustrations and accompanying descriptions follow the keys. Each species has a unique number assigned to its drawing and description.

Like the hydrophyte keys, each key consists of a series of contrasting couplets. Begin with couplet 1 and match the statements to the animal you are observing. Read both parts of the couplet to find the best fit. Going through the couplets in a key, you will eventually find a reference to one or more species (e.g., species 1 or species 2–6). Review the applicable numbered drawing(s) carefully, select the one most resembling the animal in question, then locate and read the species description in the text. Review the characteristics of the species, and pay close attention to the "Similar species" section, which discusses animals that are related or similar in appearance to the illustrated and described species. One or two of these similar species may be illustrated; they can be located in the text by following the numbering code for the principal species. For ex-

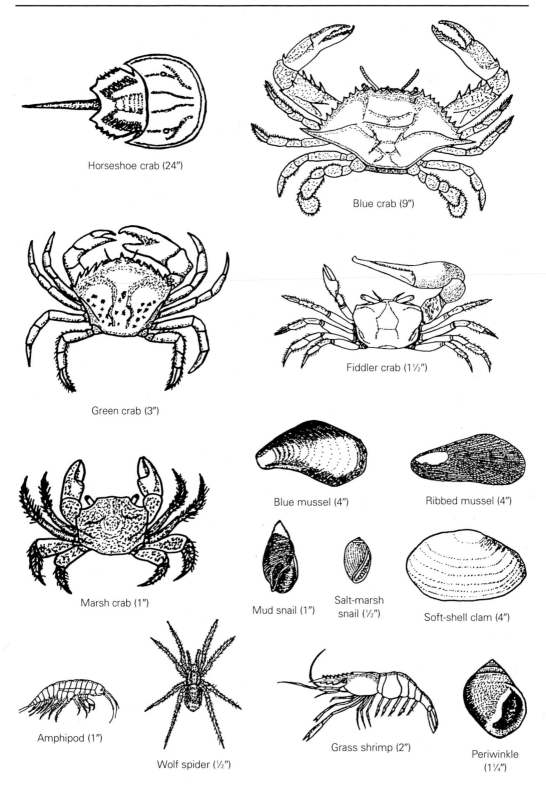

Horseshoe crab (24″)

Blue crab (9″)

Green crab (3″)

Fiddler crab (1½″)

Marsh crab (1″)

Blue mussel (4″)

Ribbed mussel (4″)

Mud snail (1″)

Salt-marsh snail (½″)

Soft-shell clam (4″)

Amphipod (1″)

Wolf spider (½″)

Grass shrimp (2″)

Periwinkle (1¼″)

Figure 12.1. Common salt- and brackish-marsh invertebrates. Approximate size is given in parentheses.

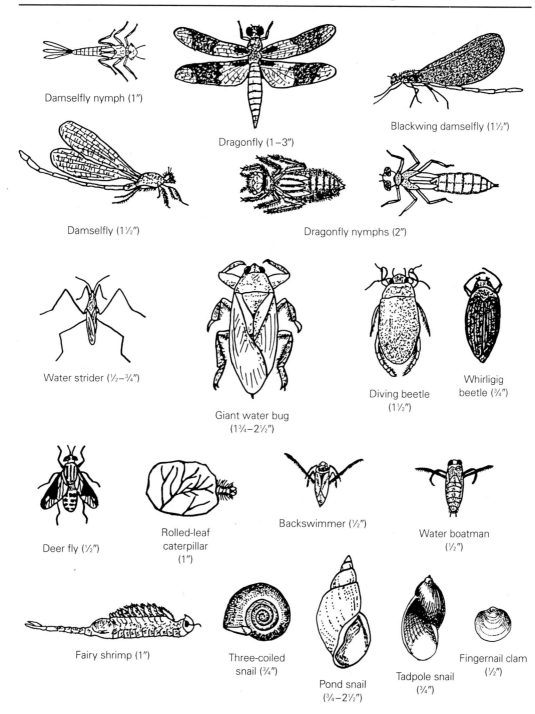

Damselfly nymph (1″)

Dragonfly (1–3″)

Blackwing damselfly (1½″)

Damselfly (1½″)

Dragonfly nymphs (2″)

Water strider (½–¾″)

Giant water bug (1¾–2½″)

Diving beetle (1½″)

Whirligig beetle (¾″)

Deer fly (½″)

Rolled-leaf caterpillar (1″)

Backswimmer (½″)

Water boatman (½″)

Fairy shrimp (1″)

Three-coiled snail (¾″)

Pond snail (¾–2½″)

Tadpole snail (¾″)

Fingernail clam (½″)

Figure 12.2. Common freshwater invertebrates. Approximate size is given in parentheses.

ample, leopard frog (species 13b), an illustrated similar species, is described under pickerel frog (species 13). If you cannot identify the animal following this procedure, review the keys again. If you still cannot find it, consult one of the other field guides listed at the end of the chapter.

More-detailed descriptions of the wetland animals can be found in the specialized field guides and other works listed at the end of the chapter. For photographs of these and other wildlife, consult the National Wildlife Federation's online field guides (http://enature.com)

Table 12.1. Butterflies found in Northeast wetlands.

Wetland Type	Associated Butterflies
Salt and brackish marshes	Aaron's Skipper, Salt Marsh Skipper, Broad-winged Skipper, and Monarch (during migration)
Inland marshes	Red Admiral, Painted Lady, Black Swallowtail, Cabbage Butterfly, Bronze Copper, Acadian Hairstreak, Regal Fritillary, Silver-bordered Fritillary, Harris Checkerspot, Pearl Crescent, Baltimore, Hop Merchant, Eyed Brown, Monarch, Persius Dusky Wing, Least Skipper, Peck's Skipper, Long Dash, Delaware Skipper, Mulberry Wing, Broad-winged Skipper, Dim Skipper, Black Dash, and Two-spotted Skipper
Wet meadows	Bronze Copper, Aphrodite Fritillary, Regal Fritillary, Silver-bordered Fritillary, Meadow Fritillary, Harris Checkerspot, Red Admiral, Eyed Brown, Delaware Skipper, Black Dash, Ocola Skipper, Two-spotted Skipper, and Mitchell's Satyr
Bogs	Bronze Copper, Brown Elfin, Silver-bordered Fritillary, Baltimore, Common Wood Nymph, Mulberry Wing, Hobomok Skipper, Dion Skipper, and Two-spotted Skipper
Shrub and forested swamps	Zebra Swallowtail, Tiger Swallowtail, Spicebush Swallowtail, Palamedes Swallowtail, Falcate Orange Tip, Harvester, Striped Hairstreak, Henry's Elfin, American Snout, Question Mark, Hop Merchant, Mourning Cloak, Viceroy, Tawny Emperor, Appalachian Eyed Brown, Confused Cloudy Wing, Horace's Dusky Wing, Clouded Skipper, Hobomok Skipper, Dion Skipper, Dun Skipper, Roadside Skipper, and Hessel's Hairstreak (Atlantic-white-cedar swamps)

Note: Common names follow Opler and Malikul (1992). Consult a field guide for identification.

Keys to Selected Wetland Vertebrates of the Northeast

1. Animal lacking hair 2
  2. Skin warty or smooth; shell, scales, or feathers lacking . . . . Key A
  2. Skin covered with shell, scales, or feathers 3
    3. Shell present . . . . Key B
    3. Shell lacking 4
      4. Body covered with scales . . . . Species 23 and 24
      4. Body covered with feathers . . . . Key C
1. Animal covered with hair . . . . Key D

Key A—Common Amphibians

1. Animal with tail 2
  2. Four toes on each foot . . . . Species 1
  2. Five toes on each foot 3
    3. Body marked with spots or blotches 4
      4. Spots present . . . . Species 2–6
      4. Blotches present . . . . Species 6
    3. Body marked by conspicuous lines . . . . Species 7
1. Animal lacking tail 5
  5. Skin warty with large parotoid glands behind eyes . . . . Species 8
  5. Skin smooth or if warty lacking parotoid glands 6
    6. Hind foot with sickle-shaped appendage ("spade") . . . . Species 9
    6. Hind foot lacking this appendage 7

    7. Feet with sticky toe pads; body small, usually 2 inches or less 8
        8. Distinctive X-mark on back . . . . Species 10
        8. Not so marked . . . . Species 11
    7. Feet lacking sticky toe pads; hind feet webbed 9
        9. Small brownish frog (less than 3½ inches
           long); head with black stripe through eye . . . . Species 12
        9. Larger frog 10
          10. Body marked with spots or blotches 11
             11. Skin brownish with dark brown squarish
                blotches; underside of legs yellowish orange . . . . Species 13
             11. Skin not so marked 12
                12. Skin brownish or green with conspicuous
                    round spots, and legs also marked with spots . . . . . Species 13b
                12. Skin green with small spots or many blotches 13
                  13. Prominent ridges extend from behind eyes . . . . Species 14
                  13. Ridges lacking . . . . Species 15
          10. Body greenish or brownish, not marked with spots
            or blotches . . . . Species 15

Key B—Common Turtles

1. Found in salt marshes 2
    2. Less than 1 foot long; shell with distinctive sculpted design,
      sides of shell with dark U- or O-shaped markings on yellow
      background . . . . Species 16
    2. Larger; rear margins of shell strongly toothed; underside of
      shell (plastron) small, with much of body exposed; be careful
      handling this one . . . . Species 17
1. Found in freshwater wetlands and bodies of water 3
    3. Rear margins of shell strongly toothed; underside of shell small,
      with much of body exposed; be careful handling this one . . . . Species 17
    3. Shell not so 4
      4. Shell marked with spots . . . . Species 18
      4. Shell not spotted 5
        5. Exudes musky odor when handled . . . . Species 19
        5. Does not do so 6
          6. Margins of shell marked with red-and-black pattern . . . . Species 20
          6. Shell not so marked; shell with sculpted pattern 7
            7. Large yellow or orange blotch behind eye . . . . Species 21
            7. No blotch on head; neck and front legs reddish orange . . . . . Species 22

Key C—Common Birds

1. Feet webbed or padded; usually observed swimming 2
    2. Waterfowl (swan, goose, duck, or merganser) . . . . Species 25–40
    2. Not waterfowl 3
      3. Aquatic bird (cormorant, grebe, moorhen, or coot) . . . . Species 41–43
      3. Sea gull . . . . Species 59–60
1. Feet not webbed or padded; usually observed walking or perching 4
    4. Very small (about 3½ inches long), with long needlelike bill
      and rapid wing beat; usually observed flying around flowers . . . . Species 67
    4. Bird not so 5
      5. Legs very long; typically wading in shallow water (wading bird) 6
        6. Body white . . . . Species 44, 45,
                                    and 50b

6. Body not white 7
  7. Body mostly grayish, bluish, or greenish; may
    have a black cap . . . . Species 46–48
                                            and 51
  7. Body brownish . . . . Species 49–50
5. Legs short 8
  8. Usually found walking along shores (at water's edge)
    or in wetlands, or resting on beach 9
    9. Usually walking along water's edge in search of
      food (shorebird) . . . . Species 51–55
    9. Found resting on beaches or in wetland vegetation 10
      10. Usually observed in wetland vegetation and
        sometimes flushed (marsh bird) . . . . Species 55–58
      10. Usually observed diving into water to catch fish
        or resting on beaches; body black, white, and gray . . . . Species 61
  8. Usually found perching on plants 12
    12. Large to medium-sized (body longer than 1 foot) 13
      13. Hawk or owl . . . . Species 62–66
      13. Woodpecker or kingfisher . . . . Species 68–69
    12. Smaller (body usually less than 1 foot long) 14
      14. Woodpecker . . . . Species 68
      14. Not woodpecker 15
        15. Usually observed feeding on flying insects
          over marshes . . . . Species 70
        15. Usually observed perching on vegetation 16
          16. Typically perched on marsh vegetation 17
            17. Body length about 6 inches or less 18
              18. Bill thin; observed mostly in
                cattail marshes and wet meadows . . . . . . Species 71
              18. Bill stout; breast streaked or not . . . . Species 72–74
                                  and 89b
            17. Body longer than 6 inches; males mostly
              black; females usually brownish . . . . Species 75–77
          16. Typically perched on woody plants 19
            19. Body brownish above, with
              streaked or spotted breast 20
              20. Body length 6 inches or less 21
                21. Bill short and stout . . . . Species 73–74
                21. Bill not noticeably short or stout . . . Species 78–80
              20. Body longer than 6 inches . . . . Species 81
            19. Body not mostly brownish above or
              breast not streaked; mostly small 22
              22. Mostly grayish above; no yellow
                markings except on tip of tail 23
                23. Bill not stout and/or no eye stripe . . . Species 82–84
                23. Bill noticeably stout;
                  dark eye stripe . . . . Species 85–86
             22. Not so colored 24
              24. Red or brown with black wings . . . . Species 87
              24. Not so colored 25
                25. Yellow body or markings 26
                  26. Body mostly yellow;
                    wings dark . . . . Species 88–89

Key D—Common Mammals

and the Patuxent Bird Identification Infocenter (http://www.mbr-pwrc.usgs.gov/Infocenter/infocenter.html). Indeed, I hope that the information here will stimulate a desire to learn more about wildlife.

Illustrations and Descriptions of Wetland Vertebrates

Selected wetland animals are briefly described and illlustrated by line drawings. Each species has been assigned a unique number that cross-references illustration and description. The descriptions include the common name of the animal, its scientific name, a brief overview of some major distinguishing characteristics (e.g., body length, color patterns, etc.), habitats, and range (in the East). Related species that may occur in wetlands are briefly described under "Similar species." More-detailed descriptions and color photographs or drawings can be found in various field guides listed at the end of the chapter. These books will also be useful for identifying animals not included in this guidebook.

Amphibians

Salamanders

1. Four-toed Salamander (*Hemidactylium scutatum*): to 4 inches long, four toes on front and hind feet, reddish brown to black back, black-speckled white belly; breeds from August into October, eggs also reported in March in Maryland; secretive (hard to find), wet moss of peat bogs, vernal pools, and slow-moving bog streams; central Maine to Virginia, in mountains to Georgia.

2. Blue-spotted Salamander (*Ambystoma laterale*): to 5 inches long, with dark blue, dark gray, or blackish back mottled with bluish white specks (body looks brown in water; colors appear only when exposed to air); larvae olive green to black, with exposed gills and a long top fin from back of head to tail; breeds in March and April in pools less than 1½ feet deep; deciduous forests and swamps; Newfoundland south to northern New Jersey and eastern Pennsylvania, in mountains to West Virginia. Similar species: Jefferson salamander (*A. jeffersonianum*) is dark brown or brownish gray, with bluish specks on lower sides and legs; southern Vermont to western Maryland, in eastern mountains to Virginia.

3. Spotted Salamander (*Ambystoma maculatum*): to nearly 10 inches long, grayish brown, bluish black, or black body with two rows of conspicuous yellow to orange spots on back; larvae greenish yellow with exposed gills; breeds from February to April, returning to the same (natal) pool to breed; deciduous or mixed mesic woods around vernal pools or streams; Nova Scotia to Georgia and Louisiana, absent from much of coastal plain. Similar species: Eastern Tiger Salamander (*A. tigrinum*), a coastal-plain species, is dark with yellowish olive spots and blotches; due to its scarcity, it is a species of concern in some states.

4. Red-spotted or Eastern Newt (*Notophthalmus viridescens*): to about 5½ inches long, larvae and adults aquatic, terrestrial "red eft" stage; larvae and adults olive green to yellowish brown with black and red spots (larvae have exposed gills), eft red, orange or reddish brown with black and red spots (lives on land for 2–7 years and grows to 3½ inches long); breeds in spring (February to May); ponds, shallows of lakes and some reservoirs, and slow-moving sections of streams (for red eft, upland forests near breeding areas); Nova Scotia to Florida.

5. Dusky Salamander (*Desmognathus fuscus*): to 5½ inches long, dark brown to blackish with mottling, lower part of sides has salt-and-pepper look, belly light gray; larvae with exposed white gills and dark skin, with pairs of whitish spots in a gray band on back; breeds from summer into spring; moist woods along streams, springs, and seeps; New Brunswick to northern New Jersey, in Piedmont and mountains to Louisiana.

6. Marbled Salamander (*Ambystoma opacum*): to 5 inches long, black to dark gray body, with distinctive whitish to silvery markings on back (juveniles with tiny white spots); breeds in September and October in Northeast, until December in Southeast; forested wetlands and adjacent woods near breeding pools; southern New Hampshire and Massachusetts to northern Florida. (Note: Known to breed in flooded sand and gravel pits in Cape May County, New Jersey, as do Tiger Salamanders).

7. Two-lined Salamander (*Eurycea bislineata*): to about 4¾ inches long, yellowish back (often with many dark specks) having two dark stripes running from eye to tail (one on each side), wiggles like snake when disturbed by overturning rocks; larvae have small exposed gills (lightly colored) and two rows of paired spots on back; breeds from fall into early spring; river swamps, floodplain forests, moist woods, rocky seeps, and shores of streams and rivers; Nova Scotia to northern Florida. Similar species: Spring Salamander (*Gyrinophilus porphyriticus*) occurs in similar habitats, but is quite different in appearance—larger (to 8½ inches long) and more colorful (pinkish orange or salmon-colored, with darker mottling pattern); Nova Scotia to northern New Jersey, in mountains to Georgia.

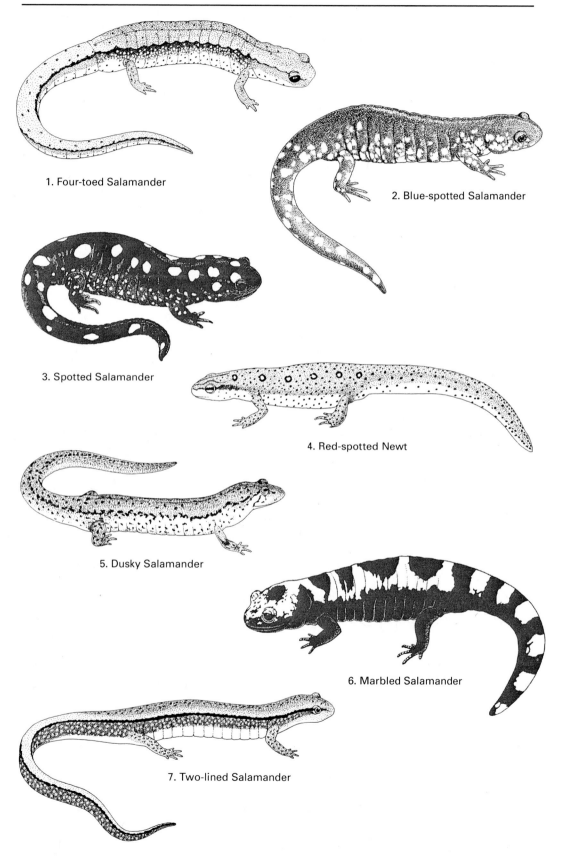

1. Four-toed Salamander

2. Blue-spotted Salamander

3. Spotted Salamander

4. Red-spotted Newt

5. Dusky Salamander

6. Marbled Salamander

7. Two-lined Salamander

Toads and Frogs

8. American Toad (*Bufo americanus*): less than 4½ inches long, brown to grayish to black skin covered with many rounded to spiny warts, two bean-shaped (parotoid) glands behind eyes (glands do not touch bony cranial crest), produces skin secretions upon handling (does not produce warts in humans); tadpoles blackish with rounded end of tail; call is high-pitched trill lasting ½ minute or longer; breeds in shallow water from March into July; all habitats, except salt marshes and estuarine waters; Newfoundland to Delaware, in eastern mountains to Georgia. Similar species: Fowler's Toad (*B. woodhousei fowleri*) is a lighter-colored toad, yellow, green or brown, with parotoid glands touching the cranial crest; its call is much different, thought by some to resemble the bleeting of a sheep.

9. Eastern Spadefoot Toad (*Scaphiopus holbrooki*): to about 3¼ inches long, eyes with vertical pupils, olive to blackish skin, mostly smooth with many scattered bumps, and hind foot with distinctive sickle-shaped spade; breeds in vernal pools in sandy and gravelly soils, lives in adjacent dry woods and thickets, also on floodplains; Massachusetts to Florida; call is a low-pitched crowlike caw.

10. Spring Peeper (*Pseudacris crucifer*, formerly *Hyla crucifer*): less than 1½ inches long, smooth skin brownish to gray, with conspicuous dark cross or X-mark on back, dark V-shaped marking between eyes, yellowish to grayish white belly, male throat yellowish or dark with yellow specks (when breeding); large toe pads; tadpoles less than 1¼ inch long, with green-orangish back, and tail covered with dark spots and blotches on outer edge; call is high-pitched peep ending on upward note, repeated as much as twenty-five times per minute; breeds from April into May; woods along ponds, marshes, and swamps, and swamps surrounding vernal pools; Nova Scotia to northern Florida.

11. Gray Tree Frog (*Hyla versicolor*): to about 2 inches long, warty skin, gray to green to light brown (can change color depending on temperature, humidity, and light), somewhat star-shaped pattern on back, belly white, males have dark throat, females and young adults have few black specks on throat; large toe pads; tadpoles dark green to black, with reddish tailfins; call is short, loud trill, usually lasting less than ½ second, sometimes to 3 seconds; breeds from May to August; shrubs and trees along shallow water, usually perched on branches; Maine to northern Florida. Similar species: Northern Cricket Frog (*Acris crepitans*), a smaller tree frog (to 1½ inches long) occurring from southern New York to Georgia, has greenish brown, yellow, red or black rough skin, without any distinctive markings on back, but has dark triangle patch on top of head between its eyes, and its legs are marked by three thick, dark bands; lacks toe pads; breeds from April to August; call is rapid, repeating clicking sound. This small frog has a distinctive zig-zagged hopping pattern that makes it difficult to catch. Green Tree Frog (*H. cinerea*), a southern-coastal-plain species (to 2½ inches long) ranging northward into Delaware, is bright green with a white stripe extending from the upper jaw along the side of the body; large toe pads; breeds April to September; call is said to sound like a cowbell from afar, but more like a "quank quank" close up. Pine Barrens Tree Frog (*H. andersonii*), a rare coastal-plain species (to 2 inches long) found in the New Jersey Pine Barrens, is also bright green but has a lavender stripe extending from snout through eye and along side; breeds from April to August; call is quiet nasal honk, repeated about every second. (See also Chorus or Swamp Cricket Frogs listed under Wood Frog, species 12.)

12. Wood Frog (*Rana sylvatica*): less than 3½ inches long, smooth skin, brownish to pinkish tan, with distinctive dark mask extending from nose through eye to back of jaw (behind eardrum), belly clear or whitish; tadpoles less than 1½ inches long, with creamy line on upper jaw and iridescent pinkish bronze belly; call is hoarse, ducklike quacking sound; breeds March and April; wooded swamps, abandoned oxbows, and upland woods near breeding pools; hibernates in rotten logs and leaf litter, and under rocks and moss in woods; Labrador to New Jersey, in eastern mountains to Georgia. Similar species: Pine Barrens Tree Frog (*Hyla andersonii*) also has a dark mask (a lavender stripe from the tip of the snout across the eyes and extending along the sides), but its body is mostly green; a rare coastal-plain species found in the New Jersey Pine Barrens, it breeds from April to August; its call is a quiet nasal honk repeated every second. Upland Chorus Frog (*Pseudacris feriarum feriarum*) and New Jersey Chorus Frog (*P. feriarum kalmi*) (also called Swamp Cricket Frogs; to 1½ inches) have smooth brown or gray

8. American Toad

9. Eastern Spadefoot Toad

10. Spring Peeper

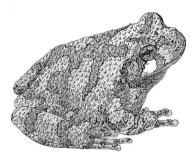

11. Gray Tree Frog

12. Wood Frog

13a. Pickerel Frog

13b. Leopard Frog

14. Green Frog

15. Bull Frog

skin, a dark stripe running through its eyes but extending to its hind legs, small toe pads, a characteristic light-colored upper lip, and either three dark stripes or rows of dark spots on the back; their calls resemble the sound made from running your fingernail slowly along the teeth of a plastic comb. The former subspecies has its back stripes broken up into dark streaks or rows of dark spots; it occurs from Pennsylvania to northern Florida. The latter subspecies, found along the coastal plain from Staten Island, New York through the Delmarva Peninsula, has thick dark stripes on its back.

13. Pickerel Frog (*Rana palustris*): less than 4 inches long, smooth brownish skin, with squarish to rectangular spots arranged in two rows on back between two bony ridges (extending along sides of back from behind eyes) and yellowish orange undersides of hind legs; call sounds like the creak of a slowly opening door; breeds March to May; wet meadows, grasslands, and moist woods; returns to water to hibernate and breed in marshes and vegetated shallow waters of slow-moving streams, ponds, and lakes; Nova Scotia to South Carolina, in mountains to Georgia and Alabama. (Note: Skin secretes irritating substance that may be toxic to other frogs.) Similar species: Leopard Frogs (*R. pipiens* and *R. sphenocephala*) are green or brown, smooth-skinned meadow frogs with two or three rows of dark irregular-shaped (roundish) spots along back between the bony ridges.

14. Green Frog (*Rana clamitans*): less than 4½ inches long, smooth skin green to greenish brown, with many dark brown to grayish spots or mottles, legs dark-striped, back with prominent raised lines extending from eye to rear of body, male eardrum larger than eye; tadpoles to about 2½ inches long, having olive green skin with dark spots, green tail with brown mottles, and iridescent cream-colored belly; call sounds like plucking a loose string of a banjo; breeds from March to August; marshes, vegetated shores of ponds, lakes, and slow-moving streams, springs, and nearby moist woods; Labrador and Nova Scotia to central Florida.

15. Bull Frog (*Rana catesbeiana*): to 8 inches long, smooth skin, greenish to brownish with or without dark specks or mottles, whitish belly, male eardrum larger than eye and same size as eye in female; tadpoles green to brown, with black-spotted tail (first year to 3½ inches, second year to over 5½ inches with hind legs);

call is a low-pitched "rahhor-room" or "jug-o-rum," repeated; breeds May to August; marshes and vegetated shores of ponds, lakes, and slow-moving streams; Nova Scotia to central Florida.

Reptiles

Turtles

16. Northern Diamondback Terrapin (*Malaclemys terrapin terrapin*): estuarine turtle to less than 9 inches long, shell with distinct keels, grooves and usually sculpted pattern on each plate (scute), and head, neck, and legs usually grayish with many small dark spots; hibernates in mud of tidal embayments in winter; salt and brackish marshes, tidal flats, and estuaries (tolerant of pollution); Cape Cod, Massachusetts, through Florida (different subspecies in southern waters). Similar species: Snapping Turtle (*Chelydra serpentina*) occasionally enters estuarine marshes and waters; it has a distinctive saw-toothed tail.

17. Snapping Turtle (*Chelydra serpentina*): weight to 70 pounds or more, shell length to 20 inches, shell blackish brown to dark reddish brown, with rear of shell strongly toothed, underside of shell (plastron) yellowish to olive and small, with much of body exposed, warty skin, and saw-toothed tail; marshes, natural and artificial ponds, lakes, and streams, including estuarine wetlands and tidal streams; Nova Scotia through Florida. (Caution: This turtle gets its name from its snapping behavior when on land. Be careful: if you must pick it up, do so by holding the tail with the plastron facing you. Also note that in water, you are not likely to be bitten.)

18. Spotted Turtle (*Clemmys guttata*): to 5 inches long, shell smooth and black, with many small yellowish spots (number of spots unique to each individual), head and neck with yellow to orange spots, males with tan chins and females with yellow chins and orangish eyes; marshes, wet meadows, ponds, forested wetlands, bogs, and slow-moving streams; central Maine to central Florida. Similar species: Blanding's Turtle (*Emydoidea blandingi*), a midwestern species, is a rare turtle in New Hampshire, Massachusetts, and eastern New York; it grows to 10½ inches and has a black shell covered with hundred of tiny spots.

19. Stinkpot or Musk Turtle (*Sternotherus odoratus*): less than 5 inches long, exuding pungent to musky odor when handled, shell oval,

16. Northern Diamondback Terrapin

17. Snapping Turtle

18. Spotted Turtle

19. Stinkpot or Musk Turtle

20. Painted Turtle

21. Bog Turtle

22. Wood Turtle

smooth, and black to brownish to grayish, plastron (underside of shell) yellowish and small, with much exposed skin (more exposed in males), two distinctive yellow stripes along neck and head, fleshy wartlike projections (barbels) on chin and throat; ponds, shallow lakes, slow-moving streams, marshes, and cranberry bogs; central Maine through Florida. Similar species: Eastern Mud Turtle (*Kinosternon subrubrum*) occurs in brackish and fresh marshes and waters from southwestern Connecticut to northern Florida; its head and neck are marked by yellowish spots or streaks and not by yellow striping, and its plastron is large and double-hinged, lacking the exposed skin.

20. Painted Turtle (*Chrysemys picta*): to about 7 inches long, shell smooth and dark brown to black, with sides marked with red-and-black pattern, margins of shell plates (scutes) yellowish, plastron yellowish, head and neck with yellow and red stripes and two distinct yellow spots behind eyes; marshes, bogs, ponds, lakes, and slow-moving streams; Nova Scotia to North Carolina, in eastern mountains to Georgia. Similar species: Red-bellied Turtle (*C. rubriventris*) has plastron with yellowish center and reddish margins, and shell marked with red mottles.

21. Bog Turtle (*Clemmys muhlenbergii*): usually less than 4 inches long, shell somewhat longer than wide, and brown with sculptured pattern in plates (scutes), head with distinct single (sometimes split into two) yellow to orange blotch behind eye; marshes, wet meadows, fens, bogs, and slow-moving streams bordered by meadows, often associated with calcareous regions; western Massachusetts and eastern New York to central Maryland, in mountains to Georgia (disjunct populations).

22. Wood Turtle (*Clemmys insulpta*): to 9 inches long, the best-climbing turtle, shell with raised plates (somewhat pyramid-shaped), underside (plastron) yellow with black blotches along margins, neck and front legs reddish orange; swamps, marshes, wet meadows, cool streams, woods, and farm fields; Nova Scotia to northern Virginia.

Snakes

23. Northern Water Snake (*Nerodia sipedon*): usually 2 to 3 feet long (to over 4 feet), body black to variably crossbanded (reddish, brown, gray, and blackish brown) with light belly (juveniles more brightly colored; older specimens usually black), crossbands appear to be widest on back and narrower on sides, head not noticeably larger than neck, eyes have round pupils; rivers, streams, lakes, marshes, and swamps; southern Maine to North Carolina. (Note: Water snakes dive underwater when startled. Regrettably, these nonpoisonous snakes are often mistaken for poisonous cottonmouths or water moccasins and killed.) Similar species: The poisonous Copperhead (*Agkistrodon contortix*) lacks rattles, but otherwise has typical pit-viper-family characteristics: 1) heat-sensing pit between eyes and nostrils, 2) eyes with vertical pupils, and 3) head much wider than neck, plus a copper- to pinkish-colored body with dark brown crossbands (noticeably narrower on back versus sides) and markings on head; uplands and swamps; active mainly at night in summer and during day in spring and fall. The Timber Rattlesnake (*Crotalus horridus*) has a large, prominent wedge-shaped head, typical viper characteristics, a rattle, a yellow, brown, gray, or black body with mottled or crossband pattern, and a black tail; mainly in upland habitats such as forests with rocky soils, but also frequents swamps and stream banks, and hibernates in Atlantic-white-cedar swamps in New Jersey Pine Barrens and rocky areas elsewhere; nocturnal during summer, and diurnal in spring and fall.

24. Eastern Ribbon Snake (*Thamnophis sauritus*): garter snake to nearly 3½ feet long, with three distinct light-colored stripes on dark back (yellow or orangish stripe) and sides (yellow stripe), side stripes on third and fourth scale rows, and long tail (about ⅓ total length); wet meadows, swamps, bogs, marshes, stream banks, and ponds; southern Maine and central New Hampshire to Florida. (Note: When disturbed near water, ribbon snakes swim and do not dive.) Similar species: Eastern or Common Garter Snake (*T. sirtalis sirtalis*): typically around 2 feet long (to over 4 feet), the most abundant snake in the Northeast, occurs in similar places; usually three stripes, but yellow side stripes are on second and third scale rows, and dark areas between stripes usually have a double row of darker spots, giving these areas a checkered appearance; can hibernate underwater beneath rocks in streambeds. The Brown Snake (*Storeria dekayi*) also frequents wetlands from southern Maine to Florida; it is a small grayish to brownish snake (less than 2 feet long) with two rows of dark spots along its back. The

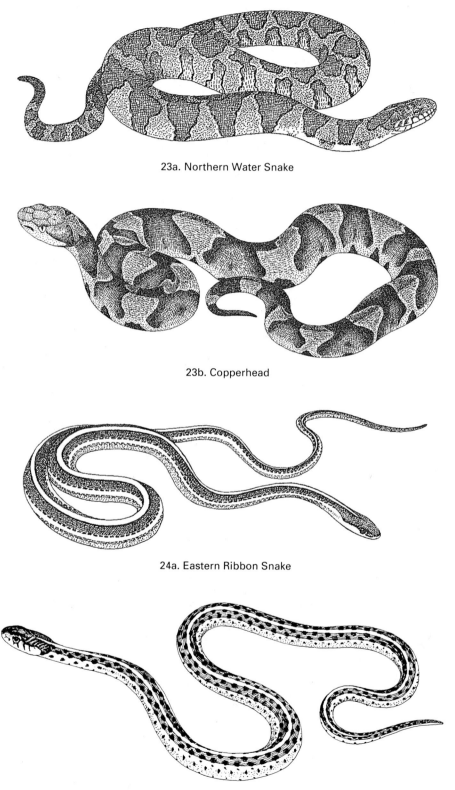

23a. Northern Water Snake

23b. Copperhead

24a. Eastern Ribbon Snake

24b. Eastern Garter Snake

Smooth Green Snake (*Opheodrys vernalis*) is another small snake that can be found in bogs and marshes to Nova Scotia.

Birds

Waterfowl

25. Mute Swan (*Cygnus olor*): over 4 feet long, white body, long neck, and orangish to pinkish bill with prominent black knob extending to forehead; feeds on plants, fish, crustaceans, and aquatic insects; year-round resident; tidal and nontidal waters (near coast), tidal and nontidal marshes, and ponds; New Hampshire to Virginia (European introduction). (Note: Very territorial, aggressive during breeding season, when it hisses at people.)

26. Snow Goose (*Chen caerulescens*): less than 3 feet long, white body, black wing tips on undersides (best seen in flight); eats plants, especially underground tubers and roots of smooth cordgrass (*Spartina alterniflora*); winter resident; tidal marshes; New Jersey to North Carolina. (Note: Winter flocks feed so heavily on cordgrass that they leave many open mud flats; quite destructive, and a problem for certain salt-marsh restoration projects.)

27. Canada Goose (*Branta canadensis*): from 2 to 4 feet long, black head, neck, and bill, white cheeks, chin, and breast, and brownish wings and body; eats wetland plants and upland lawn and park grasses (often becoming a nuisance, defecating on yards); now mostly year-round resident (almost domesticated), some populations still migratory; coastal and inland waters and marshes; Labrador to Florida. (Note: Flocks fly in characterisitc V-formations.) Similar species: Brant (*Branta bernicla*), a smaller dark goose (about 2 feet long) with black head, neck, and bill, white marking on neck, white belly and rump, and dark brown back, is a winter resident in coastal waters, where it feeds on marine and estuarine plants, especially eelgrass (*Zostera marina*) and sea lettuce (*Ulva lactuca*); flocks of hundreds are common sights.

28. Wood Duck (*Aix sponsa*): about 1½ feet long; male brightly colored, with green-and-white head bearing distinctive crest, red eyes, multicolored bill (black tip and top, white sides, and orangish to reddish base), brown-and-white checkered breast; female brownish, with dark bill and white eye ring; feeds mostly on plants (seeds of smartweeds, arrow arum, wild rice, pondweeds, and elm, as well as acorns) but also eats fish, frogs, and insects; summer resident in most of region; wooded streams, swamps (especially beaver swamps, where it nests in cavities of dead trees or in man-made nesting boxes), ponds, and inland waters; Nova Scotia to Florida (winters New Jersey south). (Note: Usually seen in mating pairs.)

29. Black Duck (*Anas rubripes*): about 2 feet long, dark brown body, lighter head and neck; male with yellowish to olive bill; female with greenish gray bill; feeds on plants (rootstocks of smooth cordgrass and pondweeds, seeds of smartweeds, wild rice, and pondweeds) and animals (snails, molluscs, crustaceans, insects, and amphibians), especially the salt-marsh snail (*Melampus bidentatus*); year-round resident, also winter resident in coastal marshes (mostly between Long Island and North Carolina); tidal and nontidal marshes and waters; Labrador to North Carolina, to Florida in winter. Similar species: Female mallard looks similar but has dark blue speculum (inner-wing patch) with a white border, a whitish tail, and an orangish and black bill, in contrast to the violet speculum and different-colored bills of the blacks.

30. Mallard (*Anas platyrhynchos*): about 2 feet long; male with conspicuous green head (giving name "greenhead"), thin white collar, yellow bill, brownish breast, and grayish body; female brownish with orangish and black bill and whitish tail; eats mostly wetland plants (e.g., smartweed, bulrush, rice cut-grass, and wild-rice seeds) but also insects and occasionally fish; year-round resident; tidal and nontidal marshes and waters; Nova Scotia to North Carolina, to Florida in winter. (Note: The most common duck in the Northeast.)

31. Blue-winged Teal (*Anas discors*): slightly more than 1 foot long; male with gray head, white crescent-shaped patch in front of face, black bill, and dark body (light blue forewing patches evident in flight); female mottled brownish color, with black bill and small light patch on face behind bill; feeds on wetland plants, especially duckweeds and bulrush and smartweed seeds; summer resident and early migrant; tidal and nontidal marshes and waters; Nova Scotia to North Carolina, to Florida in winter. Similar species: Green-winged teal (*Anas crecca*) female resembles female blue-winged, but has green speculum (wing patch) seen when flying, and head has black streak extending from bill through eye and lacks light

25. Mute Swan

26. Snow Goose

27. Canada Goose

28. Wood Duck (male, female)

29. Black Duck

30. Mallard

31. Blue-winged
Teal

patch on face; male of green-winged is distinctly different, with reddish brown head marked with large green patch extending from back of neck through eye; winter resident along coast from Massachusetts to Florida, mostly from New Jersey south, in spring migrates to breeding grounds (from northern Maine north); one of the earliest spring migrants.

32. American Wigeon (*Anas americana*): less than 2 feet long; male with grayish head and neck, white forehead and crown (giving it one of its other names, "baldpate"), green patch from back of neck to front of eye, black-tipped, light gray bill, white wing patches (seen in flight), white rump, and black tail; female with similar bill but overall brownish body, grayish head and neck, and black eye ring or dark patch around and behind eye; eats wetland plants, especially rootstocks of widgeon grass and seeds and rootstocks of pondweeds; winter resident; tidal marshes and waters, nontidal marshes, wet meadows, and waters near coast; Connecticut and Rhode Island to Florida.

33. Northern Pintail (*Anas acuta*): about 2 feet long, with conspicuous long upward-curving tail (hence the name "pintail"); male with dark brown head marked by white crescent-shaped line extending from white neck and narrowing toward back of head, black-and-gray bill, white breast, brownish gray body, and black tail; female with similar body shape, but body mottled light and dark and bill dark gray; eats plants and animals; winter resident; tidal marshes and waters; Massachusetts to Florida (mostly New Jersey south). Similar species: Oldsquaw (*Clangula hyemalis*), a sea duck with a long black tail (male only), winters along the coast from Labrador to North Carolina; male has whitish head with dark patch on upper neck, black-and-white body, and bill marked with dark and light colors.

34. Canvasback (*Aythya valisineria*): less than 2 feet long; male conspicuous with reddish brown head, long sloping forehead, long stout black bill, red eye, black breast, whitish body, and black tail; female with similar head profile and black bill, brownish head, dark eye with light eye ring, neck, and breast, grayish body, and black tail; dives for wetland plants and aquatic animals (fish, insects, and crustaceans); winter resident; tidal and nontidal marshes and waters along coast; southeastern Massachusetts to Florida (mostly from New York to Virginia). Similar species: Redhead (*Aythya americana*)

male also has a reddish head, but lacks the sloping forehead of the canvasback, and its grayish bill has a prominent black hook at the tip; female has similar distinctive bill, but brownish body and head; winter resident from New Jersey south, summer resident to the north.

35. Ring-necked Duck (*Aythya collaris*): about 1½ feet long, with distinctive white ring across gray-and-black bill; male head, neck, breast, and back black, sides white, and eyes yellowish to golden; female with gray bill having white stripe, head grayish, eye with light eye ring, light patch between eye and bill, and body brownish with darker upper wings; dives for plants and animals; winter resident mostly: tidal and nontidal marshes and waters (rarely in salt marshes and saline waters); southern New England to Florida in winter, migrates to northern New England and Newfoundland for breeding in marshes, swamps, and bogs.

36. Lesser Scaup (*Aythya affinis*): about 1½ feet long; male with blackish, purple-tinged head, neck, and tail, bluish bill (hence the name "bluebill"), yellowish eye, grayish-and-whitish back, and white sides; female dark brown, with prominent light patch at base of dark bill; dives for submerged plants and aquatic invertebrates; winter resident; tidal marshes and waters, and nontidal ponds near coast; New Jersey to Florida. Similar species: Closely resembles Greater Scaup (*Aythya marila*), a winter coastal resident from Nova Scotia to South Carolina; the male's dark head is green-tinged (not easily seen).

37. Common Goldeneye (*Bucephala clangula*): about 1½ feet long; male with blackish, green-tinged head, neck, back, and bill, head with prominent white round patch between lower base of bill and eye, eyes yellowish (hence its name); female with brown head, golden eye, black bill, white collar, and grayish body; dives for aquatic invertebrates and plants; winter resident in most of region; tidal marshes and waters, and occasionally larger rivers; Newfoundland to Florida, mostly from Long Island to North Carolina in winter, migrates in spring to northern New England and farther north for breeding.

38. Bufflehead (*Bucephala albeola*): slightly more than 1 foot long; male with black-and-white head (white behind eyes), black bill, black back, and white lower neck, breast, and sides; female has dark head marked with white stripe behind eyes, bluish gray bill, dark gray back, and whitish to grayish sides; dives for fish,

32. American Wigeon

33. Northern Pintail

34. Canvasback

35. Ring-necked Duck

36. Lesser Scaup (male, female)

38. Bufflehead

37. Common Goldeneye

invertebrates, and aquatic plants; winter resident; tidal marshes and waters, and inland lakes and rivers; Newfoundland to Florida.

39. Ruddy Duck (*Oxyura jamaicensis*): less than 1½ feet long, with conspicuous upward-pointing stiff tail and stout broad bill; male with black-and-white head (white below eyes), body grayish and bill dark in winter (body reddish brown and bill bluish in summer); female brownish overall, with a light stripe extending from base of bill (just below the eye) to back of head; dives for wetland plants and aquatic invertebrates; winter resident; tidal marshes and waters; southern New England to Florida.

40. Common Merganser (*Mergus merganser*): about 2 feet long, with long hooked, serrated bill; male with dark green head and neck, orange to red bill, white breast, and dark back; female with reddish brown, crested head, white breast, and grayish body; dives for fish, but also eats invertebrates; winter resident along coast, year-round resident in southern New England, breeds from northern New England and New York north (nests in tree cavities or nest boxes); tidal and nontidal marshes and waters (winter), inland lakes, rivers, and ponds surrounded by forests (summer); Nova Scotia to Florida. (Note: Mergansers swim lower in the water than ducks, with a good portion of their body submerged.) Similar species: Red-breasted Merganser (*M. serrator*) is a winter coastal resident in estuaries; male with dark green head having prominent open crest, white neck collar, checkered white-and-black body, and dark back; female with brown head, front of neck white, nape of neck grayish, and body grayish. Hooded Merganser (*Lophodytes cucullatus*) is much smaller (about 1½ feet long) and has shorter dark bill; male has black head with white crest (when raised); female has grayish head with brownish crest (when raised); year-round resident in much of the Northeast; summer resident in most of New York and northeastern Pennsylvania; winters mostly south of Connecticut.

Other Aquatic Birds

41. Double-crested Cormorant (*Phalacrocorax auritus*): less than 3 feet long, blackish body, orangish throat, and yellowish bill having prominent hook at end of upper beak; dives for fish; winter resident; estuarine waters; New Jersey to Florida winter, to Labrador summer. (Note: Dries its spreading wings after diving; seems to have trouble taking off from water—

must run a long distance on water's surface before becoming airborne; flocks fly in lines or V-formations.)

42. Pied-billed Grebe (*Podilymbus podiceps*): about 1 foot or more long, ducklike, with brownish body and short, stout, light-colored bill marked with a dark or black stripe near middle; dives for frogs, small fish, crayfish, and other aquatic invertebrates; year-round resident (summer inland; winter inland and coastal); tidal and nontidal marshes and waters (prefers fresh water) winter, and inland ponds and marshes summer; Nova Scotia to Florida, and southern New England south in winter. Similar species: Red-necked Grebe (*Podiceps grisegena*) overwinters in coastal bays and estuaries; its winter plumage is grayish, and its bill is long-pointed and not black-striped.

43. Common Moorhen (*Gallinula chloropus*): just over 1 foot long, ducklike, with mostly black body (some white patches near rump and on wings), black head with prominent red forehead, red bill with yellow tip, and yellow legs; feeds mostly on vegetation, but also eats snails and insects; summer resident; freshwater marshes, ponds, and lakes; southern New England (and Connecticut River valley) and New York to Florida. Similar species: American Coot or Mud Hen (*Fulica americana*) is similar sized, but is dark gray with a white forehead and bill; breeds from New York north, winters along coast from southern New Jersey south. (Note: Seems to have difficulty flying from water, as evidenced by its apparent running on the surface for some distance before becoming airborne.)

Wading Birds

44. Great Egret (*Ardea alba*): tall white heron, body over 3 feet long, wing span over 4½ feet, with black legs and feet, and yellow bill; feeds on fish, frogs, and invertebrates; summer resident; tidal and nontidal marshes and flats, river and lake shores; southern New England along coast to Florida. Similar species: Other white egrets or herons are much smaller (see Snowy Egret).

45. Snowy Egret (*Egretta thula*): small white heron, body about 2 feet long, wing span about 3 feet, with black bill and legs, and yellow feet; eats mostly small fish and aquatic and marsh invertebrates; summer resident; tidal and nontidal marshes, river and lake shores; southern New England along coast to Florida. Similar

39. Ruddy Duck

40. Common Merganser

41. Double-crested Cormorant

42. Pied-billed Grebe

43a. Common Moorhen

43b. American Coot

44. Great Egret

45. Snowy Egret

species: Immature of Little Blue Heron (*E. caerulea*) is slightly more than 2 feet long, with a grayish bill having a dark blue to blackish tip, and yellowish to greenish legs; summer resident. Cattle Egret (*Bubulcus ibis*) is less than 2 feet long, with an orange to yellowish bill, and head topped by a buff-colored patch; more typical of agricultural fields (often found following a tractor or grazing with livestock and feeding on insects); summer resident.

46. Great Blue Heron (*Ardea herodias*): tall gray-blue heron, body over 4 feet long, wing span of nearly 6 feet, with long legs and neck, whitish head with black stripe extending from over eye into long crest, and light-colored bill; feeds mostly on fish and frogs, but also eats aquatic invertebrates, reptiles, small mammals, and small birds; builds large nest platform of twigs and branches in trees, may form impressive colonies; some year-round residents, especially near coast, but mostly summer residents; tidal and nontidal marshes and flats, river and lake shores, and forested wetlands; Nova Scotia to Florida, from Massachusetts south in winter. Similar species: Other dark-colored herons are smaller (slightly more than 2 feet long). Tricolored Heron (*Egretta tricolor*) has a dark blue-gray head and upper body, white belly, and neck that is dark in back and mostly whitish in front; year-round resident New Jersey south, summer resident to Maine. Little Blue Heron (*E. caerulea*) is gray-blue to purplish (when mature) with a purplish head, light green legs, and a grayish bill having a dark blue to blackish tip; summer resident.

47. Black-crowned Night Heron (*Nycticorax nycticorax*): stocky gray, white, and black heron, body about 2 feet long, with mostly white head topped by black cap, and black bill, whitish crest, black back, grayish wings, and yellow legs that turn red when breeding; feeds on fish, frogs, small mammals, and insects; year-round resident along coast, summer resident inland; tidal and nontidal marshes, inland swamps, lakes and rivers; Nova Scotia to Florida, from southern New England south in winter. Similar species: Yellow-crowned Night Heron (*Nyctanassa violacea*) is grayish also, but its head is alternately striped with white (top and crest), black (through eyes), white (cheek stripe), and black (chin); summer resident.

48. Green-backed or Green Heron (*Butorides virescens*): stocky small heron, body about 1½ feet long, dark grayish to blackish green or grayish blue above, with yellow stripe in front of eye, dark bill, reddish brown neck, white and reddish brown–striped breast, and yellow or orange legs; eats fish, snakes, small mammals, and aquatic invertebrates; summer resident; the most common inland heron.

49. Least Bittern (*Ixobrychus exilis*): the region's smallest heron, brown-and-white body about 1 foot long, with light-striped throat and dark back, top of head (crown), and patches on outer parts of wings in males; females lack dark colors; feeds on small fish, insects, small mammals, eggs, and young birds; builds nest of sticks in dense vegetation about 1 foot above water; points bill skyward when alarmed; a weak flyer; summer resident; tidal and nontidal marshes (rarely salt marshes); New Brunswick to Florida. Similar species: American Bittern (*Botaurus lentiginosus*) is twice the size of least bittern and is light and reddish brown–streaked overall, with dark brown and white–striped throat; summer resident in most of region, winters in brackish marshes along coast from New York south; tidal and nontidal marshes.

50. Glossy Ibis (*Plegadis falcinellus*): about 2 feet long, with long, decurved stout bill, dark body with brownish head and neck, and darker grayish green wings; eats invertebrates and snakes; summer resident; tidal and nontidal marshes and flats; Maine to Florida. Similar species: White Ibis (*Eudocimus albus*) is of similar size and body shape, but is white with reddish bill and legs when mature (immature has grayish brown head, neck, and back and white belly); summer resident.

Shorebirds and Marsh Birds

51. Greater Yellowlegs (*Tringa melanoleuca*): slightly more than 1 foot long, one of the larger shorebirds, dark gray above and white below, with long yellowish to orangish legs and long black bill; feeds on worms, snails, insects, and fish; coastal migrant and winter coastal resident; tidal marshes and flats; New Jersey to Florida (winter), breeds in northern Canada. Similar species: Lesser Yellowlegs (*T. flavipes*) is smaller (less than 1 foot long) and reportedly congregates in larger flocks; its call is a two-note "too-too," versus the Greater's loud three- or four-note call, "whew-whew-whew" or "whew-whew-whew-whew."

52. Willet (*Catoptrophorus semipalmatus*): more than 1 foot long, gray body, with

46. Great Blue Heron

47. Black-crowned Night Heron

48. Green-backed or Green Heron

49a. Least Bittern

49b. American Bittern

50. Glossy Ibis

51. Greater Yellowlegs

52. Willet

long gray legs and stout long bill, best recognized by its black-and-white-striped wing tips seen in flight; eats molluscs, crustaceans (e.g., fiddler crabs), small fish, and insects, and also seeds; very noisy, with distinctive call "pill-will-willet," which gives the bird its name; summer resident; tidal marshes, flats, and beaches; New Jersey to Florida, also in Nova Scotia.

53. Least Sandpiper (*Calidris minutilla*): grayish to brownish body about ½ foot long, with white belly and somewhat streaked breast (at least upper part), black bill pointing slightly downward, and yellowish to greenish legs; feeds on worms, crustaceans, insects, and molluscs; spring and fall coastal migrant; tidal marshes and flats; winters south of region, breeds in northern Canada. Similar species: Pectoral Sandpiper (*C. melanotos*) is very similar, but is almost twice as large (somewhat less than 1 foot long) and more brownish in color, with a much-streaked breast and yellow legs. Semipalmated Sandpiper (*C. pusilla*) has dark legs and a straighter bill with a conspicuous blunt tip. (Note: Many sandpipers, plovers, and other shorebirds are observed in the Northeast only during migration.)

54. Spotted Sandpiper (*Actitis macularia*): less than 1 foot long, with grayish to brownish head and upper body, light stripe through eye in winter, partial white eye ring in summer, white belly with dark spots, and orangish bill with blackish tip; eats insects, worms, crustaceans, and small fish; summer resident; shores of fresh bodies of water, upland fields, and coastal dunes; Newfoundland to Florida. (Note: Characteristic body movement—when walking, tail bobs up and down.)

55. Killdeer (*Charadrius vociferus*): slightly less than 1 foot long, with brown back and white belly, two distinctive black rings around neck (collar) and breast, brown-and-white-striped head, and black bill; feeds on insects, crustaceans, and worms; distinctive call "kill-dee-dee," giving its name; year-round resident in much of region, summer resident to north; wet meadows, pond shores, and upland fields including mowed areas; Labrador to Florida in summer, southeastern Massachusetts south in winter. (Note: Well known for female's broken-wing act to distract predators from nest or nestlings.)

56. Common Snipe (*Gallinago gallinago*): slightly less than 1 foot long, brown, black, and white body, with very long bill and short legs; feeds on worms, insects, crayfish, and some plants; summer resident in north, winter resident in south; marshes and wet meadows; Newfoundland to northern New England in summer, southern New England to Florida in winter. (Note: Conspicuous zigzag flight pattern.)

57. American Woodcock (*Scolopax minor*): almost 1 foot long, plump brownish and grayish (well-camoflagued) body, with big head, very long bill, large dark eyes, reddish brown breast and belly; eats mostly earthworms; summer resident; alder swamps next to meadows, bottomlands, stream banks, forested wetlands and upland woods; Newfoundland to Florida in summer, southern New Jersey south in winter. (Note: Begins courtship ritual in late March and early April; male's unique courtship flight ends with a zigzag drop from a couple of hundred feet up; call a nasal "peent" sound or a high-pitched twittering while in courtship flight.)

58. Clapper Rail (*Rallus longirostris*): longer than 1 foot, the typical grayish to grayish brown rail of salt marshes, with long decurved dark bill and black (dark) and white (light) barred sides (flanks); feeds on crabs (especially fiddlers), snails, fish, and insects, also on plants (seeds and tubers); distinctive call is a clucking sound, "chit-chit" or "kek-kek"; year-round resident mostly from New Jersey south, summer resident to Massachusetts; salt marshes, especially the low marsh; Massachusetts to Florida. Similar species: King Rail (*R. elegans*) is similar sized but more brownish in color, always with a reddish brown breast and neck and an orangish bill; summer resident in brackish and tidal and nontidal fresh marshes north to southern New Hampshire, overwinters in tidal marshes north to New York. Virginia Rail (*R. limicola*) is smaller (about ¾ foot long), with reddish brown breast and mostly gray head; summer resident to Nova Scotia, winter resident north to Massachusetts; tidal and nontidal marshes (mostly fresh in summer), tidal marshes (winter). Sora (*Porzana carolina*) is a small, dark-colored rail (about ¾ feet long) with a short, stout yellow bill and a black face; year-round along coast from Long Island Sound south, summer resident north to Nova Scotia.

Gulls and Terns

59. Laughing Gull (*Larus atricilla*): slightly less than 1½ feet long, with black head (gray in winter), white neck and breast, gray wings with black tips, and orange bill; feeds on fish,

53. Least Sandpiper

54. Spotted Sandpiper

55. Killdeer

56. Common Snipe

57. American Woodcock

58. Clapper Rail

59. Laughing Gull

crustaceans, and insects; call a distinctive high-pitched "ha-ha-ha" (giving its name); mostly summer coastal resident; salt marshes and beaches; Maine to Florida.

60. Herring Gull (*Larus argentatus*): about 2 feet long, the common gray-and-white sea gull, with stout yellow bill having red spot near tip of lower bill, and pinkish to flesh-colored legs; eats fish, crabs, bivalves, small mammals, young of other birds, and garbage; year-round resident; salt marshes, beaches, lakes, rivers, and landfills; Newfoundland to Florida. Similar species: Ring-billed Gull (*L. delawarensis*) is a little smaller and has a dark stripe (ring) across its yellow bill (bill is thinner than Herring's); winter resident, but breeds in northern Maine. Great Black-backed Gull (*L. marinus*), the largest of our gulls (about 2½ feet long), with distinctive black back set against white head, neck, breast, and belly; year-round resident.

61. Common Tern (*Sterna hirundo*): slightly more than 1 foot long, with black-and-white head (black on top), orangish to reddish bill usually black-tipped in summer, gray body, orange legs, and forked tail; dives for fish; summer resident; salt marshes, beaches, and lakes; Newfoundland to Virginia in summer, winter to south. Similar species: Roseate Tern (*S. dougalli*) has more black on bill and a longer, more deeply forked tail; Long Island Sound north. Arctic Tern (*S. paradisaea*) also has a longer tail, but has a reddish bill lacking black tip. Least Tern (*S. antillarum*) is much smaller, with a black cap and black stripe through eyes, otherwise white head, white neck, and black-tipped yellow bill; southern New England south. Caspian Tern (*S. caspia*) is larger (less than 2 feet long) with white-and-black head (black from eyes above), black crest, and stout reddish bill (may have dark tip); south of New Jersey. Black Skimmer (*Rynchops niger*) is larger (about 1½ foot long) and differs in black cap, nape of neck, back, and wings (rest of bird is white) and very stout red-and-black bill with markedly longer lower bill; southern New England south.

Hawks

62. Osprey or Fish Hawk (*Pandion haliaetus*): about 2 feet long, with white head having a thick black eye stripe, body blackish to dark brownish above and white below, and conspicuous appearance in flight—wings dark on outside, white inside and body; feeds on fish; summer resident; tidal marshes, estuaries, lakes, and rivers; Newfoundland to Florida. (Note: Along coast often nests on artificial platforms constructed in salt and brackish marshes.) Similar species: Bald Eagle (*Haliaeetus leucocephalus*) is larger (about 2½ feet long), adult conspicuous with white head and tail and dark body, immature mottled dark and white, tail shorter and broader than that of osprey.

63. Northern Harrier or Marsh Hawk (*Circus cyaneus*): the common hawk of marshes, about 1½ feet long, brown or grayish body, with conspicuous white-striped rump before long, somewhat narrow tail (obvious in flight); eats rodents, birds, frogs, snakes, among others; year-round resident; tidal and nontidal marshes, wet meadows, and upland fields; Nova Scotia to Florida. (Note: Often seen gliding or cruising at low levels across marshes.)

64. Red-shouldered or Swamp Hawk (*Buteo lineatus*): the characteristic hawk of forested wetlands, about 1½ feet long, with reddish brown shoulders and barred breast, and light-and-dark banded tail evident in flight; feeds on all kinds of small vertebrates (e.g., frogs, mice, and snakes) and also on insects; year-round resident in most of region; forested wetlands and uplands, wooded edges of marshes; Maine to Florida. Similar species: Red-tailed Hawk (*B. jamaicensis*) has a whitish breast, and in flight its reddish brown tail is distinctive; year-round resident, the most common large hawk in the Northeast, and most frequently seen in open uplands.

Owls

65. Barred or Swamp Owl (*Strix varia*): the owl of forested wetlands, less than 2 feet long, with grayish-brown-streaked body and distinctive dark eyes; eats small mammals of various kinds, birds, frogs, and insects; distinctive call "hoo-hoo-hoo-hoohoo" (which has been translated into the quote "who who cooks for you?") may be heard during day; year-round resident; forested wetlands and upland woods; Newfoundland to Florida. Similar species: Great Horned Owl (*Bubo virginanus*) may also occur in swamps; it is a large gray-and-dark-brown owl (about 2 feet long) with yellow eyes, a white throat, and tufted ears. The region's smallest owl at about 8 inches long, the Saw-whet Owl (*Aegolius acadicus*) prefers evergreen forests, including Atlantic white cedar swamps. (Note: Owls can rotate their heads about 270 degrees,

60. Herring Gull

61. Common Tern

62. Osprey

63. Northern Harrier

64. Red-shouldered Hawk

65. Barred Owl

66. Short-eared Owl

which gives them the ability to see behind them without moving their bodies; owls can also see about 2½ times better than humans.)

66. Short-eared or Marsh Owl (*Asio flammeus*): the owl of marshes, slighter over 1 foot long, with brownish-streaked body and yellow eyes surrounded by black patches; eats meadow voles and other small mammals; year-round resident in most of the Northeast; marshes, fields, and dunes, and salt marshes (winter); Newfoundland to Florida, coastal Maine south in winter. (Note: Usually seen flying at low levels over marshes at dawn and dusk.)

Hummingbirds

67. Ruby-throated Hummingbird (*Archilochus colubris*): the region's smallest bird, about 3½ inches long, with needlelike bill, greenish body; male with iridescent red throat and white collar, female with white neck and breast; feeds on nectar, spiders, and insects; summer resident; forested wetlands, uplands, often near streams, and gardens; Newfoundland to Florida. (Note: Hummingbirds eat twice their body weight in food every day. Some favorite flowers are jewelweed and bee balm. The Ruby-throated Hummingbird has the fewest number of feathers of any bird.)

Woodpeckers

68. Pileated Woodpecker (*Dryocopus pileatus*): about 1½ feet long, body mostly black, with black-and-white-striped head and prominent red crest; male has red cheek stripe to bill, female has black stripe; feeds mostly on carpenter ants and insects; year-round resident; deciduous forests and swamps; Nova Scotia to Florida. (Note: Makes a loud hammering sound when probing trees for food; excavated holes are rectangular.) Similar species: Hairy Woodpecker (*Picoides villosus*) and Downy Woodpecker (*P. pubescens*) are much smaller black-and-white woodpeckers about ¾ foot and ½ foot long, respectively; year-round residents. Northern Flicker (*Colaptes auratus*) is also smaller (about 1 foot long), but is multicolored (brown, black, gray, and red, with some yellow underparts), with distinctive black-spotted breast; best field character may be its white rump seen in flight; year-round resident. Red-headed Woodpecker (*Melanerpes erythrocephalus*) has a red head and black-and-white body and wings; breeds in forested wetlands and beaver ponds from Massachusetts south. Black-

backed Woodpecker (*P. arcticus*) occurs from northern New England north; the back of its head and its back are black, and male has yellow crown patch on head, no red markings; seen feeding on insects in dead trees in beaver swamps and logged areas.

Kingfishers

69. Belted Kingfisher (*Ceryle alcyon*): about one foot long, with grayish blue head, collar, back, and wings, white neck and belly (female with reddish brown belly stripe), a distinctive crest, and a stout dark bill; eats fish, reptiles, amphibians, mice, crayfish, and insects, diving to catch fish; distinctive call given in flight is a hollow rattling sound (can often be heard along bodies of water); year-round resident in most of Northeast; lakes, rivers, streams, and ponds; Newfoundland to Florida. (Note: Nests in riverbank tunnels, usually around 5 feet but up to 15 feet long.)

Swallows and Similar Birds

70. Tree Swallow (*Tachycineta bicolor*): about ½ foot long, iridescent (metallic) blue above and white below (female with brownish forehead); feeds mostly on flying insects, bayberries, and other berries; summer resident; marshes, wet meadows, and bodies of water; Newfoundland to Virginia. (Note: Interesting migratory behavior of circling in great masses when caught in swirling winds.) Similar species: Bank Swallow (*Riparia riparia*) is mostly dark brown above and whitish below, except for brownish breast stripe or band. Northern Rough-winged Swallow (*Stelgidopteryx serripennis*) is a grayish brown swallow with a light-colored breast; it lacks white throat of Bank Swallow. Barn Swallow (*Hirundo rustica*) is dark blue above, buff below, with reddish brown chin and neck and a distinctive forked tail. Purple Martin (*Progne subis*) is slightly larger (⅔ foot long); male is shiny purplish with blackish wings, female has gray-and-white-mottled chest, dull purplish back, and black wings. Chimney Swift (*Chaetura pelagica*) is a small, dark-colored bird (less than ½ foot long) with a somewhat cigar-shaped body, large eyes, and a characteristic flight: rapid beating of stiff wings followed by a glide and repeated.

Wrens

71. Marsh Wren or Long-billed Marsh Wren (*Cistothorus palustris*): tiny bird of cattail

67. Ruby-throated Hummingbird

68a. Pileated
Woodpecker

68b. Red-headed
Woodpecker

69. Belted Kingfisher

70. Tree Swallow

71. Marsh Wren

marshes, less than ½ foot long, brownish above and lighter below, with white stripe above eye (eyebrow), cinnamon-colored sides (flanks), and blackish tail; eats mostly insects; distinctive call is either a single "check" or many, resembling a twittering sound; year-round resident from mid-Atlantic states south, summer resident farther north; tidal and nontidal marshes dominated by tall graminoids; southern New Brunswick to Florida. (Note: Male makes many dummy nests in the cattails, apparently to help protect the female's active nest.) Similar species: Sedge Wren or Short-billed Marsh Wren (*C. platensis*) is a less common summer resident of marshes and wet meadows; its head and back are streaked white and dark brown, and it lacks the prominent light eyebrow of the Marsh Wren. Winter Wren (*Troglodytes troglodytes*) frequents streams, bogs, and evergreen forests; its tail is much shorter than that of the other wrens.

Sparrows

72. Sharp-tailed Sparrow (*Ammodramus caudacutus*): less than ½ foot long, with brown-streaked back and breast, whitish belly, and orange face with dark cheek patch; eats seeds (smooth cordgrass), amphipods, and insects; year-round resident in southern half of region; salt and brackish marshes; Nova Scotia to Virginia in summer, New Jersey to Florida in winter. Similar species: Seaside Sparrow (*A. maritimus*) is more brownish gray and lacks orange face, has yellow patch in front of eyes; year-round resident of salt marshes from Massachusetts to Florida. Savannah Sparrow (*Passerculus sandwichensis*) is a black-brown-and-white-striped sparrow with yellow eyebrow; winter resident from Massachusetts south, summer resident to the north.

73. Swamp Sparrow (*Melospiza georgiana*): nearly ½ foot long, with reddish brown and dark-streaked body, white throat and whitish eye stripes (in summer), grayish face (in winter), gray back of neck and breast, reddish brown patch on top of head (crown), and faint streaking on breast; eats seeds of smartweeds, sedges, and other plants, and insects; year-round resident north to southern New England, summer resident farther north; inland marshes, shrub swamps, bogs, and shores of lakes and streams; Newfoundland to Florida.

74. Song Sparrow (*Melospiza melodia*): nearly ½ foot long, with light- and dark-brown–streaked body and wings, brown-streaked white breast, and a central dark spot on breast; feeds on seeds and insects; year-round resident from central Maine south; forested wetlands and uplands, fields, streamsides, parks, and lawns; Nova Scotia to Florida.

Blackbirds, Grackles, and Crows

75. Red-winged Blackbird (*Agelaius phoeniceus*): about ⅔ foot long; male distinctive black with yellow-fringed red shoulder patch (white-fringed orange patch in immature); female brown above and brown-and-white-streaked below, with light stripe above eye; feeds on insects and seeds, especially wild rice; year-round resident from New Jersey south, summer resident north; tidal and nontidal marshes, wet meadows, ponds, and fields; Nova Scotia to Florida. (Note: This bird is among the first spring migrants to return to Northeast marshes.)

76. Common Grackle (*Quiscalus quiscula*): about 1 foot long, iridescent purplish to greenish black bird, with yellow eyes, and fairly long wedge-shaped tail; eats insects, fish, mice, birds, and seeds; year-round resident from coast of southern New England south, summer resident north; marshes, wet meadows, swamps, farms, lawns, and fields; Newfoundland to Florida. Similar species: Boat-tailed Grackle (*Q. major*) is larger (about 1⅓ feet long), with a long tail, and is a southern coastal species (from New Jersey south); the female is brownish, with dark wings; common along the coast, especially around salt marshes. Rusty Blackbird (*Euphagus carolinus*) is a winter resident along the coast to southern New England, and summer resident from central Maine north; it has a yellow eye and a shorter tail. Common Crow (*Corvus brachyrhynchos*) and Fish Crow (*C. ossifragus*) are much larger black birds (about 1½ feet long), with the latter having a very stout bill and being restricted to the coast, where it feeds on fish; its call is a short nasal "ca, ca" versus the "caaaw" of the Common Crow.

77. Bobolink (*Dolichonyx oryzivorus*): slightly more than ½ foot long; male mostly black, with yellowish white back of head, black-and-white wings, white rump (conspicuous in flight), and black bill; female sparrowlike brownish, with light eyebrow, dark eye stripe behind eye, light- and dark-brown–striped wings, and light-colored bill; eats mostly seeds and insects (wild rice is a favorite); summer resident; marshes, wet meadows, and upland fields; Nova Scotia to New Jersey.

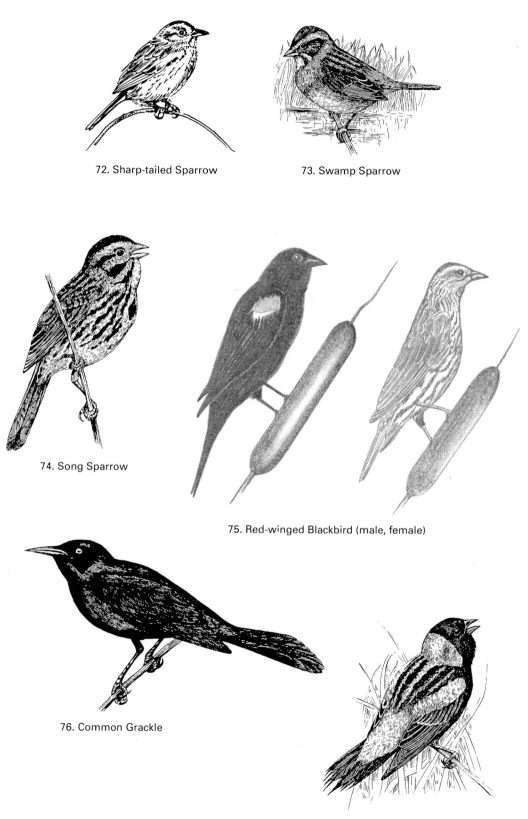

72. Sharp-tailed Sparrow

73. Swamp Sparrow

74. Song Sparrow

75. Red-winged Blackbird (male, female)

76. Common Grackle

77. Bobolink

Warblers and Other Songbirds

78. Northern Waterthrush (*Seiurus nove-boracensis*): about ½ foot long, with brown back and brown-streaked white neck, breast, and belly, distinctive yellowish eyebrow, brown eye stripe and cap, and flesh-colored legs; feeds on aquatic invertebrates, insects, and fish; summer resident; bogs, forested wetlands, and shores of lakes, ponds, and streams; Newfoundland to southeastern New York, in mountains to West Virginia. (Note: Sometimes seen wading in shallow water.) Similar species: Louisiana Waterthrush (*S. motacilla*) has a dark patch behind the eye and white eyebrow (darker in front of eye) and pinkish legs; summer resident of bottom-land forests from central New Hampshire south; seems to prefer fast-flowing wooded streams.

79. Ovenbird (*Seiurus aurocapillus*): about ½ foot long, with brownish to olive brown back, dark-streaked white breast and neck, white eye ring, orange cap (crown) bordered by dark stripes, and pinkish legs; feeds on worms, insects, and spiders on the ground; distinctive call "teacher teacher teacher"; summer resident; forested wetlands and uplands (deciduous or mixed); Labrador to Virginia, in mountains to Georgia.

80. Brown Creeper (*Certhia americana*): about ½ foot long, brown-and-white-streaked above, whitish below (well-camoflaged against bark), with conspicuous long curved bill (dark above, light below); feeds on insects and spiders on bark; year-round resident in most of region; forested wetlands and uplands; Nova Scotia to Maryland in summer, southern New England to Florida in winter. (Note: Creeps only up trunk, not up and down.)

81. Veery (*Catharus fuscescens*): about ⅔ foot long, cinnamon brown above, whitish below, with mottled reddish brown and buff breast and dark eyes; eats insects, spiders, snails, and worms on the ground, and also fruits and seeds; call is distinctive melodious, flutelike, downward-spiraling "tooreeooreeoreeoorooh"; summer resident; rich, moist forests and forested wetlands, especially along streams; Newfoundland to New Jersey, in mountains to Georgia. Similar species: Wood Thrush (*Hylocichla mustelina*) is more brownish above, white with black spots below, and head reddish brown with white eye ring; summer resident throughout region; its call is also flutelike, but rising. Brown Thrasher (*Toxostoma rufum*) is much larger (almost 1 foot long), with reddish brown head and back, reddish-brown- and white–spotted below, stout curved bill, and light-colored eye; year-round resident from New York south, summer resident north to western Maine.

82. Eastern Wood Pewee (*Contopus virens*): about ½ foot long, grayish above and lighter below, with a distinctive-shaped head, almost crestlike, and a short bill (dark above, orangish below); eats flying insects and caterpillars; call sounds like its name, "pee-wee"; summer resident; forested wetlands along rivers and upland forests; Nova Scotia to Florida. Similar species: Eastern Phoebe (*Sayornis phoebe*) is more-brownish gray, has head more rounded, black bill, and a habit of bobbing or wagging its tail when perched.

83. Alder Flycatcher (*Empindonax traillii*): about ½ foot long, olive gray above, white throat, and light below, with light-colored ring around eye; eats airborne insects; summer resident; alder swamps, shrub swamps, and shrubby wetland or lake borders; Newfoundland to eastern Pennsylvania, in mountains to West Virginia. Similar species: Other similar-looking small flycatchers that frequent wetlands include Acadian Flycatcher (*E. virescens*) and Willow Flycatcher (*E. alnorum*). The former occurs from the southern New England coast south, the latter north to southwestern Maine and central New Hampshire and Vermont. Eastern Phoebe (*Sayornis phoebe*) is slightly more than ½ foot long, lacks an eye ring, and has grayish brown head and upper body, whitish neck, breast, and belly, and black bill; its habit of tail bobbing is quite distinctive. Great Crested Flycatcher (*Myiarchus crinitus*), a forest species, is larger (about ¾ foot long) with gray-crested head, brownish back, yellowish belly, white-and-black wing bars (seen at rest), and reddish brown wings and tail.

84. Eastern Kingbird (*Tyrannus tyrannus*): about ¾ foot long, dark gray above and light below, with black head, bill, eyes, and conspicuous white-tipped black tail; feeds on insects and berries; summer resident; fields, marshes, forest and swamp edges, and streams; New Brunswick to Florida.

85. Red-eyed Vireo (*Vireo olivaceus*): about ½ foot long, grayish to olive green above and whitish below, with gray stripe through red eye, whitish stripe (with black upper border) above eye, gray cap, and dark bill; eats mostly insects, but also fruits; summer resident;

78. Northern Waterthrush

79. Ovenbird

80. Brown Creeper

81. Veery

82. Eastern Wood Pewee

83. Alder Flycatcher

84. Eastern Kingbird

85. Red-eyed Vireo

deciduous forested wetlands and uplands; Newfoundland to Florida. Similar species: Yellow-throated Vireo (*V. flavifrons*) has a yellowish head, neck, and breast, dark eyes, a light eye ring, and distinctive white wing bars. Warbling Vireo (*V. gilvus*) is also grayish to olive green above and white below, but has dark eyes and lacks the prominent striping of the Red-eyed Vireo. White-eyed Vireo (*V. griseus*) is yellow and gray above, whitish below, with yellow sides, a yellow eye ring, and whitish eyes; summer resident north to southern New England.

86. Cedar Waxwing (*Bombycilla cedrorum*): slightly more than ½ foot long, grayish brown body, with distinctive crest, black eye stripe (mask), yellow belly, small red markings on wings, and yellow band at tip of tail; feeds on fruits and insects; year-round resident in most of Northeast, except northern New England; open forested wetlands and uplands, orchards, farms, and urban areas; Nova Scotia to Georgia. (Note: Typically observed in flocks.)

87. Scarlet Tanager (*Piranaga olivacea*): slightly more than ½ foot long; male red, with black wings and tail and light-colored bill; female yellowish brown, with dark wings; eats insects and fruits; summer resident; deciduous and mixed forested wetlands and uplands, including larch swamps; Nova Scotia to Virginia, in mountains to Georgia.

88. Yellow Warbler (*Dendroica petechia*): less than ½ foot long, yellow body, with dark wings, tail with yellow spots, light-colored legs; male has reddish-brown-streaked breast; eats insects, including caterpillars; summer resident; call is distinctive "sweet-sweet-I'm so-sweet" or "sweet-sweet-sweet, sweeter than sweet"; shrub swamps, shrubby borders of streams and lakes, farms, orchards, and garden shrubs; Newfoundland to Georgia. Similar species: Pine Warbler (*D. pinus*) is a mostly yellow bird of pine and mixed pine-hardwood forests and swamps; it has gray-and-white wings and black legs; summer resident from New Jersey north, year-round farther south.

89. Prothonotary Warbler (*Protonotaria citrea*): about ½ foot long, yellow body, with blackish to grayish wings and tail, dark bill, and dark-colored legs; eats insects, spiders, snails, and other small aquatic invertebrates; summer resident; forested wetlands and bottomlands, stream borders, and pond shores; central New York to Florida, rarely in Connecticut and Rhode Island. Similar species: American Gold-finch (*Carduelis tristis*) is similarly colored, but has white markings on wings, orangish bill (summer), and light-colored legs; male has a black cap; seen mostly in open fields and shrub thickets; year-round.

90. Common Yellowthroat (*Geothlypis trichas*): less than ½ foot long; male with conspicuous black mask, white eyebrow, and yellow throat and breast; female lacking mask, but has prominent white eye ring and yellowish throat and breast, light-colored legs; feeds on insects, spiders, and seeds; distinctive calls include "witchity witchity," "which is it, which is it, which is it," or "your money, your money, your money"; summer resident; shrub swamps, shrubby borders of streams and ponds, tidal and nontidal marshes, pastures, and edges of upland woods; Newfoundland to Florida.

91. Hooded Warbler (*Wilsonia citrina*): about ½ foot long, olive above and yellow below, with yellow face almost surrounded by black hood (immature female lacking black hood), light-colored legs; feeds on insects and spiders; call is "tawee-tawee-tawee-tee-o"; summer resident; Connecticut and New York to Florida.

92. Northern Parula (*Parula americana*): less than ½ foot long; male with gray-blue head and back, white upper and lower eye markings, yellow throat, yellow patch on back, reddish brown markings on yellow breast, and white belly; female more grayish, lacking streaked breast; eats insects and spiders; summer resident; bogs and forested wetlands, especially evergreens with old man's beard lichens; Nova Scotia to Florida (common in Maine).

93. Black-throated Green Warbler (*Dendroica virens*): less than ½ foot long, olive above, with mostly yellow head, olive eye stripe and cap, black throat and breast (male), dark-streaked breast and belly (female), and dark legs; eats insects and fruits; summer resident; evergreen forested wetlands and uplands, especially hemlock; Newfoundland to New Jersey, in mountains to Georgia.

94. Canada Warbler (*Wilsonia canadensis*): about ½ foot long, mostly yellow and gray; male bluish gray, with yellow eye ring, throat, neck, and belly, yellow patch above bill, black-streaked yellow breast, and light-colored legs; female breast not as prominently streaked, and back gray; eats insects and spiders; summer resident; forested wetlands and uplands, northern-cedar swamps, and shrubby stream

86. Cedar Waxwing

87. Scarlet Tanager

88. Yellow Warbler

89. Prothonotary Warbler

90. Common Yellowthroat (male)

91. Hooded Warbler

92. Northern Parula

93. Black-throated Green Warbler

94. Canada Warbler

95. Black-and-White Warbler

96. Rufous-sided Towhee

borders; Nova Scotia to New Jersey, in mountains to Georgia. (Note: Has a twitchy tail.)

95. Black-and-white Warbler (*Mniotilta varia*): about ½ foot long, black and white striped body, male with black throat, female more grayish with white throat; feeds on insects; summer resident; deciduous or mixed forested wetlands and uplands; Newfoundland to Florida.

96. Rufous-sided Towhee (*Piplio erythrophthalmus*): about ⅔ foot long; male with black head, back, wings, neck, and breast, white belly, and cinnamon brown (rufous) sides; female brown instead of black; feeds on seeds, fruits, insects, spiders, and even snakes and lizards; year-round resident; call is distinctive "drink your tea"; year-round resident in most of region; edges of forested wetlands and uplands, shrub thickets, pastures, and clearings; central Maine to Florida in summer, from southern New England south in winter.

Mammals

Small Mammals

97. Star-nosed Mole (*Condylura cristata*): about 5 inches long, with conspicous star-shaped nose (composed of twenty-two fleshy appendages), blackish to dark brown body, and hairy tail (to 3 inches long); active all day long; feeds on worms, aquatic invertebrates, and insects; marshes and wet meadows; Labrador to northeastern North Carolina, in eastern mountains to South Carolina. Similar species: Eastern Mole (*Scalopus aquaticus*) lacks the star-shaped nose and has a naked tail (about 1 inch long); wet meadows and moist, open grasslands, lawns, and golf courses; southern New Hampshire to Florida.

98. Short-tailed Shrew (*Blarina brevicauda*): about 4 inches long, with short tail (about 1 inch or less), body dark gray above, lighter below, velvety fur, external ears hidden in fur, and minute eyes; active year-round and all day long; eats insects, worms, and other invertebrates, as well as plants; marshes, swamps, upland woods and fields; New Brunswick to Florida. (Note: Saliva reportedly is poisonous, and shrew has gland capable of exuding a foul odor, presumably for defense.)

99. Masked Shrew (*Sorex cinereus*): about 2 inches long, body grayish brown, with tail less than or nearly equal to body length, light below,

darker above; active all day long; eats mostly insects, but also worms, other invertebrates, and salamanders (reportedly eats more than its weight daily, up to 3½ times its weight in captivity); forested wetlands and shrub swamps, moist woods, fields, and thickets; Labrador to Maryland, in eastern mountains to Tennessee. Similar species: Least Shrew (*Cryptotis parva*) is reddish brown to cinnamon-colored, with a short tail (less than 1 inch); western Connecticut and Long Island south to Florida. Northern Water Shrew (*Sorex palustris*) is an excellent swimmer and a bit larger (about 3 inches long), with body dark gray above, light gray to whitish below; feeds on aquatic invertebrates; Labrador to Long Island, in eastern mountains to North Carolina.

100. Red-backed Vole (*Clethrionomys gapperi*): almost 5 inches long, with tail less than half body length, reddish back, gray sides, and ears appressed to head; active all day long; feeds on herbaceous vegetation, also seeds, nuts, and fungi; mature Atlantic-white-cedar swamps (including logged sites), hardwood swamps, and salt marshes; Labrador to Delmarva Peninsula, in eastern mountains to South Carolina.

101. Meadow Vole (*Microtus pennsylvanicus*): up to 5 inches long, dark brown to grayish brown, with tail less than half body length, dark above and light below, small eyes, and ears appressed to head; active all day long; eats graminoids, seeds, bark, and insects; tidal and nontidal marshes (including salt marshes), wet meadows, bogs, hardwood swamps, logged Atlantic-white-cedar swamps, grasslands near bodies of water, fields, and pastures; Labrador to Georgia.

102. Southern Bog Lemming (*Synaptomys cooperi*): up to 4½ inches long, brownish gray, with short tail (less than 1 inch), and ears appressed to head (difficult to see); active all day long; eats vegetation and seeds; marshes, wet meadows, bogs, and moist deciduous forests; southern Canada to eastern North Carolina, in eastern mountains to Georgia.

103. Meadow Jumping Mouse (*Zapus hudsonius*): about 3 inches long, brownish body, with prominent erect ears, big white hind feet (for jumping), and a long tail (much longer than body, to 6 inches); mostly nocturnal, occassionally seen leaping through wet meadows in daylight; hibernates in winter; eats invertebrates, seeds, fruits, and nuts; marshes, wet meadows, shrub swamps, and transitional areas between

97. Star-nosed Mole

98. Short-tailed Shrew

99. Masked Shrew

100. Red-backed Vole

101. Meadow Vole

103. Meadow Jumping Mouse

102. Southern Bog Lemming

swamps and uplands; Labrador to North Carolina, in eastern mountains to Alabama. Similar species: Woodland Jumping Mouse (*Napaeozapus insignis*) is brighter-colored, with brown back, yellowish sides, and white belly; forests and shrub thickets (near water), and bogs; Labrador to northern New Jersey (from Massachusetts south occurs away from coast), in eastern mountains to Alabama.

104. White-footed Mouse (*Peromyscus leucopus*): about 4 inches long, brownish to reddish brown, with tail usually less than body length, white belly and feet, large eyes, and prominent erect ears; eats seeds, acorns, nuts, and insects; woods, thickets, fields, and houses (in winter); central Maine to North Carolina, in interior to Georgia and Louisiana in the Southeast. Similar species: Deer Mouse (*P. maniculatus*) is very similar (grayish to reddish brown, with white belly), but usually has a two-colored tail (dark above and light below); Labrador to northern New Jersey, inland in eastern mountains to Georgia. Rice Rat (*Oryzomys palustris*), a grayish brown southern species (from southern New Jersey south) of marshes and rice fields, has soft velvety fur, a scaly tail nearly as long as or longer than body, a gray or reddish brown belly, and white feet.

105. Red Squirrel (*Tamiasciurus hudsonicus*): about ⅔ foot long, reddish brown, with bushy tail (up to 6 inches), white belly (grayish white in winter), and tufted ears (in winter); active year-round, mostly during the day; feeds on nuts, seeds (especially from pines), mushrooms, and bird eggs; evergreen forests and swamps; Labrador to Virginia, in mountains to South Carolina.

106. Southern Flying Squirrel (*Glaucomys volans*): less than ½ foot long, olive brown, with white belly and skin folded on sides (expands when gliding); active at night; eats acorns, seeds, nuts, insects, and bird eggs; forested wetlands, including Atlantic-white-cedar swamps, and upland forests; southern Maine to central Florida. Similar species: Northern Flying Squirrel (*G. sabrinus*) occurs from northern New Jersey and Georgia (in mountains) north; it is somewhat larger (about ½ foot) and has white-tipped gray belly hairs.

Medium-sized Mammals

107. Beaver (*Castor canadensis*): over 2 feet long, with dark brown fur and paddlelike tail (less than 1 foot long); active year-round, most active at night beginning at dusk; feeds on bark, twigs, and branches of willows, birch, alders, maples, and aspens, in summer eats water lilies and spatterdock; builds dams and conical lodges of sticks and mud; mountain streams, ponds, lakes, marshes, and shrub swamps where there is an ample supply of its woody foods; Labrador to Florida. (Note: Slaps tail on water repeatedly as warning signal.)

108. Muskrat (*Ondatra zibethicus*): about 1 foot long, with naked, scaly, laterally flattened black tail (less than 1 foot long), brown fur, and silvery belly; marshes (including brackish and cattail marshes) and streams; feeds on cattail rootstocks, young stems, and leaves, rice cutgrass, bur reeds, and other aquatic vegetation, and also on fish, frogs, mussels, and clams; Labrador to North Carolina, in eastern interior to Georgia, Alabama, Mississippi, and Louisiana; builds cone-shaped lodges of herbaceous plants or lives in bank burrows. Similar species: Nutria (*Myocastor coypus*), a South American mammal first introduced into Louisiana, is grayish brown to yellowish brown and larger (to 2 feet long), with a long, round, scaly tail; Maryland and southern New Jersey.

109. Eastern Cottontail (*Sylvilagus floridanus*): slightly more than 1 foot long, with brown or grayish fur, white (cotton) tail, whitish feet, and reddish brown at back of neck; most active in evening and morning; eats herbaceous plants in summer and bark and twigs of red maple and other species in winter; wetland and upland shrub thickets, fields, and swamps; Massachusetts to Florida. Similar species: New England Cottontail (*S. transitionalis*), a less-common species, may frequent swamps in mountainous areas; fur is more reddish in summer and reddish gray marked with white in winter, lacking the reddish brown nape of the Eastern Cottontail; Vermont and southern Maine to New Jersey, in mountains to Alabama.

110. Snowshoe Hare (*Lepus americanus*): to 1½ feet long, with fur that is dark brown in summer and turns white in winter, black-tipped ears, and large hind legs and feet; forested wetlands and upland forests and thickets; Labrador to northern New Jersey and Pennsylvania, in eastern mountains to North Carolina and Tennessee.

111. Opossum (*Didelphus marsupialis*): North America's only marsupial, about 1½ feet long, with fur whitish to light grayish, ratlike prehensile tail (used to grasp branches), and

104. White-footed Mouse

105. Red Squirrel

106. Flying Squirrel

107. Beaver

108. Muskrat

109. Eastern Cottontail

110. Snowshoe Hare

111. Opossum

thin black ears; bears very undeveloped young (up to twenty, each the size of a housefly), which then go to mother's pouch for further development; active at night; omnivorous; woods and streams; Massachusetts to Florida. (Note: It has been referred to as a living fossil, since evidence of the opossum can be found from the age of dinosaurs, over 100 million years ago: the Cretaceous Period).

112. Raccoon (*Procyon lotor*): about 2 feet long, with gray-and-black fur, black mask, and ringed tail (less than 1 foot long); most active at night; eats plants and animals (frogs, crayfish, molluscs, and insects); marshes, swamps, and upland habitats; New Brunswick and Nova Scotia to Florida.

113. Gray Fox (*Urocyon cineroargenteus*): about 2 feet long or more, with fur mottled black and gray, long bushy gray tail (to almost 1½ feet long) with a black stripe down top, reddish brown feet and legs, and reddish brown markings on face, ears, and sides (flanks); active mostly at night; eats mainly small mammals (mice, voles, etc.), but also eggs, insects, and plants; swamps, upland woods, and edges of marshes; western Maine and New Hampshire to Florida. Similar species: Red Fox (*Vulpes fulva*) is mostly reddish-colored, with black legs and a white-tipped bushy tail; Labrador to Virginia, in eastern interior to Georgia, Alabama, Mississippi, and Louisiana. Coyote (*Canis latrans*) is now in the Northeast; it is about 3 feet long, grayish above and whitish below, and lacks the reddish brown markings of the gray fox.

114. River Otter (*Lutra canadensis*): 2–3 feet long, with long thick tail (to 1½ feet), fur dark brown above, whitish chin and front neck, and webbed feet; eats fish, molluscs, frogs, crayfish, aquatic invertebrates, and even birds and young muskrats and beavers; river and stream banks and lake shores; Labrador to Florida. (Note: Its presence may be an indicator of good water quality.) Similar species: Mink (*Mustela vison*) is a smaller, dark brown weasel (about 1–1½ feet long) with white chin and long neck, and lacking webbed feet; active mostly at night; feeds on small mammals (especially muskrats), fish, birds (and eggs), frogs, young snapping turtles, and crayfish; river and stream banks and lake shores; Labrador to Florida. Long-tailed Weasel (*Mustela frenata*) is much smaller (less than 1 foot long), with body distinctly brown above and whitish below, noticeably long neck, and black-tipped tail (changes

color in winter in northern part of range: white, except for black-tipped tail); mostly nocturnal; eats mostly small mammals; forested wetlands, woods, and other habitats near water; western New Brunswick to Florida.

115. Fox Squirrel (*Sciurus niger*): 1 foot or longer, with gray fur and bushy tail (around a foot long); eats acorns, nuts, seeds, buds, bark, mushrooms, and bird eggs; forested wetlands and upland forests, including pine forests; Delmarva Peninsula (endangered) and southern Pennsylvania to Florida. Similar species: Eastern Gray Squirrel (*S. carolinensis*) is smaller (less than 1 foot long), gray often with a brownish tinge, and its tail has white-tipped hairs (summer); forested floodplains and deciduous upland forests; Maine and New Brunswick to Florida; it makes leaf nests in trees. (Note: Fox squirrel spends more time on ground than gray squirrel.)

Large Mammals

116. White-tailed Deer (*Odocoileus virginianus*): to 6 feet long and 4 feet high, weighing to more than 400 pounds, with fur reddish brown in summer, grayish in winter, white belly and tail (raised when startled); fawns are white-spotted; feeds on various plant parts—twigs, bark, shrubs, acorns, mushrooms, and leaves and stems; yards in groups of two dozen or more in winter; critical winter habitat in northern areas is evergreen forested wetlands, especially Atlantic-white-cedar swamps along coast and northern-white-cedar swamps in New England; summer habitat is forested wetlands, upland forests, and thickets; eastern Canada to Florida. (Note: Can run about 30 miles per hour and jump over 8 feet high or more than 30 feet in distance.)

117. Moose (*Alces alces*): to 9 feet long and 6–7 feet tall, weighing over 1,000 pounds, with dark brown fur, humpback, hanging "bell" from throat, grayish legs, and short tail (3 inches); male with distinctive antlers; eats wetland plants, especially spatterdock, in summer, and twigs, bark, and other woody parts in winter; male sheds antlers in winter; boreal forests, forested wetlands, and bodies of water; Labrador to Maine and northern New England, occasionally south into Massachusetts. (Note: May run at nearly 15 miles per hour.)

118. Black Bear (*Ursus americanus*): up to 6 feet long, weighing to 500 pounds or more, face brown, often with white chest; active

112. Raccoon

113. Gray Fox

114. River Otter

115. Fox Squirrel

116. White-tailed Deer

117. Moose

118. Black Bear

119a. Eastern Pipestrelle

119b. Silver-haired Bat

mostly at night; eats fish, small mammals, berries, vegetables, and skunk cabbage; wooded swamps and upland woods; Labrador to northern New Jersey and Pennsylvania, in eastern mountains to Georgia, also in Dismal Swamp (eastern Virginia and North Carolina). (Note: Sense of smell is excellent, eyesight poor, and hearing average; for short distances, can run more than 30 miles per hour.)

Flying Mammals

119. Bats (several species): nocturnal, often seen at dusk flying over marshes, feeding on flying insects; some species migrate south; distinguishing among them usually requires close observation, often looking at ears. Eastern Pipestrelle (*Pipistrellus subflavus*) and Little Brown Bat (*Myotis lucifugus*) are small brownish bats, with the former sometimes being a yellow-brown color. Silver-haired Bat (*Lasionycteris noctivagans*) is blackish, with white-tipped (silvery) hairs; it feeds on flying insects in forests. (Note: One bat can catch about six hundred mosquitoes in an hour.)

Additional Field Guides

Invertebrates

Borror, D. J., and R. E. White. 1970. *A Field Guide to the Insects of America North of Mexico.* Houghton Mifflin Company, Boston.

Chu, H. F. 1949. *How to Know the Immature Insects.* Wm. C. Brown Company Publishers, Dubuque, IA.

Jaques, H. E. 1947. *How to Know the Insects.* Wm. C. Brown Company Publishers, Dubuque, IA.

Milne, L., and M. Milne. 1995. *National Audubon Society Field Guide to North American Insects and Spiders.* Alfred A. Knopf, Inc., New York.

Opler, P. A., and V. Malikul. 1992. *A Field Guide to Eastern Butterflies.* Houghton Mifflin Company, Boston.

Pollock, L. W. 1996. *Practical Guide to the Marine Animals of Northeastern North America.* Rutgers University Press, New Brunswick, NJ.

Pyle, R. M. 1995. *National Audubon Society Field Guide to North American Butterflies.* Alfred A. Knopf, Inc., New York.

Vertebrates

Behler, J. L., and F. W. King. 1995. *National Audubon Society Field Guide to North American Reptiles and Amphibians.* Alfred A. Knopf, Inc., New York.

Bull, J., and J. Farrand, Jr. 1994. *National Audubon Society Field Guide to North American Birds: Eastern Region.* Alfred A. Knopf, Inc., New York.

Burt, W. H., and R. P. Grossenheider. 1964. *A Field Guide to the Mammals.* Houghton Mifflin Company, Boston.

Conant, R. 1958. *A Field Guide to Reptiles and Amphibians of the United States and Canada East of the 100th Meridian.* Houghton Mifflin Company, Boston.

DeGraaf, R. M., and D. D. Rudis. 1983. *Amphibians and Reptiles of New England: Habitats and Natural History.* University of Massachusetts Press, Amherst, MA.

Hunter, M. L., Jr., J. Albright, and J. Arbuckle. 1992. *The Amphibians and Reptiles of Maine.* Maine Agricultural Experiment Station, Orono, ME. Bulletin 838.

Klemens, M. W. 1993. *Amphibians and Reptiles of Connecticut and Adjacent Regions.* State Geological and Natural History Survey of Connecticut, Department of Environmental Protection, Hartford, CT. Bulletin No. 112.

Peterson, R. T. 1947. *A Field Guide to the Birds: Eastern Land and Water Birds.* Houghton Mifflin Company, Boston. (More recent edition available)

Stokes, D., and L. Stokes. 1996. *Stokes Field Guide to Birds: Eastern Region.* Little, Brown and Company, Boston.

Thomson, K. S., W. H. Weed III, and A. G. Taruski. 1971. *Saltwater Fishes of Connecticut.* State Geological and Natural History Survey of Connecticut, Department of Environmental Protection, Hartford, CT. Bulletin 105.

Whitworth, W. R., P. L. Berrien, and W. T. Keller. 1968. *Freshwater Fishes of Connecticut.* State Geological and Natural History Survey of Connecticut, Connecticut Department of Environmental Protection, Hartford, CT. Bulletin 101.

General

Niering, W. A. 1985. *Wetlands.* Alfred A. Knopf, Inc., New York.

Reid, G. K. 1987. *Pond Life.* Golden Press, New York.

CHAPTER 13

Finding Wetlands and Their Boundaries
Wetland Identification and Delineation

Positive identification of wetlands and accurate delineation of their boundaries require both reviewing existing information on the project location and conducting field investigations at the site. Existing information (e.g., wetland maps and soil survey maps) should be reviewed prior to examining a wetland in the field. Use of these maps and general field procedures to identify and delineate wetlands are discussed below. The field procedures have been intentionally simplified for the nonspecialist from those published in official wetland delineation manuals. The approach described here does not require documentation of three criteria (hydrophytic vegetation, hydric soil, and wetland hydrology) as the manuals do, but permits identification by a single characteristic unique to wetlands, as recently recommended by the National Research Council. More-technical guidance for identifying regulated wetlands is available in the official manuals, which are from the appropriate regulatory authorities.

Printed Maps: Gospel or Guidance?

Available wetland information varies from place to place, but the two primary sources are 1) National Wetlands Inventory (NWI) maps produced by the U.S. Fish and Wildlife Service and 2) county soil-survey reports published by the USDA Natural Resources Conservation Service. These sources are available for many areas in the country. Some states have produced their own wetland maps. All Northeast states have maps of coastal or tidal wetlands, and many have published or are producing maps showing inland wetlands for at least some portion of the state. Most Great Lakes states have relied on the NWI maps for their wetland inventory, but Wisconsin and Ohio have produced their own wetland maps. Although wetland maps provide useful information, all maps have limitations on the uses

to which they may be put. Contact the appropriate agency for specifics.

National Wetlands Inventory Maps

The U.S. Fish and Wildlife Service (FWS), through its National Wetlands Inventory project, is producing a series of large-scale (1:24,000) maps that show the location, size, and type of wetlands within defined geographical areas for the entire country. An example of a 1:24,000 NWI map is shown in Figure 13.1. Call 1-800-USA-MAPS for information on current map availability for your area. Many maps are available in digital form through the Internet (http://www.nwi.fws.gov) for computer (geographic information system) applications.

NWI maps are useful for identifying the likely presence of wetland in a given area. The target mapping unit (i.e., the smallest area consistently mapped) varies with the scale of the photography used to prepare the map. Source photography is designated on the legend of each map. The NWI maps show the following:

- wetlands generally 1 acre in size and larger where 1:40,000 color-infrared photography was interpreted;
- wetlands 1 to 3 acres and larger where 1:58,000 color-infrared photography was used;
- wetlands 3 to 5 acres and larger where 1:80,000 black-and-white photography was interpreted;
- the location and shape of wetlands;
- the type of wetland based on vegetation (or substrate, where vegetation is absent); and
- water regime, salinity (for tidal areas), and other characteristics.

An alpha-numeric code designates the wetland type according to the FWS's *Classification of Wetlands and Deepwater Habitats of the United States*, published in 1979 (see Chapter 14 and Figure 13.2 for overview). For example, the

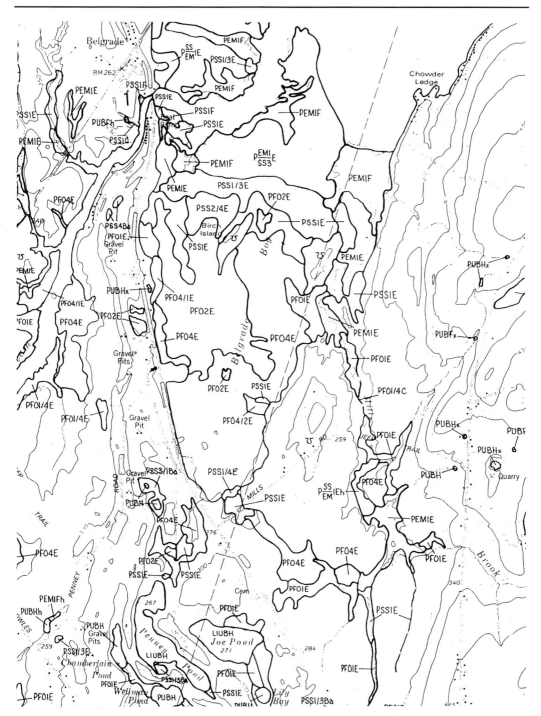

Figure 13.1. A National Wetlands Inventory map (Belgrade, Maine). Alpha-numeric codes signify different types of wetlands and deepwater habitats. For example, PFO4E is a palustrine forested wetland (needle-leaved evergreen), seasonally flooded/saturated (e.g., a black-spruce bog); PFO2E is a palustrine forested wetland (needle-leaved deciduous), seasonally flooded/saturated (i.e., a larch swamp); PFO1E is a broad-leaved deciduous hardwood swamp; PSS1E is a deciduous shrub swamp; PEM1F is a palustrine emergent wetland, semipermanently flooded (e.g., a cattail marsh); PUBH is a pond; L1UBH is a lake; U is upland. Compare this map to a soil map for the same area (Figure 13.2). (Source: U.S. Fish and Wildlife Service)

code PFO1E, representing a common nontidal wetland type in the Northeast, can be broken down as follows: P—palustrine (system); FO—forested wetland (class); 1—broad-leaved deciduous (subclass); and E—seasonally flooded/saturated (water regime). This type includes many red-maple swamps. A legend at the bottom of each NWI map explains the alpha-numeric code.

While the NWI maps give the location of a large number of wetlands, not all wetlands are shown. Since the maps were prepared through aerial-photointerpretation techniques with spot field checking, there is an inherent margin of error. Some limitations are listed below.

- Wetlands smaller than one acre and those occurring as a narrow fringing band (as wide as 100 feet) along watercourses are usually not shown. (Note: Be mindful of the target mapping unit listed in an earlier paragraph: the older NWI maps, including those originally produced for New Jersey, Massachusetts, Connecticut, Rhode Island, Pennsylvania, and parts of New York and Virginia, have target mapping units ranging from 3 to 5 acres, so many wetlands 3 to 5 acres and smaller are not designated.)
- Farmed wetlands are generally not mapped in the Northeast, with the exception of cranberry bogs. More-recent maps show depressional (potholelike) farmed wetlands that were flooded at the time the aerial photographs were taken.
- The aerial photographs reflect conditions during the specific year and season when they were taken: if taken during a dry season or a dry year, many wetlands would be missed and not mapped.
- Many forested wetlands are conservatively mapped. Some evergreen forested wetlands escaped detection due to full-canopy cover that prevents observation of wet soil conditions.
- Drier-end wetlands (e.g., seasonally saturated and temporarily flooded) are conservatively mapped. These wetlands are the most difficult to identify on the ground as well as through remote sensing.
- The mapped boundaries may be somewhat different from what they would be if based on detailed field observations, especially in areas with subtle changes in topography (e.g., pitch-pine lowlands or flatwoods on the coastal plain).

- The activities of humans (e.g., filling and drainage) or beavers may have caused changes in wetlands since the aerial photos were taken. So pay close attention to the date of the photograph used to prepare the maps; it is shown on the legend.

NWI maps do not display all wetlands, and they tend to err more by omission than commission. This means that if an area is designated as wetland on an NWI map, it is wetland most of the time. If the map does not show a wetland in an area, it is usually not, but there is a chance that a wetland may be there, especially if the area occupies a favorable landscape position (e.g., depression, or flat along a stream). It is important, therefore, that you be aware of these limitations and use the NWI maps to establish the presence and general configuration of a wetland in a given location. Soil survey maps should also be consulted to aid in wetland interpretation. A site inspection should *always* be conducted in order to accurately delineate the boundary of mapped wetlands and to identify any missed wetlands for a specific project site.

The NWI Program also has developed a Web-based tool where users can view NWI map information for areas where NWI data have been digitized. The "wetlands mapper" allows users to produce maps and do some basic analysis for specific geographic areas of interest. The mapper is posted at http://wetlands.fws.gov.

County Soil Surveys

The Natural Resources Conservation Service is continually conducting soil surveys throughout the country for natural-resource planning, especially agriculture. In the field, soil scientists correlate differences in soils with changes in landscape and then draw the boundaries of individual soil types on aerial photographs. Soil maps are subsequently prepared from these data. When the survey is completed for a given county or a major portion of a county, a soil survey report is published. The report contains invaluable information about the county and its soils (e.g., climate, soil series and map units, use and management of soils, and formation and classification of soils) plus large-scale (often 1:20,000) photo-based maps showing the location and configuration of individual soil-map units. Soil-map units represent mapped areas of various soil types (soil series and land types)

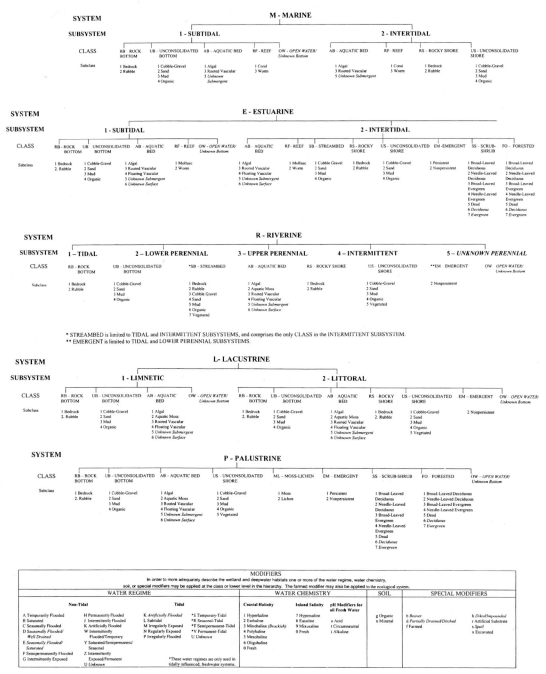

Figure 13.2. Legend for a National Wetlands Inventory map showing classification categories and map codes.

designated by an alpha code or number code on the maps. An example of a soil survey map is presented in Figure 13.3.

When used in conjunction with a list of hydric soil-map units (available from the Nat-

ural Resources Conservation Service's county offices), the soil-survey maps can help identify the likely presence of wetland in a given area. Yet, as with any map, there are limitations, including the following:

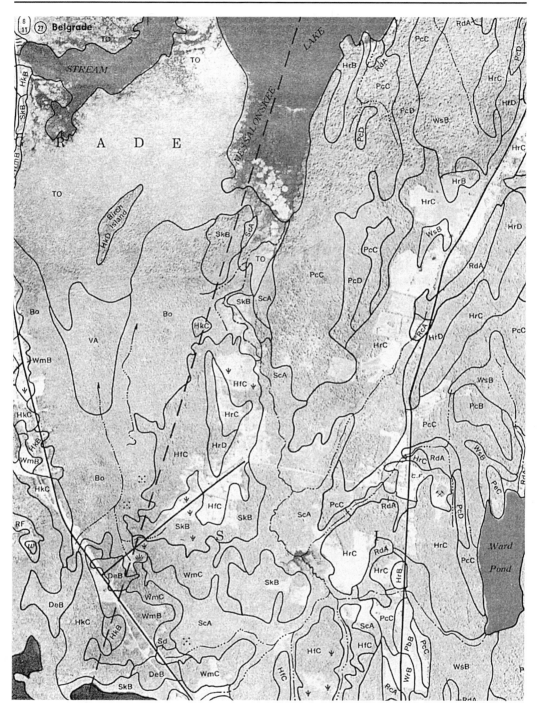

Figure 13.3. A soil survey map (Belgrade, Maine). Alpha codes represent different types of soil-map units. Hydric soil-map units are TO (Togus peat), VA (Vassalboro peat), Bo (Biddeford mucky peat), and ScA (Scantic silt loam). (Source: USDA Natural Resources Conservation Service)

- In general, the minimum map unit of soil types ranges from 1.5 to 10 acres, depending on landscape diversity and survey objectives. For a 1:24,000 map, the minimum size is 5.7 acres.

- Soil-map units often have inclusions of other soils, so that a large area mapped as nonhydric (upland) soil may, in fact, have substantial acres of hydric soil (wetland) within its borders, and vice versa. (As much as 40 percent of a map unit may contain other soils.)

- Some obvious wet spots may be identified by a "wet" symbol within areas of upland soil.
- Soil maps usually do not differentiate between undrained hydric soils and drained or recently filled hydric soils, thus making it virtually impossible to separate current wetlands from historic wetlands, except in uncommon instances where a flooded phase or undrained phase or made land (fill) is mapped.
- Some members of hydric soil series do not possess hydric properties and are not associated with wetlands, while other members are hydric; these "facultative hydric soils" may or may not be hydric, depending largely on landscape position (those in depressions are usually hydric, but those on slopes are often not); soil maps do not distinguish between them.

Failing to recognize these points, some people have unfortunately used the acreage of hydric soil-map units to estimate the acreage of land subject to wetland regulations. This has resulted in grossly exaggerated reports on acres of wetlands subject to government regulation across the country. The soil surveys and acreage of hydric soil-map units should not, therefore, be the sole sources used to identify the presence of wetlands or to determine the acreage of land subject to regulation. Soil maps should be used in conjunction with the NWI maps, as well as other information (e.g., current aerial photos), to identify the likely presence of wetland in a given area and soils that may be encountered during field inspections. After reviewing these materials, on-site inspections should be performed to verify the presence of wetland and determine its boundaries.

Seeking Swampland: How to Find Wetlands in the Field

Regardless of the wetland definition chosen, it is important to recognize that wetland identification on the ground will depend on the indicators used. The choice of indicators is the key to deciding whether a particular parcel of land, mud, or water will be called a wetland by scientists and regulators. Much of this book is devoted to explaining how the critical indicators of distinctive vegetation, soils, and signs of wetness vary with the wetland type, how to recognize them in the field, and how to use the indicators for wetland identification and delineation.

Traditionally, wetlands were recognized by their characteristic vegetation. However, upon closer examination of the diverse types of wetlands, it became evident that not all wetlands possessed unique plant communities. As we have seen, many drier-end wetlands and the borders of most wetlands are colonized by plant species that may live in both wetlands and the drier uplands and as individual plants that are virtually indistinguishable from one another from the vegetation standpoint (see Chapters 5 and 10). For this reason, scientists have turned to soils for additional help in identifying wetlands. In most landscapes, soil properties change significantly in response to prolonged wetness (see Chapters 4 and 11). Various combinations of plants, soils, and other signs of wetness are used to identify wetlands and delineate their boundaries.

To date, animals have not been used for wetland identification, although it is acknowledged that certain wildlife are specific to wetlands and that wildlife habitat is an important wetland function (see Chapters 6 and 12). Most animals tend to be too mobile or secretive, making their use impractical for wetland-identification purposes, with some exceptions (e.g., aquatic invertebrates). The presence of certain species is a useful indicator of wetland (e.g., muskrat, beaver, or mallard duck), but they do frequent uplands at times, so their presence at a given site at a moment in time does not conclusively mean that the area is a wetland. Plants and soils remain in place, and therefore serve as the principal indicators for identifying wetlands and their boundaries.

To accurately locate wetlands on your property or elsewhere, you have to examine vegetation, soils, and other signs on the ground. Available maps may provide some valuable insight and may adequately depict the wetland boundary in areas with abrupt topographic changes, but field inspection is a must if you want to really know the location and extent of wetland on your property. This book attempts to give you a good foundation for knowing what constitutes a wetland and how to recognize wetlands, mainly by their vegetation (hydrophytes) and soils (hydric soils). With this knowledge, you should be able to determine if a wetland is present on your land, and if present, the wetland boundary or upper limit for typical wetlands.

When trying to identify wetland on a site, first walk the entire area to learn its topography and determine where wetlands probably exist. Pay attention to areas with different plant communities and to changes in microtopography (slight changes in elevation can make a big difference in hydrology). For example, in open meadows, the presence of yellowish green "grasses" (often sedges) in low spots or along sloping drainageways, surrounded by green upland grasses, indicates possible wetlands. Similarly, during droughts or in late summer, the occurrence of lush green patches in a mostly brown field or pasture, where the majority of the vegetation has succumbed to the heat and accompanying lack of water, may also indicate the presence of wetlands. These rather simple "windshield" observations are often surprisingly reliable wetland indicators. More thorough examination of vegetation and soils will usually verify such interpretations.

Plants usually reflect changes along environmental gradients, especially changes in soil moisture with topographic variations. Most wetlands can be identified by a prevalence of hydrophytes, a predominance of undrained hydric soils, and direct or indirect evidence of flooding, ponding, or soil saturation during the growing season. You should have little difficulty identifying a wetland where you get your feet wet in late spring and summer. In these cases, OBL and FACW plants (see Chapter 10), characteristic of seasonally flooded and wetter wetlands, often predominate. These plants growing together are reliable indicators of wetlands.

Since not all wetlands look alike (due mainly to differences in hydrology, vegetation, and soils), it is not always easy to recognize wetlands on the ground. Remembering, however, that wetlands occur along a natural wetness (soil-moisture) gradient, you should be able to recognize the range of wetlands associated with various water regimes, from permanently flooded wetlands in shallow water to seasonally saturated wetlands that are only briefly saturated, usually early in the growing season. These latter wetlands pose the greatest problem for wetland recognition, since most, if not all, of the plants growing here can also be found on uplands. In fact, many of the plants in these drier wetlands are FAC and FACU species. Thus, in these wetlands, vegetation alone will not provide the answer to the question, Is this wetland? Since saturation is short-lived and surface water is not present, water will be absent from the upper soil most of the time, especially in summer. There may, however, be indirect evidence of inundation in nearby depressions, such as water-stained leaves or stems (Plate 38a,c) or signs of soil saturation (e.g., oxidized rhizospheres; Plate 1e) in nonflooded microsites. These are perhaps the best of several indicators of wetland hydrology used by government regulators in wetland delineation. Other hydrology indicators are sediment deposits (Plate 38f), silt marks on vegetation (Plate 38b), evidence of scouring (on floodplains), drift lines (Plate 38d,e), presence of peat moss (Figure 7.4), algae deposits (Plate 38g), lichen lines, and aquatic invertebrates (Plate 38h), plus morphological plant adaptations (Plate 38i, also see Tables 2.3 and 5.1 and Plate 1).

You still need to confirm that these hydrology signs are indicative of recurring prolonged wetness. This is done by examining the soil for hydric properties within 1.5 feet of the surface. Examining the soil provides verification of the presence or absence of wetland, for soil properties usually reflect long-term hydrology. If indicators of undrained hydric soils (poorly and very poorly drained soils) are present, you should be confident that the area is wetland and can proceed to mark the upper boundary at the project site. If you find major drainage ditches or know of activities (e.g., groundwater withdrawals) that may have significantly altered the area's hydrology, it is best to call in an expert to assess the current hydrology.

To achieve the best results in the field, inspect the site in question at two times during the year—at the beginning and the peak of the growing season, if possible. Avoid or at least be aware of extreme flooding conditions or periods immediately following heavy rains, because low-lying uplands could be flooded at these times or soils could be wet only at the surface. While early spring may be the best time from a hydrologic standpoint for making a wetland determination in the eastern United States, it is not the optimal time for identifying vegetation, since most herbaceous plants are not recognizable at this time. Woody plants, however, can be identified in all seasons. While both spring and summer visits are recommended, experienced wetland specialists can usually make a wetland determination by a single site visit. After using this book for a while, you too might be able to do this for most sites.

You should also be aware that small wetlands may be scattered throughout a large upland tract of land and that these small areas should be treated separately for wetland-determination purposes. The converse is also true: small upland islands may exist within large wetland complexes. Be sure to identify homogeneous areas having similar vegetation, topography, and other characteristics within the project site for separate evaluation. An initial walk through the entire project area should reveal any significant differences.

For regulatory purposes, wetlands are currently identified in the field by the presence of hydrophytic vegetation, hydric soils, and signs of wetland hydrology. Briefly described in Chapter 1, this approach is commonly called the three-parameter approach. In order to validate the occurrence of wetland, positive indicators of all three criteria must be recorded in most circumstances. Positive indicators of hydrophytic vegetation are generally OBL, FACW, and FAC plants, while positive indicators of hydric soils are the soil properties listed in Chapter 11. Wetland-hydrology indicators are listed above (see Table 2.3). Refer to specific manuals for details, as there are differences in the interpretation of these features.

If you are simply interested in seeing if there are wetlands on your property, the first thing to consider is the landscape position. Wetlands are typically found in depressions, along watercourses and waterbodies, and in drainageways or seepage areas on slopes or at the toes of slopes. Recognizing these landscapes allows you to separate areas with high probability of wetlands from areas with low probability. After this initial overview, continue the determination process by using the following rapid-assessment technique. It is derived from a method I proposed for identifying wetlands that are not significantly hydrologically modified—in other words, wetlands lacking extensive drainage structures or other hydrologic manipulation. Called the "Primary Indicators Method," it uses for wetland identification the presence of vegetation, soil characteristics, or other unique features that occur only in wetlands (see Table 13.1 above for examples). In *Wetlands: Characteristics and*

Table 13.1. Unique characteristics of Northeast wetlands that are reliable wetland indicators.

Vegetation Indicators

1. OBL species constitute more than 50% of the common species.
2. OBL and FACW species constitute more than 50% of the common species.
3. OBL species represent at least 10% of the plant community and are evenly distributed throughout the community.
4. At least one common species has one of the following morphological adaptations: hypertrophied lenticels, buttressed stems or trunks, and floating leaves. (Note: These features must be observed on the majority of individuals.)
5. Surface incrustations of algae are materially present.
6. Significant patches of peat moss (*Sphagnum* spp.) are present. Reliable in most areas, except higher elevations in Adirondacks and New England mountains.

Soil Indicators

1. More than 8 inches of organic material (peat or muck) present (Plates 24–26).
2. Strong smell of rotten eggs (hydrogen-sulfide gas) detected within upper 12 inches of the soil.
3. For nonsandy soils, gleyed subsoil, usually with (but sometimes without) orange, yellow, or reddish brown mottles conspicuous and plentiful, and other hydric soil properties within 12 inches of the surface (see Chapter 10 and Plates 26–30; Plate 34; Plate 37).
4. For sandy soils, black sandy topsoil, blotchy or mottled gray sandy subsoils, and other hydric soil properties within 6 inches of the surface (see Chapter 10 and Plates 32 and 35).
5. Remains of aquatic invertebrates (e.g., snails, clams, or dragonfly nymphs) on the soil surface or within 12 inches of it.

The presence of any one of these typically identifies a wetland, provided that the area is not subjected to significant hydrologic modification, such as drainage. Common species are plants representing more than 20% of the plant community's cover.

Boundaries, the National Research Council concluded that this approach was supported by the scientific literature.

From a vegetation standpoint, OBL species are emphasized because they are the best plant indicators of wetland. When OBL species are dominant, abundant, or present in modest numbers but scattered throughout the site, the area should typically be a wetland. For example, when skunk cabbage covers the forest floor in the spring, you can be pretty certain that the area is a forested wetland. It's nice to know the associated plants, but for wetland identification, you already know that the area is a wetland based on the predominance of skunk cabbage. When OBL species are absent, even the predominance of FACW species does not unequivocally indicate wetlands, so you must find other signs—other primary indicators to verify the presence of wetland. In these cases, unless you know the hydrology of the site, the soils must be examined for hydric properties (Chapter 11; Plates 24–37). If these properties are present, the area should be a wetland (unless, of course, the area is effectively drained).

Drawing a Line in the Dirt: Delineating the Boundary

Establishing the wetland-upland boundary is of critical importance, especially if you plan to do some type of construction work in or near the wetland. To draw this line, you must make astute observations, correlating differences in vegetation and soils with topographic changes.

Where the change in slope is abrupt, a marked change in the plant community usually occurs as wetland plants quickly give rise to UPL plants or FAC and FACU plants growing under upland conditions. Even though these differences are apparent, it is still worth examining the soil along the gradient to ensure that observed vegetation patterns are the result of a change in soil moisture—that is, hydric soil to nonhydric soil—and not the result of other phenomena, such as soil fertility, soil chemistry, or human disturbance.

In many situations, changes in topographic relief are more subtle, and the obvious differences in vegetation patterns are lacking. This happens throughout the coastal plain, along floodplains, where elevations are low and topographic changes are gradual, and also on gentle slopes in groundwater-seepage areas. In these areas, and along the upper edges of many wetlands, OBL and FACW plants may gradually intermix with FAC and FACU plants to form what is sometimes called a "transition zone." Flooding or soil saturation may not be particularly evident here, since these wetlands may be only temporarily flooded (e.g., have water ponded on the surface only briefly during the growing season) or seasonally saturated (i.e., have soils saturated for only the early part of

the season). In these cases, where hydrology is not apparent and vegetation is inconclusive, the soils need to be examined, for they reflect the long-term hydrology.

For these sites, you should set up a transect (a straight survey line) to evaluate subtle changes in vegetation and soils. The transect should go downhill, for this insures that you will be evaluating changes along the soil-moisture gradient (Figure 13.4).

To identify a wetland boundary, follow these steps:

(1) First walk downslope into the area to a point where OBL plants or other "primary indicators" of wetland are clearly present.

(2) Then walk in a straight line upslope toward the perceived boundary, evaluating changes in vegetation and periodically examining the soil to see if hydric properties are present (especially where OBL species are absent).

(3) When hydric soil properties or other "primary indicators" are no longer present, mark that point with flagging tape on a nearby tree or shrub. You have just identified one boundary point. The relationship between wetland and nonwetland may then be seen by looking at subtle differences in vegetation on each side of the boundary. In many cases, the boundary will follow a certain elevation (topographic contour). The boundary line for the rest of the wetland may simply follow this contour, or you can establish it by repeating the transect procedure at other locations and connecting the points together (Figure 13.4).

If you're doing a delineation for regulatory purposes, you must follow agency guidelines specified in wetland-delineation manuals. Be sure to take thorough notes on your observations, because good documentation is crucial for explaining your findings to others. Also remember that even when soils are examined, the wetland boundary may not be clear-cut, yet by examining vegetation, soils, and landscape position and considering possible hydrology as necessary, you will be able to draw the best line possible during a single site visit.

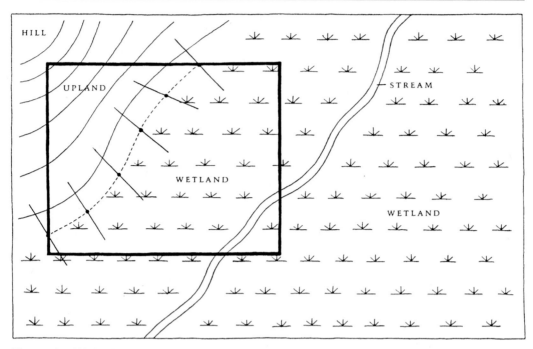

Figure 13.4. Wetland boundaries are often determined by evaluating vegetation and soil changes along a slope. Transects running downslope are established for this purpose. Wetland indicators reveal the presence of wetland. Transects may be formal, with specific data recorded, or informal, simply for making observations of changing vegetation and soils. The wetland boundary is usually marked with flagging tape in a number of locations to show where wetland indicators are no longer present. (Source: Tiner 1988)

Additional Readings

Environmental Laboratory. 1987. *Army Corps of Engineers Wetland Delineation Manual.* U.S. Army Corps of Engineers, Waterways Experiment Station, Vicksburg, MS. Technical Report Y-87-1.

Federal Interagency Committee for Wetland Delineation. 1989. *Federal Manual for Identifying and Delineating Jurisdictional Wetlands.* U.S. Army Corps of Engineers, U.S. Environmental Protection Agency, U.S. Fish and Wildlife Service, and USDA Soil Conservation Service, Washington, DC.

Jackson, S. 1995. *Delineating Bordering Vegetated Wetlands under the Massachusetts Wetlands Protection Act.* Massachusetts Department of Environmental Protection, Division of Wetlands and Waterways, Boston.

Sipple, W. S. 1987. *Wetland Identification and Delineation Manual.* Volumes I and II. U.S. Environmental Protection Agency, Office of Wetland Protection, Washington, DC.

Tiner, R. W., Jr. 1988. *Field Guide to Nontidal Wetland Identification.* U.S. Fish and Wildlife Service, Newton Corner, MA, and Maryland Department of Natural Resources, Water Resources Administration, Annapolis, MD. (Out of print, but copies available for purchase from IWEER, P.O. Box 288, Leverett, MA 01054.)

Tiner, R. W. 1993. The primary indicators method—a practical approach to wetland recognition and delineation in the United States. *Wetlands* 13: 50–64.

Tiner, R. W. 1995. Practical considerations for wetland identification and boundary delineation. In *Wetlands: Environmental Gradients, Boundaries, and Buffers,* ed. G. Mulamoottil, B. G. Warner, and E. A. McBean, pp. 113–137. CRC Press, Lewis Publishers, Boca Raton, FL.

Tiner, R. W. 1999. *Wetland Indicators: A Guide to Wetland Identification, Delineation, Classification, and Mapping.* CRC Press, Lewis Publishers, Boca Raton, FL.

CHAPTER 14

Typing and Evaluating Wetlands
Wetland Classification and Functional Assessment

After identifying an area as a wetland you might be interested in classifying it and getting an idea of its functions. In reading the wetland primer section of this book, you've learned much about the diversity of wetlands and their functions. Since wetlands have formed in different climates, geologic settings, landscapes, and soils, and under different hydrologic regimes and water chemistries, all wetlands are not alike in either form or function. Scientists studying wetlands have used many terms to describe these differences. Over time some of these terms have been applied in two basic ways—they may simply refer to a wetland of any kind in common language or have a more precise meaning to scientists. For example, terms like marsh, bog, and swamp have been widely used by average citizens to refer to any land that is waterlogged or periodically flooded. On the other hand, most scientists in the United States generally equate marsh to a wetland dominated by herbs (non-woody plants) growing in shallow water or frequently flooded sites, bog to a wetland formed on peat soils, and swamp to a wetland dominated by shrubs and/or trees growing on wet mineral or mucky soils. These terms actually have more specific definitions than presented here.

Classification from the natural resource standpoint is the grouping of habitats or natural features into categories with similar characteristics, properties, and/or functions. The unifying properties vary according to the needs of the classifier. For example, wetlands may be classified biologically, physically, chemically, hydrogeomorphically, and in other ways depending on the discipline and interests of the classifier. Most of the classification has focused on the form of the wetland type rather than on the function. Recent attention has begun to emphasize the latter due to the need to evaluate wetland functions when assessing the impact of proposed projects subject to federal, state, and local wetland regulations and ordinances.

Wetland classification is important because it usually provides a visual picture of the wetland (its physiognomy) and an idea of what its functions may be. The more detailed the classification, the more information one has to evaluate and the more likely it will accurately describe the wetland and its functions. Classification also allows researchers to identify similar areas for comparative studies and to conduct wetland inventories that can be used in wetland management and conservation.

Defining Properties

Formal classification systems have been developed by scientists with a particular expertise (e.g., ecologists, botanists, foresters, wildlife biologists, soil scientists, hydrologists, and natural resource planners) and the distinguishing features used to characterize wetlands are based largely on their area of interest. Some of the more commonly used characteristics are vegetation, hydrology, water chemistry, soil type, landscape position, and landform. The following discussion emphasizes some of the more commonly used terms; for more technical terminology consult *Wetland Indicators: A Guide to Wetland Identification, Delineation, Classification, and Mapping* (Tiner 1999).

Vegetation

Vegetation has been the focal point of most wetland classification systems, since plant community differences are obvious and most classifiers have had an interest in plant ecology, forest management, wildlife habitat management, or natural resource conservation. Some useful vegetative features considered in wetland classification are: life form (tree, shrub, herb, moss/lichen, aquatic), leaf type (needle-leaved, broad-leaved), leaf persistence (deciduous, evergreen, persistent herb, non-persistent herb), percent cover by life forms, and community type (dominant plant species). Because of the great interest in wetlands by plant ecologists and wildlife managers, predominant species and plant associations (e.g., salt-hay grass, salt grass marsh or red maple, green ash swamp) have been used to characterize wetlands, especially for local or site-specific studies. This level of detail is less useful for state and regional inventories (mapping efforts) since it cannot often be detected through remote sensing technology (e.g., aerial photointerpretation or satellite image analysis). Classification systems designed for mapping purposes usually focus on vegetation life form and other factors (e.g., degree of wetness and water chemistry), although conspicuous dominant species that are readily detected through remote sensing may be identified (e.g., Atlantic white cedar or common reed). Consequently, wetland types in many classifications describe assemblages of plants such as salt marsh, wet meadow, shallow marsh, deep marsh, shrub swamp, shrub bog, and wooded swamp or forested wetland. Species composition of these types varies regionally.

Peatland scientists have described a wide range of peatlands, especially for northern climates. They usually identify communities based on a combination of species associations and the physiognomy of the vegetation. Typical physiognomic types include hummock communities (growing on mounds of peat), lawn communities (firm turflike vegetation), carpet communities (softer than lawns, including quaking mats), and mud bottom communities (frequently submerged muddy areas).

Hydrology

Hydrologic characteristics are important descriptors of wetlands because hydrology is the forcing function that creates and maintains wetlands. Differences in flooding and soil saturation create varied environmental conditions that affect plant composition, wildlife use, soil development, and other properties as well as wetland functions.

From a hydrologic perspective, wetlands subject to tidal influence are separated from nontidal wetlands. Then the frequency and duration of flooding or soil saturation is used to separate other types (e.g., temporarily flooded, seasonally flooded, and semipermanently flooded, see

Table 2.6). Water depth has been used to identify different types of basin or depressional wetlands (e.g., shallow marsh from deep marsh). Directional flow of water (inflow, outflow, throughflow, bidirectional, or isolated) is important in assessing whether a wetland behaves as a source, sink, or pass-through system. Hydrodynamics may also be characterized to separate wetlands with fluctuating water tables from wetlands subject to one-directional flow (i.e., downstream or downhill) and from wetlands with bidirectional flow (e.g., tides and lake seiches). Wetlands may also be classified by water source (e.g., precipitation, snowmelt, groundwater, and surface-water, including tides). Unfortunately, most classification systems offer only a few broad hydrologic categories because there is a general lack of detailed, long-term studies of wetland hydrology.

Water Chemistry

The chemical composition of water is important to wetlands everywhere as it has great impact on plant composition. Four major areas where it is particularly useful for classifying wetlands are: 1) estuaries and marine shores where tidal waters with ocean-derived salts (mostly sodium chloride) create saline soils, 2) arid and semiarid regions where inland salts (calcium, magnesium, sodium, potassium, chloride, bicarbonates, and sulfates) accumulate forming alkaline wetlands (e.g., wetlands along the Great Salt Lake, Utah), 3) boreal and arctic regions where nutrient-poor, acidic bogs (pH < 4.5) and nutrient-rich, circumneutral to basic fens (pH 5.5–9.0) develop depending on groundwater interactions, and 4) calcareous, limestone regions where alkaline wetlands establish. Water chemistry descriptors include salinity (inland salts), halinity (coastal salts), and pH (hydrogen ion concentration). The latter provides a measure of acidity (ranging from acidic to alkaline). Terms like ombrotrophic (typically nutrient-poor wetland dependent on rainwater) and minerotrophic (wetland fed by mineral-rich groundwater) have been used to describe peatlands.

Soil Types

Wetlands may be separated into those with organic soils (peats and mucks) and those with mineral soils (sands, silts, clays, and various mixtures). Peatlands are commonly culled out from the mucklands due to their abundance in northern regions (e.g., boreal climates and higher elevations in temperate climates). Specific soil types (i.e., soil series) may be classified and mapped by soil scientists during soil surveys, but the mineral soil wetlands tend to be classified more by vegetation and hydrology than by soil texture (e.g., silt clay, loam, sandy loam, and fine sand) or other soil properties in various wetland classification systems.

Landscape Position and Landform

Wetlands typically form in several important hydrogeologic settings: topographic depressions, slope breaks, areas of stratigraphic change, permafrost areas, and paludified landscapes (Figure 14.1). Landscape position and landform are important geomorphologic features that relate to wetland functions and aid in grouping wetlands by function. Landscape positions may include marine (ocean shores), estuarine (estuary shores), riverine or lotic (river and stream

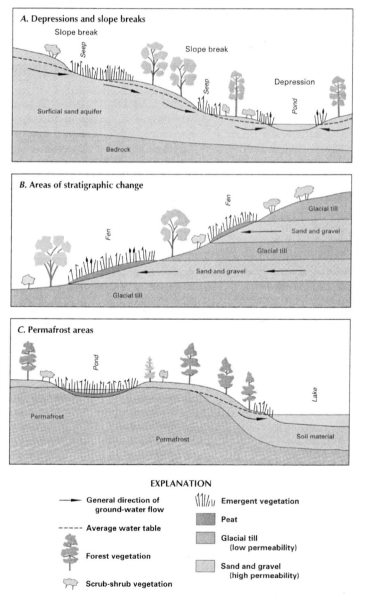

Figure 14.1. Three main hydrogeologic settings favoring wetland formation. (Source: Carter 1996)

shores), lacustrine or lentic (lake shores), and terrene (surrounded by upland), for example.

Name that Wetland: Classifying Wetlands

There are two basic approaches to classification: horizontal and vertical. The former lists a finite number of types based on a few major characteristics, whereas the latter provides a stepwise approach starting with general characteristics and then further subdividing the groups into smaller and smaller groups based on a larger suite of properties. Examples of these two basic approaches follow, emphasizing U.S. systems. By design, this is an overview and not an in-depth analysis of the history of wetland classification. For information on other classification systems, readers should consult *Wetland Indicators: A Guide to Wetland Identification, Delineation, Classification, and Mapping.*

Horizontal Classifications

Horizontal classifications tend to be very generalized out of the necessity to limit the number of wetland types. These classifications have produced terms like marsh, swamp, and bog that are familiar to the public. Unfortunately, many of these common terms lack universally accepted definitions and have been inconsistently used. Despite this, many scientists, including myself, still use these terms when referring to certain types of wetlands as they remain useful descriptors of broad categories of wetlands. For example, in Chapter 5 of this book, I described seventeen major types of wetlands found in the Northeast and Great Lakes region: 1) salt marshes, 2) brackish marshes, 3) tidal fresh marshes, 4) tidal swamps, 5) tidal submerged aquatic beds, 6) nontidal marshes (including Great Lakes coastal marshes), 7) savannas (including Carolina bays), 8) wet meadows, 9) fens, 10) bogs (including shrub and forested types), 11) shrub swamps, 12) hardwood swamps (including northern hardwood swamps and coastal plain swamps), 13) larch swamps, 14) floodplain forested wetlands, 15) evergreen forested swamps (including cedar swamps, hemlock swamps, and pine swamps, lowlands, and flatwoods), 16) vernal pools, and 17) nontidal aquatic beds.

In one of the earliest plant ecology textbooks, *Oecology of Plants,* the Dutch plant ecologist Eugenius Warming (1909) identified two major types of wetland plant communities: saline swamps (with halophytic vegetation) and freshwater swamps. Freshwater swamps included three types: reed swamp, bush swamp, and forest swamp. The former type is dominated by herbs, whereas the latter types are characterized by shrubs and trees, respectively. Among terms used in the earliest classification systems for inventorying wetlands nationwide were river swamps, lake swamps, upland swamps (wet woods and climbing bogs), permanent swamps, wet grazing land, periodically overflowed land, and periodically swampy land. They were used to identify "wastelands" that were targeted for drainage to support agriculture, given the strong farming interests of a growing country. A 1922 federal survey of wetlands based on soil surveys, topographic maps, and other sources divided wetlands into several types including tidal marsh, inland marsh, swamp and timbered overflow lands, and very deep peat.

In the 1950s, the U.S. Fish and Wildlife Service developed a classification for inventorying wetlands important for waterfowl. This system identified twenty types (Table 14.1) of which four (types 5, 11, 14, and 19) are mostly shallow open waterbodies. Putting all of the U.S. wetlands into just twenty types meant significantly different wetlands were included in the same class (e.g., black spruce forests and cypress swamps in wooded swamps). Inadequate definition of types caused inconsistencies in application across the country.

Vertical Classifications

With more precise definitions and more properties considered, a vertical or hierarchical approach will likely produce more consistent mapping and more detailed characterization than a horizontal approach. Lower levels share more generalized characteristics, such as landscape position and water source, while higher levels are based on more detailed and specific properties like vegetation life form (including dominant species), substrate type, and water level fluctuations. By providing more descriptive characterizations of wetlands, a hierarchical classification system can be used to better predict the potential functions of individual wetlands than a horizontal classification offers.

Table 14.1. Wetland types used by the U.S. Fish and Wildlife Service to inventory wetlands from the 1950s to the mid-1970s

Type	Name	Definition
1	Seasonally flooded basins or flats	Soil covered with water and waterlogged during variable seasonal periods; usually well drained during much of the growing season; located on river bottoms, along borders of drawn-down reservoirs, and in "dry lakes," shallow potholes, and other shallow upland depressions; vegetation includes bottomland woods and herbaceous areas
2	Inland fresh meadows	Soil without standing water but waterlogged within at least a few inches of its surface during the growing season; located mostly in glaciated areas, the Nebraska sandhills, and in Florida, or may border marshes on landward side; vegetation includes grasses, sedges, rushes, and broad-leaved species
3	Inland shallow fresh marshes	Soil normally waterlogged during the growing season; often covered by as much as 6 inches of water; locally mostly in glaciated areas, the Nebraska sandhills, and in Florida, or may border deep marshes on landward side, or adjoin irrigation systems; vegetation includes grasses, bulrushes, spikerushes, cattails, arrowheads, pickerelweed, smartweeds, reed, whitetop, rice cutgrass, bur-reeds, maidencane, sawgrass, and Baker cordgrass
4	Inland deep marshes	Soil covered with 0.5 to 3.0 feet of water during the growing season; located mostly in glaciated areas, the Nebraska sandhills, and in Florida where they occur in and along margins of basins, potholes, limestone sinks, and sloughs; vegetation includes cattails, reed, bulrushes, spikerushes, wild rice, and various aquatic plants
5	Inland open fresh water	Shallow ponds and reservoirs; water usually less than 10 feet deep and fringed by a border of emergent vegetation; vegetation includes pondweeds, naiads, wild celery, coontail, water milfoils, muskgrasses, water lilies, spatterdocks, and (in the South) water hyacinth
6	Shrub swamps	Soil normally waterlogged during the growing season, often covered with up to 6 inches of water; located mostly in the eastern United States, also in Pacific Northwest; typically occur along sluggish streams and occasionally on floodplains; vegetation includes alders, willows, buttonbush, dogwoods, and swamp privet
7	Wooded swamps	Soil waterlogged at least to within a few inches of its surface during the growing season, often covered with up to 1 foot of water; located mostly in the eastern United States, also in the Pacific Northwest; occurs mostly along sluggish streams, on floodplains, on flat uplands, and in shallow lake basins and potholes; vegetation includes tamarack, white cedars, black spruce, balsam fir, red maple, black ash, western hemlock, red alder, and willows
8	Bogs	Soil usually waterlogged, generally blanketed with a spongy covering of mosses; located in glaciated areas, western mountains, and along the Atlantic and Gulf coastal plains; mostly in shallow lake basins and potholes, on flat uplands, and along sluggish streams; vegetation includes heath shrubs (leatherleaf, labrador tea, cranberries, Cyrilla, Persea, and Gordonia), black spruce, tamarack, pitcher plants, sphagnum moss, and sedges
9	Inland saline flats	Soil without standing water, but waterlogged within at least a few inches of the surface during the growing season; located in the Great Basin, the northern Great Plains, and other arid regions; mostly in shallow lake basins; vegetation includes salt-tolerant plants such as sea blite, salt grass, Nevada bulrush, saltbush, and burroweed
10	Inland saline marshes	Soil normally waterlogged during the growing season, often flooded with as much as 2 feet of water; located in the Great Basin, the northern Great Plains, and other arid regions; mostly in shallow lake basins; vegetation includes alkali or hardstem bulrush, widgeon-grass, and sago pondweed
11	Inland open saline water	More permanent areas of shallow saline water often closely associated with types 9 and 10; water depth is variable; vegetation mainly at depths less than 6 feet includes sago pondweed, widgeon-grass, and muskgrasses

(continued)

Table 14.1. (continued)

Type	Name	Definition
12	Coastal shallow fresh marshes	Soil always waterlogged during the growing season, may be covered at high tide with up to 6 inches of water; located along the Atlantic, Gulf, and Pacific coasts, on the landward side of deep marshes along rivers, sounds, and deltas; vegetation includes grasses (e.g., big cordgrass, reed, giant cutgrass, and maidencane), sedges (e.g., *Carex*, spikerushes, three-squares, and sawgrass), cattails, arrowheads, smartweeds, and arrow arum
13	Coastal deep fresh marshes	Soil covered at average high tide with 0.5 to 3 feet of water during the growing season; located along tidal rivers and elsewhere on the Atlantic and Gulf coasts; vegetation includes cattails, wild rice, pickerelweed, spatterdock, alligatorweed, water hyacinth, water lettuce, and aquatics
14	Coastal open fresh water	Shallow portions of open water along fresh tidal rivers and sounds; vegetation at depths of less than 6 feet includes pondweeds, naiads, wild celery, coontail, waterweeds, water milfoils, and muskgrasses and in the Gulf, some mats of water hyacinth
15	Coastal salt flats	Soil almost always waterlogged during the growing season, varying from areas submerged only by occasional wind tides to areas that are fairly regularly flooded with a few inches of water; located along the Atlantic, Gulf, and Pacific coasts, on the landward side of, or as islands or basins within, salt meadows and salt marshes; vegetation is sparse or patchy, including glassworts, sea blite, salt grass, saltflat grass, and saltwort
16	Coastal salt meadows	Soil always waterlogged during the growing season, rarely flooded; located along the Atlantic, Gulf, and Pacific coasts, on the landward side of salt marshes or bordering open water; vegetation includes saltmeadow cordgrass, salt grass, black rush, Olney three-square, salt marsh fleabane, *Carex*, hairgrass, and *Jaumea*
17	Irregularly flooded salt meadows	Soil covered by wind tides at irregular intervals during the growing season; located on the Atlantic coast from Maryland south and on the Gulf coast, along the shores of nearly enclosed bays, sounds and rivers; vegetation includes black needlerush
18	Regularly flooded salt marshes	Soil covered at average high tide with 0.5 feet or more of water during the growing season; located along the Atlantic, Gulf, and Pacific coasts, mostly along sounds, but also along the open ocean in places; vegetation includes salt marsh cordgrass, alkali bulrush, glassworts, and arrowgrass
19	Sounds and bays	Portions shallow enough to be filled and diked including all waters landward of the low-tide line; vegetation includes eelgrass, widgeon-grass, sago pondweed, muskgrasses, and in the Southeast, shoalgrass, manateegrass, and turtlegrass
20	Mangrove swamps	Soil covered at average high tide with 0.5 to 2 feet of water during the growing season; located along the southern half of Florida; vegetation is mostly red mangrove and some black mangrove

U.S. Fish and Wildlife Service Wetland Classification

In 1974, the U.S. Fish and Wildlife Service (FWS) began planning for a new national wetlands inventory. The decision was made that the new system should be hierarchical in structure, so that the users could select a level of detail appropriate to their needs. In addition, the new system had three primary objectives: 1) to group ecologically similar habitats so that value judgments can be made, 2) to furnish habitat units for inventory and mapping, and 3) to provide uniformity in concepts and terminology throughout the entire United States. This system, *Classification of Wetlands and Deepwater Habitats of the United States*, has been used to map wetlands across the country (National Wetlands Inventory maps; see Chapter 13) and to report on wetland status and trends. In the 1990s, it was adopted by the federal government as the national data standard for wetland classification for government data collection and reporting on wetland status and trends.

The FWS's wetland classification system is hierarchical, proceeding from general to specific as illustrated in Figure 14.2. Wetlands are first defined at a rather broad level—the *system*. The

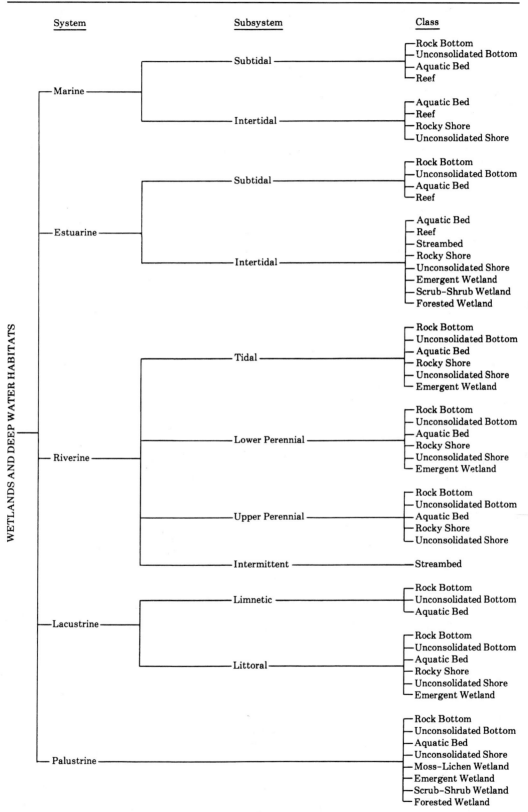

Figure 14.2. Hierarchy of wetlands and deepwater habitats following the U.S. Fish and Wildlife Service's official classification system. (Source: Cowardin et al. 1979)

term *system* represents "a complex of wetlands and deepwater habitats that share the influence of similar hydrologic, geomorphologic, chemical, or biological factors." Five systems are defined: marine, estuarine, riverine, lacustrine, and palustrine. The marine system generally consists of the open ocean and its associated high-energy coastline, while the estuarine system encompasses salt and brackish marshes, nonvegetated tidal shores, and brackish waters of coastal rivers and embayments. Freshwater wetlands and deepwater habitats fall into one of the other three systems: riverine (rivers and streams), lacustrine (lakes, reservoirs and large ponds), or palustrine (e.g., marshes, bogs, swamps and small shallow ponds). Thus, at the most general level, wetlands can be defined as either marine, estuarine, riverine, lacustrine or palustrine (Figure 14.3; see Table 14.2 for dichotomous key).

Each system, with the exception of the palustrine, is further subdivided into *subsystems*. The marine and estuarine systems both have the same two subsystems, which are defined by tidal water levels: 1) subtidal (continuously submerged areas) and 2) intertidal (areas alternately flooded by tides and exposed to air). Similarly, the Lacustrine System is separated into two systems based on water depth: 1) littoral (wetlands extending from the lake shore to a depth of 6.6 feet below low water or to the extent of nonpersistent emergents such as arrowheads, pickerelweed, or spatterdock if they grow beyond that depth), and 2) limnetic (deepwater habitats lying below 6.6 feet at low water). By contrast, the riverine system is further defined by four subsystems that represent different reaches of a flowing freshwater or lotic system: 1) tidal (water levels subject to tidal fluctuations for at least part of the growing season), 2) lower perennial (permanent, flowing waters with a well-developed floodplain), 3) upper perennial (typically high-gradient, permanent flowing water with very little or no floodplain development), and 4) intermittent (channel containing nontidal flowing water for only part of the year).

The next level—*class*—describes the general appearance of the wetland or deepwater habitat in terms of the dominant vegetative life form or the nature of the substrate where vegetative cover is less than 30 percent (Table 14.3). Of the eleven classes, five refer to areas where vegetation covers 30 percent or more of the surface: aquatic bed, moss-lichen, emergent, scrub-shrub, and forested. The remaining six classes represent areas generally lacking vegetation, where the composition of the substrate and

Table 14.2. Key to wetland ecological system following the official federal wetland classification system

1. Wetland is under tidal influence 2
1. Wetland is not affected by tides 3
  2. Water is fresh (less than 0.5 ppt at low water) 3
  2. Water is salt or brackish 7
    3. Wetland is vegetated 4
    3. Wetland is nonvegetated 5
      4. Wetland is colonized by aquatic bed species or nonpersistent emergents 5
      4. Wetland is colonized by woody plants or persistent herbaceous plants .... Palustrine
        5. Wetland is within the banks of a river or stream .... Riverine
        5. Wetland is along the shores of a lake or pond 6
          6. Wetland is along the shores of a lake (generally >20 acres and deeper than 6.6 feet at low water) .... Lacustrine
          6. Wetland forms a pond (generally <20 acres and less than 6.6 feet at low water) or along a pond .... Palustrine
      7. Wetland is within an estuary, not forming an ocean shoreline .... Estuarine
      7. Wetland forms the shoreline of an ocean .... Marine

Note: For classification of class, subclass, water regime, and other modifiers, see tables and text. (Source: Cowardin et al. 1979.)

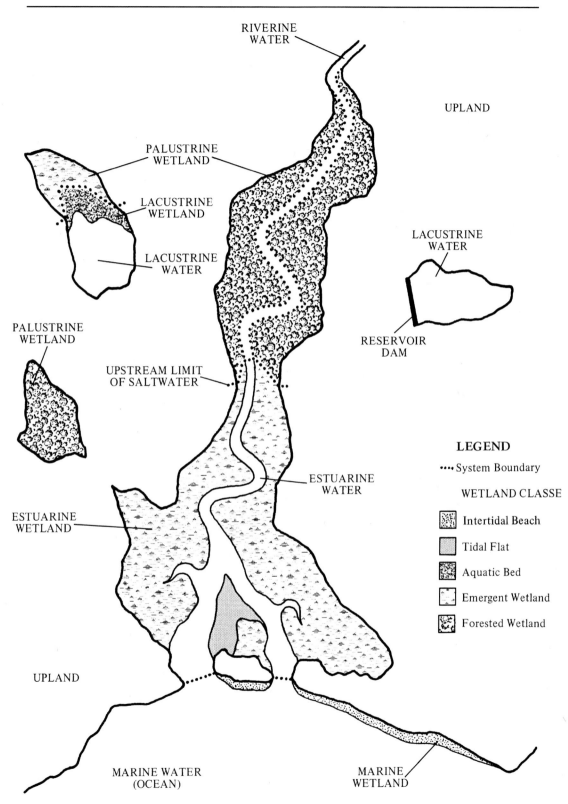

Figure 14.3. Diagram showing major wetland and deepwater habitat systems on the landscape. (Source: Tiner and Burke 1995)

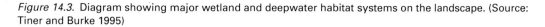

Table 14.3. Classes and subclasses of wetlands and deepwater habitats.

Class	Brief Description	Subclasses
Rock Bottom	Generally permanently flooded areas with bottom substrates consisting of at least 75% stones and boulders and less than 30% vegetative cover.	Bedrock; Rubble
Unconsolidated Bottom	Generally permanently flooded areas with bottom substrates consisting of at least 25% particles smaller than stones and less than 30% vegetative cover.	Cobble-gravel; Sand; Mud; Organic
Aquatic Bed	Generally permanently flooded areas vegetated by plants growing principally on or below the water surface line.	Algal; Aquatic Moss; Rooted Vascular; Floating Vascular
Reef	Ridge-like or mound-like structures formed by the colonization and growth of sedentary invertebrates.	Coral; Mollusk; Worm
Streambed	Channel whose bottom is completely dewatered at low water periods.	Bedrock; Rubble; Cobble-gravel; Sand; Mud; Organic; Vegetated
Rocky Shore	Wetlands characterized by bedrock, stones or boulders with areal coverage of 75% or more and with less than 30% coverage by vegetation.	Bedrock; Rubble
*Unconsolidated Shore	Wetlands having unconsolidated substrates with less than 75% coverage by stone, boulders and bedrock and less than 30% vegetative cover, except by pioneer plants.	Cobble-gravel; Sand; Mud; Organic; Vegetated
Moss-Lichen Wetland	Wetlands dominated by mosses or lichens where other plants have less than 30% coverage.	Moss; Lichen
Emergent Wetland	Wetlands dominated by erect, rooted, herbaceous hydrophytes.	Persistent; Nonpersistent
Scrub-Shrub Wetland	Wetlands dominated by woody vegetation less than 20 feet (6 m) tall.	Broad-leaved Deciduous; Needle-leaved Deciduous; Broad-leaved Evergreen; Needle-leaved Evergreen; Dead
Forested Wetland	Wetlands dominated by woody vegetation 20 feet (6 m) or taller.	Broad-leaved Deciduous; Needle-leaved Deciduous; Broad-leaved Evergreen; Needle-leaved Evergreen; Dead

*Note: This class combines two classes of the 1977 operational draft system—Beach/Bar and Flat. (Source: Cowardin et al. 1979.)

degree of flooding distinguish classes. Permanently flooded nonvegetated areas are classified as either rock bottom or unconsolidated bottom, while exposed areas are typed as streambed, rocky shore, or unconsolidated shore. Invertebrate reefs are found in both permanently flooded and periodically exposed areas.

Each class is further divided into *subclasses* to better define the type of substrate in nonvegetated areas (e.g., bedrock, rubble, cobble-gravel, mud, sand, and organic) or the type of dominant vegetation (e.g., persistent or nonpersistent emergents, moss, lichen, or broad-leaved deciduous, needle-leaved deciduous, broad-leaved evergreen, needle-leaved evergreen, and dead woody plants). Below the subclass level, *dominance type* can be applied to specify the predominant plant or animal in the community.

To describe hydrologic, chemical, and soil characteristics and human impacts, the classification system contains four types of specific modifiers: 1) water regime, 2) water chemistry, 3) soil, and 4) special. These modifiers may be applied to class and lower levels of the classification hierarchy.

Water regime modifiers define flooding or soil saturation conditions and are divided into two main groups: tidal and nontidal (Table 14.4). Tidal regimes can be subdivided into two general categories, one for salt and brackish water tidal areas and another for freshwater tidal areas. This distinction is needed because of the special importance of seasonal river overflow and groundwater inflows in freshwater tidal areas. By contrast, nontidal modifiers define conditions where surface water runoff, ground-water discharge, and/or wind effects (e.g., lake seiches) cause water level changes.

Water chemistry modifiers are divided into two categories which describe the water's salinity or hydrogen ion concentration (pH): 1) salinity modifiers, and 2) pH modifiers. Like water regimes, salinity modifiers are further subdivided into two groups: halinity modifiers for tidal areas and salinity modifiers for nonti-

dal areas (Table 14.5). Estuarine and marine waters are dominated by sodium chloride which is gradually diluted by fresh water as one moves upstream in coastal rivers. On the other hand, the salinity of inland waters is dominated by four major cations (i.e., calcium, magnesium, sodium, and potassium) and three major anions (i.e., carbonate, sulfate, and chloride). Interactions between precipitation, surface runoff, groundwater flow, evaporation, and sometimes plant transpiration form inland salts. This is most common in arid and semiarid regions. The pH modifiers are used to identify acid (pH < 5.5), circumneutral (5.5–7.4) and alkaline (pH > 7.4) conditions.

Soil modifiers are presented because the nature of the soil exerts strong influences on plant growth and reproduction as well as on the animals living in it. Two soil modifiers are given: mineral and organic. In general, if a soil has 20 percent or more organic matter by weight in the upper 16 inches, it is considered an organic soil, whereas if it has less than this amount, it is a mineral soil.

The final set of modifiers—special modifiers

Table 14.4. Water regime modifiers, both tidal and nontidal groups.

Group	Type of Water	Water Regime	Definition
Tidal	Saltwater and brackish areas	Subtidal	Permanently flooded tidal waters
		Irregularly exposed	Exposed less often than daily by tides
		Regularly flooded	Daily tidal flooding and exposure to air
		Irregularly flooded	Flooded less often than daily and typically exposed to air
	Freshwater	Permanently flooded–tidal	Permanently flooded by tides and river or exposed irregularly by tides
		Semipermanently flooded–tidal	Flooded for most of the growing season by river overflow but with tidal fluctuation in water levels
		Regularly flooded	Daily tidal flooding and exposure to air
		Seasonally flooded–tidal	Flooded irregularly by tides and seasonally by river overflow
		Temporarily flooded–tidal	Flooded irregularly by tides and for brief periods during growing season by river overflow

(continued)

Table 14.4. (*continued*)

Group	Type of Water	Water Regime	Definition
Nontidal	Inland freshwater and saline areas	Permanently flooded	Flooded throughout the year in all years
		Intermittently exposed	Flooded year-round except during extreme droughts
		Semipermanently flooded	Flooded throughout the growing season in most years
		Seasonally flooded	Flooded for extended periods in growing season, but surface water is usually absent by end of growing season
		Saturated	Surface water is seldom present, but substrate is saturated to the surface for most of the season
		Temporarily flooded	Flooded for only brief periods during growing season, with water table usually well below the soil surface for most of the season
		Intermittently flooded	Substrate is usually exposed and only flooded for variable periods without detectable seasonal periodicity (not always wetland; may be upland in some situations)
		Artificially flooded	Duration and amount of flooding is controlled by means of pumps or siphons in combination with dikes or dams

(Source: Cowardin et al. 1979)

Table 14.5. Salinity modifiers for coastal and inland areas.

Coastal Modifiers[1]	Inland Modifiers[2]	Salinity (‰)	Approximate Specific Conductance (Mhos at 25° C)
Hyperhaline	Hypersaline	>40	>60,000
Euhaline	Eusaline	30–40	45,000–60,000
Mixohaline (Brackish)	Mixosaline[3]	0.5–30	800–45,000
Polyhaline	Polysaline	18–30	30,000–45,000
Mesohaline	Mesosaline	5–18	8,000–30,000
Oligohaline	Oligosaline	0.5–5	800–8,000
Fresh	Fresh	<0.5	<800

[1] Coastal modifiers are employed in the Marine and Estuarine Systems.
[2] Inland modifiers are employed in the Riverine, Lacustrine and Palustrine Systems.
[3] The term "brackish" should not be used for inland wetlands or deepwater habitats.
(Source: Cowardin et al. 1979)

—were established to describe the activities of people or beavers affecting wetlands and deepwater habitats. These modifiers include: excavated, impounded (i.e., to obstruct outflow of water), diked (i.e., to obstruct inflow of water), partly drained, farmed, and artificial (i.e., materials deposited to create or modify a wetland or deepwater habitat).

Hydrogeomorphic Classification

In 1993, the wetland ecologist Mark Brinson developed a hydrogeomorphic-based (HGM) wetland classification to aid functional assessment of wetlands. He recognized a major shortcoming of the FWS wetland classification system—it did not address abiotic features, namely hydrogeomorphology, that are directly linked to many wetland functions. The HGM system is actually more of an approach to provide a framework for wetland evaluation rather than a classification system for mapping wetlands. The HGM system identified seven basic hydromorphic classes: 1) depressional wetlands (within topographic depressions, e.g., kettles, potholes, vernal pools, playas, and Carolina bays), 2) organic soil flats (extensive peatlands, e.g., bogs and pocosins), 3) mineral soil flats (broad, nearly level wetlands with inorganic soils, e.g., flatwoods), 4) riverine wetlands (along rivers and streams, e.g., floodplains and riparian areas), 5) slope wetlands, 6) lacustrine fringe (lakeshore wetlands), and 7) estuarine fringe (tidal wetlands). Riverine wetlands are further divided into three gradients that reflect stream flow and fluvial processes: 1) high gradient (flow likely continuous, but mostly flashy; wetland on coarse substrate maintained by upslope groundwater source; unstable substrate colonized by pioneer species; streamside vegetation contributes to nutrient supply), 2) middle gradient (flow likely continuous; channel processes establish variable topography/hydroperiod/habitat interspersion on floodplain; alluvium is renewed by surface accretion and point bar deposition; interspersion of plant communities contributes to species migration), and 3) low gradient (flow continuous with cool season flooding; high suspended sediments; flood storage; conserves groundwater discharge; major fish and wildlife habitat and biodiversity; strong biogeochemical activity and nutrient retention). Examples of these hydrogeomorphic classes are given in Table 14.6.

Landscape Position, Landform, Water Flow Path, and Waterbody Classification

Recognizing the need to expand its characterization of wetlands to help map users better understand the functions of individual wetlands, the FWS began creating a set of HGM-type descriptors for the Northeast in 1994. In developing the HGM system, Brinson's used several key terms found in the FWS's classification system, but defined them differently. This eliminated the possibility of a simply adding HGM terminology to FWS wetlands inventory data. For example, the HGM riverine class includes both riverine and palustrine wetlands as defined by the FWS. Likewise, the HGM lacustrine class includes more than the lacustrine wetlands of the FWS. Consequently, new terminology had to be developed that could be easily integrated into the FWS system for mapping purposes. The FWS developed a set of descriptors for landscape position, landform, water flow path, and waterbody type (LLWW descriptors) that could

Table 14.6. Examples of hydrogeomorphic (HGM) classes of wetlands.

Hydrogeomorphic Class	Examples of Wetland Types
Riverine	Floodplain wetlands, bottomland hardwood swamps, riparian wetlands, willow streambank thickets, streamside marshes
Depressional	Bogs (small), woodland vernal pools, fens, basin marshes, prairie potholes, Carolina bays, isolated swamps, wet meadows, farm ponds, playas (small), and West Coast vernal pools, sinkhole wetlands
Slope	Seepage wetlands, snow-meltwater wetlands, fens, wet meadows, blanket bogs, avalanche chutes
Mineral soil flat	Coastal plain and glaciolacustrine plain flatwoods, playas
Organic soil flat	Bogs, pocosins
Lacustrine fringe	Great Lakes marshes, Flathead Lake marshes, Great Salt Lake marshes, lakeshore marshes and aquatic beds
Estuarine fringe	Salt and brackish tidal marshes, mangrove swamps

Note that some wetland types can occur in more than one class. (Adapted from Brinson 1993.)

be added to its original classification or used independently. The LLWW descriptors attempt to link the FWS system with the HGM system in a way that can be easily applied to existing and future National Wetlands Inventory (NWI) maps and databases. Since 1994, these descriptors have been added to NWI digital data in the Northeast where updating has been performed and they have been used to prepare preliminary assessments of wetland functions for several watersheds. The most recent version of this classification is *Dichotomous Keys and Mapping Codes to Landscape Position, Landform, Water Flow Path, and Waterbody Type Descriptors for U.S. Wetlands and Waters.*

Landscape position describes the linkage between the wetland and a waterbody. Five landscape position descriptors are defined (Figure 14.4): marine (ocean), estuarine (estuaries including salt and brackish embayments and rivers), lotic (rivers and streams), lentic (lakes, reservoirs, and large deep ponds), and terrene (headwater wetlands [sources of streams], wetlands hydrologically decoupled from adjacent streams, and geographically isolated wetlands [surrounded by upland]). The lotic landscape position is further divided into five gradients (high, middle, and low following HGM, plus intermittent and tidal) and rivers are separated from streams (simply based on width with streams being linear map features and rivers polygonal features). Table 14.7 provides a dichotomous key to wetland landscape position, landform, and water flow path.

After identifying landscape position, wetlands are described by their landforms and water flow paths. Landform defines the shape of the wetland. Seven major types of landforms are identified for inland areas: slope, island, fringe, floodplain, interfluve, basin, and flat. The type of island wetland (delta, river, stream, lake, pond, or floating mat) and former floodplains may also be designated. For coastal wetlands, three landforms are defined: island, fringe, and basin (formed by impoundment or occurring behind a road, railroad bed, or other human-built obstruction). Coastal landforms may be further defined by their position in the estuary (e.g., barrier island, barrier beach, bay, coastal pond, river, headland, or ocean).

Each landform is further characterized by water flow path. For inland wetlands, typical water flow paths include isolated, throughflow, inflow, outflow, or bidirectional, with surface water connections usually emphasized in practice. Some may be even further classified by the nature of the flow (what is upslope and downslope from the wetland) or by paludification. For coastal wetlands, the water flow is bidirectional-tidal but may be tidally restricted or controlled.

This classification system also has provisions for identifying headwater wetlands, drainage-divide wetlands, plus an endless array of defined regional types of individual wetland landforms (e.g., basin types like prairie potholes, West Coast vernal pools, playas, woodland vernal pools, pocosins, sinkholes, and cypress domes).

What Good is That Wetland?: Assessing Wetlands

Knowing the location of wetlands is important for determining the limits of environmentally sensitive areas that may be subject to government regulation. Once a wetland is found on a particular property, other questions arise including: 1) what type of wetland is it? and 2) what are its functions? Background for answering the first question has been provided above (wetland classification). While an overview of wetland functions has been provided in Chapter 7, the following sections will address ways of evaluating relationships between wetland characteristics and wetland function. It is not an attempt to review various wetland assessment methodologies as such topic would require a technical book in itself. For such a review, consult *A Comprehensive Review of Wetland Assessment Procedures: A Guide for Wetland Practitioners* (Bartoldus 1999).

More than forty wetland assessment methods have been developed in recent years. In general, they can be divided into three categories: 1) inventory-based methods, 2) rapid field assessments, and 3) model-based field assessments. Some methods focus on functions, a few on values, and others on a combination of functions and values ("functional values"). Best professional judgment may be used to develop the assessment or data-driven models may form the basis for evaluation.

Inventory-based Methods

These types of assessment methods focus on wetland classification and inventory. They are

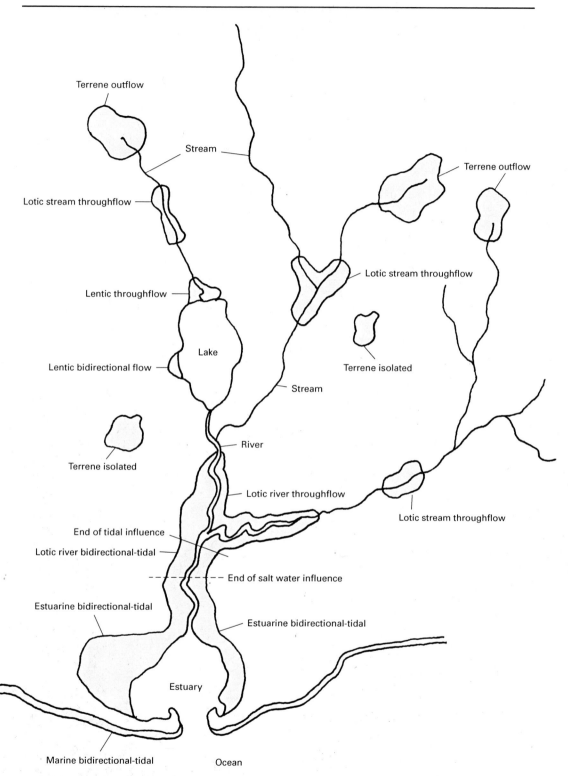

Figure 14.4. Typical wetland landscape positions and water flow paths in the eastern United States.

Table 14.7. Simplified keys for classifying wetlands by landscape position, landform, and water flow path.

Landscape Position

---

1. Wetland borders a river, stream, in-stream pond, lake, reservoir, estuary, or ocean 2

1. Wetland does not border one of these waterbodies; it is completely surrounded by
   upland or borders a pond surrounded by upland . . . . Terrene

   2. Wetland lies along an ocean shore and is subject to tidal flooding . . . . Marine

   2. Wetland does not lie along an ocean shore 3

      3. Wetland lies along an estuary (salt to brackish tidal waters) and is subject to tidal
         flooding . . . . Estuarine

      3. Wetland does not lie along an estuary or if so, it is not subject to tidal flooding 4

         4. Wetland lies along a lake or reservoir or within its basin . . . . Lentic

         4. Wetland does not occur in these locations 5

            5. Wetland is the source of a river or stream and this watercourse does not
               extend through the wetland . . . . Terrene

            5. Wetland is not the source of a river or stream 6

               6. Wetland is periodically flooded by river or stream . . . . Lotic[1]

               6. Wetland is not periodically flooded by the river or stream or by tides
                  (episodic flooding may occur) . . . . Terrene

Landform

---

1. Wetland occurs on a slope >2% . . . . Slope

1. Wetland does not occur on a slope >2% 2

   2. Wetland forms an island completely surrounded by water (at least seasonally) . . . . Island

   2. Wetland does not form an island 3

      3. Wetland occurs in the shallow water zone of a permanent nontidal water body, the
         intertidal zone of an estuary with unrestricted tidal flow, or the regularly flooded
         (daily tidal inundation) zone of freshwater tidal wetlands . . . . Fringe

      3. Wetland does not occur in these waters or in intertidal zones with unrestricted
         tidal flow 4

         4. Wetland occurs in a portion of an estuary with restricted tidal flow due to
            tide gates, undersized culverts, dikes, causeways, or similar obstructions . . . . Basin

         4. Wetland does not occur in such location 5

            5. Wetland forms a nonvegetated bank or is within the banks of a river or stream . . . Fringe

            5. Wetland is not a nonvegetated riverbank or streambank or within the
               banks of a river or stream 6

               6. Wetland occurs on an active alluvial plain . . . . Floodplain*

               6. Wetland does not occur on an active floodplain 7

                  7. Wetland occurs on a broad interstream divide (including headwater positions)
                     associated with coastal or glaciolacustrine plains or similar plains . . . . Interfluve*

                  7. Wetland does not occur on such a landform 8

                     8. Wetland occurs in a distinct depression . . . . Basin

                     8. Wetland occurs on a nearly level landform . . . . Flat

(continued)

Table 14.7. (continued)

Water Flow Path[2]

1. Wetland is typically surrounded by upland (nonhydric soil); receives precipitation and runoff from adjacent areas with no channelized inflow and no apparent outflow . . . . Isolated**

1. Wetland is not geographically isolated 2

   2. Wetland is a sink receiving water from a river, stream, or other surface water source, lacking surface water outflow . . . . Inflow

   2. Wetland is not a sink; surface water flows through or out of the wetland 3

      3. Wetland is subjected to tidal flooding . . . . Bidirectional-Tidal

      3. Wetland is not tidally influenced 4

         4. Water flows out of the wetland, but does not flow into this wetland from another source . . . . Outflow

         4. Water flows in and out of the wetland 5

            5. Water flows through the wetland, often coming from upstream or uphill sources (typically wetlands along rivers and streams) . . . . Throughflow

            5. Wetland is along a lake or reservoir and its water levels are subjected to the rise and fall of this waterbody . . . . Bidirectional-Nontidal

*basin and flat sub-landforms can be identified within these landforms when desirable.
**wetland is geographically isolated; hydrological relationship to other wetlands and watercourses is usually more complex than can be determined by simple visual assessment of surface water conditions.
[1] Lotic wetlands are separated into river and stream sections (based on watercourse width at map scale of 1:24,000—polygon = lotic river vs. linear = lotic stream) and then divided into one of five gradients: 1) high (e.g., shallow mountain streams on steep slopes), 2) middle (e.g., streams with moderate slopes), 3) low (e.g., mainstem rivers with considerable floodplain development), 4) intermittent (subject to periodic flows), and 5) tidal (hydrology under influence of the tides).
[2] Surface water connections are emphasized because they are more readily identified than groundwater linkages.
(Adapted from Tiner 2003a)

usually applied to large geographic areas such as watersheds. The underlying principle is that by describing certain properties of wetlands, one can make some judgments as to the likely functions the wetlands will perform. These methods usually require application of remote sensing techniques, perhaps coupled with map interpretation to develop a set of variables that relate to wetland function. The result is a landscape-level assessment. The FWS developed this type of assessment using its NWI data and adding additional descriptors for landscape position, landform, water flow path, and waterbody type (LLWW descriptors). Various combinations of these properties are correlated to ten functions: surface water detention, streamflow maintenance, sediment and other particulate retention, nutrient transformation, coastal storm surge detention, shoreline stabilization, provision of fish and shellfish habitat, provision of waterfowl and waterbird habitat, provision of other wildlife habitat, and conservation of biodiversity.

Table 14.8 outlines these relationships for Northeast wetlands. Correlations were based on best professional judgment supported by the published literature and a review by regional wetland experts. The correlations can also be used to identify functions for individual wetlands based on their classification through field inspections. The FWS has used this combination of classification and inventory and geographic information system technology to produce wetland characterizations for several watersheds in the Northeast. Examples are posted on the web at http://wetlands.fws.gov (look under publications for a report on two Maryland watersheds—the Nanticoke River and the Coastal Bays watersheds) and at http://library.fws.gov (the FWS's National Conservation Library; use keyword search for "wetlands" to find a list of relevant publications).

Table 14.8. Correlations between wetland functions and characteristics based on the Northeast.

Function	Potential	Characteristics of Wetlands Performing Functions at Significant Levels
Surface water detention	High	Estuarine, marine unconsolidated shore, lentic, lotic floodplain, lotic basin, terrene basin (throughflow), ponds (throughflow, bi-directional, bidirectional-tidal)
	Moderate	Lotic flat, interfluve, other terrene basin, other ponds
Coastal storm surge detention	High	Estuarine, lotic tidal, marine unconsolidated shore (flat/beach)
Streamflow maintenance	High	Headwater (not ditched)
	Moderate	Headwater-ditched, lotic floodplain, pond (outflow) (Note: Outflow must be associated with stream)
Nutrient transformation	High	Seasonally flooded and wetter vegetated
	Moderate	Saturated and temporarily flooded vegetated
Sediment retention	High	Estuarine vegetated, lotic floodplain, lotic basin, lotic fringe (vegetated), lentic basin, lentic fringe (vegetated), pond (throughflow, outflow, and bidirectional)
	Moderate	Estuarine nonvegetated (excluding rocky shore), lotic flat, lotic tidal fringe, marine unconsolidated shore (Note: Excludes bogs)
Shoreline stabilization	High	Vegetated and along waterbodies (excluding ponds)
	Moderate	Vegetated and along ponds
Fish and shellfish habitat	High	Semipermanently and permanently flooded, estuarine emergent, estuarine unconsolidated shore, estuarine reef, estuarine aquatic bed, marine aquatic bed, marine reef, marine unconsolidated shore, palustrine tidal emergent (not *Phragmites*), riverine tidal unconsolidated shore, riverine tidal emergent
	Moderate	Lentic and palustrine emergent seasonally flooded/saturated, lotic and palustrine emergent seasonally flooded/saturated, semipermanently flooded *Phragmites* contiguous with a waterbody, ponds (except industrial, sewage treatment, stormwater, and other developed ponds)
	Stream shading	Lotic stream (not intermittent) and forested, lotic stream (not intermittent) and scrub-shrub (not bog)
Waterfowl and waterbird habitat	High	Semipermanently and permanently flooded wetlands, estuarine emergent (not *Phragmites*), estuarine unconsolidated shore, estuarine reef, estuarine aquatic bed, estuarine rocky shore, marine aquatic bed, marine reef, marine unconsolidated shore, marine rocky shore, palustrine tidal emergent (not *Phragmites*), riverine tidal unconsolidated shore, riverine tidal emergent, ponds associated with semipermanently flooded vegetated wetlands, lotic stream and palustrine emergent seasonally flooded/saturated, lotic river and palustrine emergent seasonally flooded/saturated, ponds associated with semipermanently flooded vegetated wetlands
	Moderate	Semipermanently flooded phragmites and contiguous with a waterbody (not intermittent), estuarine *Phragmites* and contiguous with waterbody, other ponds (except industrial, sewage treatment, stormwater, and other developed ponds), other palustrine emergent associated with waterbody

(continued)

Table 14.8. (continued)

Function	Potential	Characteristics of Wetlands Performing Functions at Significant Levels
	Wood Duck	Lotic stream (not upper perennial or intermittent) and forested or scrub-shrub, lotic river (not upper perennial or intermittent) and forested or scrub-shrub
Other wildlife habitat	High	Any wetland complex >20 acres, wetlands 10–20 acres with two or more classes (excluding *Phragmites*), narrow wetlands connecting large wetlands, small isolated wetlands in dense clusters within a forest matrix
	Moderate	Other vegetated wetlands
Conservation of biodiversity	Regionally significant*	Slightly brackish to tidal fresh emergent and scrub-shrub, Atlantic white cedar swamps, calcareous fens, bald cypress swamps, eelgrass beds, woodland vernal pools, headwater seepage swamps
	Locally significant*	Urban wetlands, shrub bogs at southern end of range, larch swamps at southern end of range, oyster and mussel reefs, northern white cedar swamps at southern end of range, hemlock swamps, salt marshes

*Listing is not complete; examples only. (Source: Tiner 2003b.)

Rapid Field Assessments

In this approach, wetlands are examined quickly in the field to identify characteristics associated with certain functions. This method often requires preparatory work such as reviewing maps or doing an inventory-based assessment. The New England Corps of Engineers developed a qualitative assessment method originally for evaluating highway alternatives, but now recommends its use for all permit applications. It considers thirteen wetland functions and values (Table 14.9) and presents numerous features or conditions to verify the suitability for each function/value. Most methods involve scoring a number of variables. Some methods recommend evaluating reference wetlands beforehand to develop a standard for comparison, whereas the Ohio Environmental Protection Agency maintains information (calibrated scores) for various wetland types for use with its rapid assessment method. Some methods were designed for general planning purposes, while many were created for site-specific wetland assessment.

The "New Hampshire Method" is an example of one designed for planning, inventory, and educational purposes. It employs a combination of reviewing published information, compiling a wetland map, answering questions, and field evaluation. Fourteen "functional values" are addressed including both wetland functions and values. The authors of this method estimate that 8–24 hours are required for map compilation, 4 hours per wetland for field investigations, and 1 hour per wetland to analyze the information and prepare the findings. Wetland value units are calculated for each functional value to determine a score for each wetland. This type of assessment can be done for an entire town or a watershed.

The Ohio Rapid Assessment Method is designed for evaluating individual wetlands. It begins with a narrative rating based on the answers to several questions, then specific properties of the wetland and adjacent land are assessed and rated. Ratings for individual wetlands are compared to a calibrated score for the wetland type (based on Ohio EPA investigations) and the assessment concludes with an overall category rating ranging from low quality (1) to very high quality (3). The "Wisconsin Method" analyzes eight "functional values" and may require an estimated 4 hours for evaluating a 1-acre site. The "Minnesota Rapid Assessment Method" focuses on twelve variables and may require 2–3 hours for evaluating a small wetland (<10 acres). Most, if not all, of these rapid assessment methods recognize unique wetlands containing plant communities on natural heritage lists, habitats for rare and endangered species, wilderness areas, and designated wetlands of high significance for providing certain functions. These

Table 14.9. Examples of functions and/or values considered in several rapid assessment methods for wetlands.

Method (Source)	Functions and/or Values Evaluated
New Hampshire Method (Ammann and Stone 1991)	Ecological integrity, wetland wildlife habitat, finfish habitat, educational potential, visual/aesthetic quality, water-based recreation, flood control potential, groundwater use potential, sediment trapping, nutrient attenuation, shoreline anchoring and dissipation of erosive forces, urban quality of life, historical site potential, and noteworthiness
Highway Methodology (U.S. Army Corps of Engineers 1995)	Groundwater recharge/discharge, sediment/shoreline stabilization, floodflow alteration (storage and desynchronization), wildlife habitat, fish and shellfish habitat, recreation (consumptive and nonconsumptive), sediment/toxicant retention, educational/scientific value, nutrient removal/retention/transformation, uniqueness/heritage, production export (nutrient), visual quality/aesthetics, and endangered species
Method for the Assessment of Wetland Function (Fugro East, Inc. 1995; Maryland Dept. of the Environment 1997)	Groundwater discharge, flood flow attenuation, modification of water quality, sediment stabilization, aquatic diversity/abundance, and wildlife diversity/abundance
Ohio Rapid Assessment Method for Wetlands (Mack 2001)	Wetland size, buffer and surrounding land use, hydrology (source of water, hydrologic connectivity, maximum water depth, duration of inundation/saturation, hydrologic modification), habitat alteration/disturbance (substrate/soil disturbance, habitat development, habitat alteration), special wetland communities (e.g., bog, fen, old growth forest, unrestricted coastal wetland), and vegetation, interspersion and microtopography
Minnesota Rapid Assessment Method for Evaluating Wetland Functions (Minnesota Dept. of Natural Resources 2004)	Vegetation diversity/integrity, maintenance of characteristic hydrologic regime, flood/storm water attenuation, downstream water quality, maintenance of wetland water quality, shoreline protection, groundwater interaction, maintenance of characteristic wildlife habitat structure, maintenance of characteristic fish habitat, maintenance of characteristic amphibian habitat, aesthetics/recreational/educational/cultural, and commercial uses
Wisconsin Rapid Assessment Methodology (Wisconsin Dept. of Natural Resources 1992)	Floral diversity, wildlife habitat, fishery habitat, flood/stormwater attenuation, water quality protection, shoreline protection, groundwater, and aesthetics/recreation/education

methods can usually be applied with minimal training.

Model-based Field Assessments

Model-based field assessments require collecting information on a fairly large number of existing wetlands or waterbodies before this method can be employed. These data serve as the foundation for models that can be used to assess a variety of wetland functions against a reference standard. These methods require considerable upfront time and effort to develop the models, which need to be developed for specific geographic areas or watersheds. Two common model-based field assessment methods are the "Hydrogeomorphic (HGM) Approach" and the "Indices of Biological Integrity (IBI) Method." Some rapid assessment methods involve preliminary data collection but the level of effort is typically much less than required for model development for the HGM and IBI methods.

The HGM method was developed specifically for evaluating wetlands for federal permit reviews and related planning. The Corps of Engineers has taken the lead role in developing this methodology at the national level. Models (HGM regional guidebooks) are prepared by an interdisciplinary team for certain HGM wetland types in specific geographic areas. This involves first specifying the type of wetland and the relevant geographic region ("reference domain").

Field investigations are then conducted to examine "reference wetlands" representing the range of conditions in the reference domain and to determine properties associated with wetlands representing the highest level of functioning for all functions ("reference standard"). The regional guidebooks serve as a foundation for determining the type of information that needs to be collected in the field for specific projects and provides the framework for assessment relative to the reference standard. Development of a guidebook may take a year or more of effort, but once published, it provides a field-validated assessment procedure that may require as little as 1–2 hours of work per wetland assessment area. Regional guidebooks are posted on the web at: http://www.wes.army.mil/el/wetlands/guidebooks.html.

The IBI method also requires establishing reference standards based on field investigations. The concept was initially created to assess water quality of streams and lakes from the biological perspective. Various organisms (macroinvertebrates, fishes, algae, etc.) are used as biological indicators of water quality. More recently, the techniques have been applied to evaluate wetland quality. Biological integrity is defined as the ability to support and maintain a balanced, integrated, adaptive biological system that has a full range of elements and processes expected in a region's natural habitat. Like the HGM method, IBI requires developing models based on data derived from field sampling within a given suite of similar habitats. Reference standards are established for each metric with a range of values associated with varied levels of human impact assigned different scores. The IBI for each site is calculated by adding the individual scores. Model development may take nearly a year to accomplish. Application of the model may require 8 hours per site (4 hours of field work and 4 hours of lab analysis). The U.S. EPA has organized a national interagency group called the Biological Assessment of Wetlands Working Group to pursue the application of IBI to wetlands. Information on this approach can be found at: http://www.epa.gov/owow/wetlands/monitor/#bio.

Selecting a Method

Given the plethora of wetland assessment methods, one is left with the ultimate question—which one should I use? To answer this question, several other questions must be answered including:

What is the purpose of the assessment?

Is it to assess wetlands across a large geographic area for educational or natural resource planning purposes?

Is the assessment needed to determine lost functions and to design appropriate mitigation?

Do you feel comfortable using a method based on best professional judgment or do you want to use a mechanistic model based on field-validated information?

How much time and effort do you want to spend on this assessment?

With your knowledge, skill, and available technology, what method is best suited for your use?

Are there specific functions that you are interested in evaluating or do you want to consider all possible functions?

Are you interested in wetland values (what wetlands do for people)?

Do you want to compare your findings to other types of wetlands or wetlands in other regions?

Is there a regional HGM guidebook available for your area of interest?

The answers to these and other questions should help you decide what method to use. Clearly, you will have much homework to do making this decision. In all likelihood, if you are involved in natural resource planning, you'll probably choose an inventory-based method, whereas if you're doing the assessment for regulatory purposes you'll choose one of the other approaches. In many cases, the applicable regulatory agency will probably dictate the method to use as several methods have been developed by government agencies specifically for regulatory purposes.

Additional Readings

Bartoldus, C. C. 1999. *A Comprehensive Review of Wetland Assessment Procedures: A Guide for Wetland Practitioners*. Environmental Concern, Inc., St. Michaels, MD.

Brinson, M. M. 1993. *A Hydrogeomorphic Classification of Wetlands*. U.S. Army Engineer Waterways Experiment Station, Vicksburg, MS. Wetlands Research Program Tech. Rep. WRP-DE-4.

Cowardin, L. M., V. Carter, F. C. Golet, and E. T. LaRoe. 1979. *Classification of Wetlands and Deepwater Habitats of the United States.* U.S. Fish and Wildlife Service, Washington, DC. FWS/OBS-79/31.

Karr, J. R., and E. W. Chu. 1997. *Biological Monitoring and Assessment: Using Multimetric Indexes Effectively.* University of Washington, Seattle, WA. EPA 235-R97-001.

Martin, A. C., N. Hotchkiss, F. M. Uhler, and W. S. Bourn. 1953. *Classification of Wetlands of the United States.* U.S. Fish and Wildlife Service, Washington, DC. Spec. Sci. Rep.: Wildlife No. 20.

Smith, R. D., A. Ammann, C. Bartoldus, and M. M. Brinson. 1995. *An Approach to Assessing Wetland Functions Using Hydrogeomorphic Classification, Reference Wetlands, and Functional Indices.* U.S. Army Corps of Engineers, Waterways Experiment Station, Vicksburg, MS. Wetlands Research Program Tech. Rep. WRP-DE-9.

Tiner, R. W. 1999. *Wetland Indicators: A Guide to Wetland Identification, Delineation, Classification, and Mapping.* Lewis Publishers, CRC Press, Boca Raton, FL.

Tiner, R. W. 2003a. *Dichotomous Keys and Mapping Codes to Landscape Position, Landform, Water Flow Path, and Waterbody Type Descriptors for U.S. Wetlands and Waters.* U.S. Fish and Wildlife Service, Ecological Services, National Wetlands Inventory Program, Hadley, MA.

Tiner, R. W. 2003b. *Correlating Enhanced National Wetlands Inventory Data with Wetland Functions for Watershed-wide Assessments: A Rationale for Northeastern U.S. Wetlands.* U.S. Fish and Wildlife Service, National Wetlands Inventory Program, Hadley, MA.

A Call to Action
SOS—Saving Our Swampland

How many clean rivers do you want? How much open space is enough? What level of water quality do you want: fishable, swimmable, or drinkable? How much wetland wildlife do we want to share the earth with us? To maintain and improve the quality of the natural environment and the quantity of plants and animals comes at some expense, and the costs—just like taxes—are not necessarily borne equally by all. In a cooperative, responsible society, sacrifices are made for the public good. Most companies, farmers, and others did not reduce their environmental impacts of their own volition. It took the support of the majority to convince legislators and congressmen to pass laws requiring that sacrifices of private land use be made.

Government agencies promulgated regulations to provide the necessary safeguards to control water pollution and exploitation of wetlands, and to prevent the further despoliation of our natural resources. These regulations, although not perfect in scope or implementation, have provided incentives and controls needed to encourage more environmentally responsible use of our land, air, and water resources. The costs of cleaning up a polluted planet far exceed the cost of prevention. Just look at the costs of environmental remediation at a single Superfund site—they're staggering! Preventive maintenance through regulating (not necessarily prohibiting) uses of the remaining wetlands, as well as restoring damaged and lost wetlands, is the direction we should be going.

From the late 1960s to 1990, the government was on the right track. With every decade, wetlands were getting more protection, but in the 1990s, something happened. Many people, including numerous members of Congress, felt that government regulations had gone too far and had taken away their private-property rights. While admittedly government has, on a few occasions, overstepped its bounds, overall, federal and state governments have acted responsibly in the public interest to improve environmental quality for us, our children, and future generations. Government is acting on behalf of the people, and the positive results in the natural environment are a matter of record. A few excesses should not cause us to eliminate environmental protection as we know it. Instead, some fine tuning should be performed; you don't throw out your television because of bad reception, but you try to fix the antenna or adjust the reception in other ways.

All of us, landowners or not, should be as concerned about the natural environment—its current condition and future status—as we are about our jobs and families. The quality of the world around us determines the quality of our lives. Wetlands are a vital component of the natural landscape and a resource providing many valuable functions. Public-opinion polls show that the majority of Americans support efforts to strengthen wetland protection. Most people would probably like to have all bodies of water at least fishable and swimmable.

Since wetlands will continue to be under pressure for development or exploitation as our population increases (today's U.S. population is 263 million vs. a projected 400 million by 2050), it is important for all citizens to gain a better understanding of wetlands and their values, so that everyone can make informed decisions on the fate of these natural resources. All of us, regardless of whether we own land or not, have a stake in wetlands and benefit substantially from their existence.

With less than half of the country's original wetlands remaining and many of those left degraded, it is important to take action to protect, restore, and enhance America's wetlands. In doing this, wetland functions such as water-quality renovation, shoreline stabilization, and floodwater storage, which benefit all citizens, will be maintained and hopefully improved. At the same time, habitats for many unique and interesting forms of wildlife and plantlife, vital components of the earth's biodiversity, will be conserved and restored, thereby allowing the more-visible species to continue to provide pleasure for Americans of all ages. Wetland protection could be a lasting gift from all of us to future generations of Americans. And it sure beats the inheritance of the national debt and the bill to clean up a polluted environment!

The following is an action agenda for people

298 / Appendix A

interested in taking a direct hand in improving wetland protection. It suggests many things that the average citizen can do to benefit wetlands now and in the future. The suggestions are presented as a starting point for citizen involvement.

The Most Important Thing You Can Do

Regardless of your energy level or available time or whether you own land or not, there is one simple thing that you can do to help save America's remaining wetlands. It will take only a few minutes of your time and a few postage stamps once a year, but it really is the most important thing that you can do and it's virtually effortless. **Write your elected officials.** Tell them that wetlands are important to you and that the functions wetlands perform must be protected and preserved for us and for future Americans as well as for wildlife. Express your support for wetland protection through regulation, acquisition, and restoration. Politicians react to current crises and concerns expressed by voters. All too often, the only letters politicians receive about wetlands are from irate citizens upset that they can't fill or alter a wetland or from special-interest groups advocating private-property rights over environmental regulations. Your voice counts, and a hundred or a thousand voices get attention.

The Next Step

The next step to take is to get involved in local wetland-protection efforts. This can mean participating in a variety of activities, ranging from attending local planning and environmental commission meetings and voicing your support for wetland protection to becoming a commission member or an activist supporting wetland acquisition and restoration. Small changes in local ordinances and bylaws can make a big difference in the future of wetlands.

Development is rapidly changing the landscape and the character of many communities. Wetlands as well as other forms of open space are being gobbled up or degraded by development. Many of the larger swamps and marshes, especially those most vital to migratory waterfowl, will likely be preserved or protected through easements by federal and state governments, but these wetlands represent only a small portion of the nation's wetlands. Only about 25 percent of the wetlands in the coterminous United States are in public

ownership. The majority of wetlands, therefore, are privately owned.

We need to encourage our local, state, and federal governments to acquire more wetlands and other open space for the future. In all likelihood, the costs of developing wetlands for residential housing would exceed the taxes that might be generated. With the influx of new residents, schools would undoubtedly need to be expanded. This would increase the costs of town services and, most likely, lead to an increase in taxes for all residents. The mayor of Stafford Township, New Jersey, saw the value of the state preserving local wetlands known as the Oxly Tract. Although the township would lose between $500,000 and $1 million in ratables (taxable properties), he recognized that the land would never require any services. Moreover, establishment of the state preserve would give the township some money in lieu of taxes over the next thirteen years.

Establishing public land trusts is a way to preserve local wetlands, and it is probably the only way that many of these important resources will be truly protected in perpetuity With increasing population growth, there will be even greater pressure brought to bear on those who seek to protect wetlands through regulations. Regulation can and will change. Adding more wetlands to public lands will help save them, although there will always be concern about their degradation from adjacent or upstream pollution sources. The cost of land rarely gets cheaper than it is in the Northeast today, so wetland acquisition should be accelerated immediately.

Encourage your town to establish a land trust supported by adding a surcharge to real-estate transactions or by dedicating a percentage of annual taxes to a land-acquisition fund. Nantucket Island and Martha's Vineyard, Massachusetts, have financed enormously successful public land trusts through placing a 2 percent surcharge on all real-estate transactions. Block Island, Rhode Island, has also done this. The Cape May County Farmland and Open Space Trust Fund is financed by one penny for every hundred dollars of assessed land value, approved in a referendum by a two-to-one margin. The county is using this fund to purchase farmland development rights, but in the future, the fund may be used for wetland and open-space acquisition. Through such acquisition, property values are maintained and even enhanced by the abundance of open space, and natural areas can be protected forever. As noted above land preservation may actually help minimize future tax increases. Open space provides people with opportunities to explore and enjoy nature, which can add to the attractiveness of communities to the point that would-be residents are willing to pay more for the existing homes, thereby increasing local property values. Acquisition protects land from development and is more effective than attempting to protect wetlands through regulations.

Other Steps to Take

There are many other things that you can do to help improve the status of wetlands. One of the first things is to learn how to recognize wetlands. This book should give you sufficient background knowledge and many of the tools you'll need to identify wetland plants, soils, and boundaries, and to interpret existing wetland and soil maps. Knowing where wetlands are on your property should help you practice wise resource management—land stewardship—as well as know when you see wetland destruction or degradation during your travels. Such knowledge could also save you money, by helping you avoid buying a piece of undevelopable property or assess the limitations of available lots.

Some of the many other actions you can take are listed below. They are separated into three categories: wetland management, action groups, and going farther.

Wetland Management

If you own land, there are many things that you can do in terms of managing it.

1. Be a good land steward. Protect your wetlands; don't use them for trash disposal. Compost your lawn clippings and leaves and chip your brush.

2. Put a conservation easement or deed restriction on the wetland portion of your property. This should not deter any person with a positive environmental ethic from purchasing your property, and it will allow that wetland to continue performing valuable functions in perpetuity.

3. If you've got a stream or river running through your property, maintain or restore a forested border. Tree-lined streams help moderate water temperatures, thereby improving fish habitat. The streamside forest will also serve as a buffer to maintain or improve water quality by trapping sediments and nutrients from adjacent uplands. Vegetated buffers around wetlands are beneficial. If you want a view of the wetland, don't clearcut the area, but instead do some selective pruning. Be sure to check with regulatory agencies, since removal of wetland vegetation (cutting and mowing) may require a permit.

4. If your lawn extends into a wetland, consider establishing a vegetated buffer. Let part of your lawn go native: maintain a wildflower meadow rather than a manicured lawn. This is aesthetically pleasing and will help buffer the wetland from lawn runoff (e.g., fertilizers), which could increase nutrients in the wetland, alter native vegetation, and degrade water quality. The need for buffers is especially great in wetlands with acidic soils, like the New Jersey Pine Barrens, and in nutrient-poor wetlands such as bogs. To enhance the buffer for wildlife, plant some woody plants to attract them, and enjoy bird-watching throughout the year. The book *Trees, Shrubs, and Vines for Attracting Birds* (DeGraff and Witman 1979) provides helpful tips.

5. If you are using wet meadows or marshes to feed livestock, light grazing is preferable to heavy grazing, which can compact soils and lower plant productivity. Also, try keeping the animals out when the area is very wet, since more damage can be done at such times. Consider delaying grazing or mowing hay until wetland birds have completed nesting. Most ducks and wetland-nesting birds should be finished nesting by mid-July, but consult your state fish and wildlife department for specifics.

6. If you have large landholdings that contain a significant amount of wetlands, consider donating or selling the wetlands to the government or a nonprofit organization to establish a refuge or nature preserve. There should be some significant tax benefits from this. Landowners in Harding Township, New Jersey, have donated conservation easements to the Harding Land Trust to preserve wetlands and open space.

7. If you are developing land with wetlands, avoid wetland and floodplain alterations wherever possible, and consider setting aside wetlands as open space to provide passive recreational use (nature observation, walking trails, etc.) for residents. Make wetlands a benefit for your planned development rather than a liability. If you want to cross a wetland to provide access to an upland island or simply want a means of walking through a swamp without getting your feet wet, consider building a boardwalk. Encourage your town to do this at public parks, with the addition of interpretive signs.

8. Manage your wetland to improve fish and wildlife habitat. If you have a pond, put up nesting boxes for wood ducks and plant native species that provide food and will attract wetland wildlife. If you live on a tidal marsh, consider erecting a nesting platform for ospreys; be sure to contact your town's environmental officer and the state

wildlife department before doing this. Learn to recognize exotic and other invasive species (e.g., common reed and purple loosestrife), and help restore native plant communities to your wetland (technical or financial support may be available from government agencies).

9. If you have land with drained wetlands that are idle or not producing good crop yields, seek to restore the wetlands by rejuvenating the hydrology. Several agencies are actively looking for landowners with an interest in restoring wetlands on their property. Contact the U.S. Fish and Wildlife Service (Partners for Fish and Wildlife program), the USDA Natural Resources Conservation Service or your state wetland agency for assistance. A farmer in Leigh County, Pennsylvania, worked with these agencies to restore two wetlands on his farm. Now he enjoys seeing wood ducks and mallards. He also has invited local school classes to visit and experience the wonders of restored wetlands.

10. Build a wetland on your property. Construct a pond with a marsh–wet meadow complex and a shrub border. Any size will do, but the larger and more diverse, the better for attracting more animals.

11. Take cuttings in the spring before budbreak (mid-March to mid-April in New Jersey) from the best blueberries in a swamp, plant them in a peat-sand mix, water sufficiently, then transplant them in your wetland in the fall. Later, when berries are produced, you can practice your pickin' skills: pick with your thumbs and let the berries roll off into a cup!

12. If you harvest timber from wetlands, follow the best management practices; contact your state forester for the latest guidance. Consider staying out of the wettest swamps; you'll probably cause lasting damage moving heavy equipment over their unstable soils. (See pages 138–140.)

Action Groups

If you have time to dedicate to the cause, there are many opportunities awaiting you.

1. Join a watershed association and become an active participant, advocating wetland protection, restoration, and acquisition.

2. Start an adopt-a-wetland group. Begin with a survey of wetlands in your neighborhood. If your neighborhood wetlands don't have names, name them; it's easier to relate to something with a name. When you conduct annual cleanups, keep an eye out for pollution and other degradation. Notify the local press of your various activities; publicize your successes and your concerns. The book *Living Waters* (Owens 1993) discusses how

to build a neighborhood or watershed group based on the experiences of local stream-restoration groups.

3. Join or start an organization dedicated to preserving wetlands at any level: local, state, federal, or international.

4. Encourage your hunting or fishing club or your school's environmental club to get actively involved by supporting wetland protection, acquisition, and restoration. A school group in Reading, Massachusetts, has been very active in building support for vernal-pool protection in the state, even publishing an excellent guide, *Wicked Big Puddles* (Kenney 1994). Elsewhere in Massachusetts, the environmental club at the Greater New Bedford Technical High School has adopted a stream—the Paskamanset River—and is monitoring water quality and conducting river cleanups. The same can be done for wetlands.

Going Farther

1. Build an environmental ethic in your family—become an advocate for wetland protection. Take your children on nature walks that traverse wetlands (see Appendix B for locations) and educate them on the values of wetlands. Talk to your friends about these issues, and have them join you in your swamp treks.

2. Use your vote to support wetland protection, acquisition, and restoration initiatives.

3. Donate time, money, or other support to wetland acquisition and restoration. Buy federal and state duck stamps: the revenue from stamp sales goes toward acquiring wetlands for the public.

4. Encourage your local schools (elementary and secondary) to add wetland studies to the existing curriculum. If you're a teacher, start an environmental-education program if one doesn't exist. Wetlands are excellent outdoor laboratories for teaching science. Places like Broadmoor Sanctuary (Natick, Massachusetts), the Somerset County Environmental Center in the Great Swamp (Basking Ridge, New Jersey), Jug Bay Wetlands Sanctuary (Anne Arundel County, Maryland), or National Wildlife Refuges are great for class field trips.

5. Review your company's environmental policy, and encourage it to adopt a policy to develop land responsibly, including avoidance of wetland alteration wherever possible. If you're an officer, encourage your company to contribute to wetland acquisition, restoration, and education projects. If there are wetlands on your company's grounds, consider designing interpretive walking

trails to educate co-workers on the values of wetlands and encourage public use, if possible. There may even be restorable wetlands on company property.

6. Be a wetland watchdog—report any observed fill in wetlands to appropriate regulatory agencies at all levels of government. With shrinking budgets, they need all the help they can get in locating unauthorized activities.

7. Be a good wildlife Samaritan—remove critters from harm's way. Help turtles or ducklings cross the road. Beware of the snappers; they have a nasty disposition, but need your help too! Fishermen and crabbers: use by-catch-reduction apparatus to keep diamond-backed terrapins from drowning in your nets or traps. When driving along causeways crossing marshlands, slow down if you see a heron or egret flying low across the road. If there's a spring migration of salamanders to their breeding pools in your neighborhood that involves crossing roads, try to organize a program to restrict night traffic during this brief period. Get your neighbors involved; kids love to help. Several Massachusetts communities do this. Some towns in Germany have an elaborate system of traffic signs and lights to warm motorists and protect salamanders. Salamander migration usually takes place during the first warm, rainy nights in spring (when temperatures are above 42°). Your local herpetological society or environmental center may be able to provide some assistance.

8. Encourage nurseries to stop selling plants that are known to be invasive in wetlands, especially purple loosestrife. There are many other pretty plants available for home gardens.

9. Don't dispose of unwanted aquarium plants by dumping them in the nearest river. Many of these species, such as Eurasian milfoil and hydrilla, have replaced native aquatic plants and disrupted local ecosystems. Compost these aquarium plants, or simply put them in the garbage.

10. Invite a wetland scientist to give a lecture at your local library. This can increase public awareness of wetland issues and build broader support for wetland protection.

11. Consider organizing a road race, walkathon, or similar event to raise funds to support wetland restoration. These funds can be matched with federal or state funding, especially the U.S. Fish and Wildlife Service's Partners for Fish and Wildlife program.

Wetlands to Explore

The following is a simple listing of sites arranged by state and town. It provides examples of federally-owned or state-owned wetlands that are open to the public, plus other large wetlands that may be accessible from roads. For more information on these and other wetland sites, contact the local parks or conservation (environmental) commission, state parks, forests, or wildlife departments, or the U.S. Fish and Wildlife Service or National Park Service. Ask if wetlands are accessible on designated trails, or are at least open for public use. That way you should identify many wetlands in any geographic area of interest. Reference to National Wetlands Inventory maps will also provide general information on the types of wetlands that are present (e.g., salt marshes, freshwater marshes, shrub swamps, bogs, forested wetlands, etc.).

Maine

Bath—Merrymeeting Bay
Beddington—Bog Brook
Belgrade—Belgrade Bog
Brownsfield—Brownsfield Bog
Calais—Moosehorn National Wildlife Refuge
Edmunds—Cobscook Bay State Park
Freeport—Wolf Neck State Park
Georgetown—Reid Beach State Park
Lee—Dwinal Flowage
Lovell—Kezar Outlet Fen
Lubec—Quoddy Head State Park
Magalloway—Lake Umbagog
Mount Desert Island—Acadia National Park
    (numerous sites)
Orono—Caribou Bog
Phippsburg—Popham Beach State Park
Pierce Pond—Black Brook Pond
Scarborough—Scarborough Marsh
Schoodic—Schoodic Point (Acadia National
    Park)
Steuben—Petit Manan National Wildlife Refuge
Unity—Fowler Bog
Wells—Wells National Estuarine Sanctuary,
    Rachel Carson National Wildlife Refuge

New Hampshire

Adams Point—Jackson Estuarine Laboratory
Cambridge—Lake Umbagog National Wildlife
    Refuge
Center Ossipee—Heath Pond Bog
Deerfield—Deerfield Black Gum Swamp
Dover—Great Bay Marshes
Durham—Great Bay National Wildlife Refuge
Fremont—Spruce Swamp
Greenland—Great Bay Marshes
Hampton—Hampton Harbor Marshes
Hillsborough—Fox State Park
Hopkinton—Smith Pond Bog
Kingston—Cedar Swamp Pond Bog and Swamp
Manchester—Manchester Cedar Swamp
Nashua—Ponemah Bog
Portsmouth—Great and Packer Bogs
Rochester—Rochester Heath Bog
Whitefield—Pondicherry Wildlife Refuge

Vermont

Addison—Dead Creek Marshes
Alburg—Palmer Swamp, Kelly Bay Marshes
Bloomfield—Yellow Bogs
Brandon—Scanlon Bog
Brighton—Clyde River Marshes
Castleton—Lake Bomoseen Marsh
Colchester—Winooski River Marshes
Cornwall—Cornwall Swamp, Otter Creek
    Marshes
Coventry—Black River Marsh, South Lake
    Memphremagog Wetlands
Dorset—Dorset Marsh
Fairfield—Fairfield Swamp
Ferdinand—Ferdinand Bog
Ferrisburg—Little Otter Creek Marshes, Lewis
    Creek Marshes
Franklin—Lake Carmi Bog
Granby—Victory Bog
Island Pond—Mollie Beattie Bog
Leicester—Leicester River Swamp
Marshfield—Peacham Bog
Orwell—East Creek Marshes

Proctor—Otter Creek Marshes
Shelburne—LaPlatte River Marshes
Swanton—Missisquoi National Wildlife Refuge
Weybridge—Cranberry Bog
Wolcott—Bear Swamp

Massachusetts

Andover—C. W. Ward Reservation
Barnstable—Barnstable Marshes
Brewster—Nickerson State Park, Cape Cod
    Museum of Natural History
Canton—Ponkapoag Pond Bog
Carver—Myles Standish State Park
Chappaquiddick Island—Poucha Pond Preserve
Chatham—Monomoy National Wildlife Refuge
Dartmouth—Allens Pond, Demarest Lloyd State
    Park, Lloyd Environmental Center
Duxbury—Duxbury Marshes
Eastham—Cape Cod National Seashore,
    Cape Cod Rail Trail
Easthampton—Arcadia Wildlife Sanctuary
Edgartown—Caroline Tuthill Preserve, Sheriff's
    Meadow Sanctuary, Cape Poge Wildlife
    Refuge, Wasque Reservation, Felix Neck
    Wildlife Sanctuary
Hingham—World's End Reservation
Ipswich—Crane Reservation
Manchester—Agassiz Rock and Swamp Rock
Marshfield—South River Marshes
Mashpee—South Cape State Park
Medfield—Henry Shattuck Reservation, Noon
    Hill Reservation
Nantucket—Eel Point Reservation (Madaket),
    Cranberry Bog
Natick—Broadmoor Wildlife Sanctuary
Newburyport—Merrimack River Marshes
Norwell—Black Pond Nature Preserve, North
    River Marshes
Plum Island—Parker River National Wildlife
    Refuge, Plum Island (Sandy Point) State
    Reservation
Rowley—Parker River Marshes
Salisbury—Salisbury Beach State Park
Sudbury—Great Meadows National Wildlife
    Refuge
Taunton—Hockomock Swamp
Tisbury—Cranberry Acres
Topsfield—Ipswich River Wildlife Sanctuary
Truro—Cape Cod National Seashore
Wellfleet—Cape Cod National Seashore,
    Wellfleet Bay Wildlife Sanctuary
Westborough—Great Cedar Swamp
Westport—Horseneck Beach State Park
West Tisbury—Wompesket Preserve

Connecticut

Canaan—Robbins Swamp
Cromwell—Dead Man's Swamp
Deep River—Post and Pratt Coves
East Lyme—Rocky Neck State Park
Glastonbury—Wangunk Meadows
Goshen—Black Spruce Bog
Greenwich—Audubon Fairchild Garden
Groton—Bluff Point State Park
Guilford—Westwoods Trails
Litchfield—Cranberry Pond Bog
Lyme—Selden Creek
Madison—Hammonasset Beach State Park
Middletown—Pecausett Meadows
Milford—Milford Point Sanctuary
New Haven—Quinnipiac Meadows
Norfolk—Beckley Bog
Norwalk—Manresa Island Marshes
Stonington—Barn Island Wildlife Management
    Area
Stratford—Stratford Great Meadows
Voluntown—Pachaug State Forest
Weston—Devil's Den Preserve
Westport—Sherwood Island State Park

Rhode Island

Charlestown—Ninigret National Wildlife
    Refuge
Coventry—Mishnock Swamp
Hopkinton—Long and Ell Ponds Natural Areas
Lonsdale—Lonsdale Marsh
Middletown—Norman Bird Sanctuary
South Kingston—Trustom Pond National Wild-
    life Refuge
Westerly—Chapman Swamp
West Kingston—Great Swamp
West Exeter—Arcadia Management Area
Woodville—Diamond Bog

New York State

Alabama—Iroquois National Wildlife Refuge
Alexandria Bay—Black Ash Swamp
Amherst—Great Baehre Swamp
Ancram—Drowned Lands Swamp
Argyle—Tamarack Swamp
Beekmantown—Montys Bay
Bergen—Bergen Swamp
Bridgehampton—Sagg Swamp
Canton—Beaver Creek
Catskill—Great Vly Marsh
Champlain—Kings Bay
Chelsea—Sawmill Creek Marsh

Cicero—Muskrat Bay, Cicero Bog
Conesus—Conesus Lake State Wildlife Management Area
Constantia—Big Bay Wetland
Cuba—Vanderlinden Marsh
DePeyster—Black Lake Wetlands
Dewitt—White Lake Swamp
Diana—Bonaparte Swamp
East Fishkill—Swamp River, Great Swamp
East Hampton—Merrill Lake Nature Trail (part of the Accabonac Harbor Preserve)
East Norwich—Muttontown Preserve
Ellisburg—Black Pond Marsh, Little Stony Creek Marsh
Flanders—Hubbard Creek Marsh and Penny Pond
Glen Cove—Welwyn Preserve
Guilderland—Black Creek Marsh
Hammond—Black Creek Marsh, Chippewa Creek
Horseheads—Horseheads Marsh
Jamaica Bay—Jamaica Bay Wildlife Refuge
Lisbon—Brandy Brook Wetlands
Lloyd Neck—Caumsett State Park
McLean—Lloyd-Cornell McLean Wildlife Reservation
Mendon—Mendon Ponds Park
Moravia—Owasco Inlet
Montauk—Montauk Point State Park
Newcomb—Wolf Pond Fen
Northeast—Millerton Bog
Orient—Orient Beach State Park
Oswego—Snake Creek Marsh
Palermo—Lot Ten Swamp
Pawling—Great Swamp
Quogue—Quogue Wildlife Refuge
Randolph—Randolph Swamp
Richfield Springs—Maumee Swamp
Richland—Brennan Beach Fen
Rossville—Isle of Meadows, Fresh Kills
Rye—Marshlands Conservancy
Sag Harbor—Morton National Wildlife Refuge
Sailor's Haven—Fire Island National Seashore
Sandy Creek—Rainbow Shores Bog
Sangerfield—Sangerfield (Ninemile) Swamp
Schodack—Papscanee Marsh
Schroeppel—Peter Scott Swamp
Seneca Falls—Montezuma National Wildlife Refuge
Shelter Island—Mashomack Preserve
Sherman—Alder Bottom
Smith Point—Fire Island National Seashore
Smithtown—Nissequoque River State Park
Southampton—Quogue Wildlife Refuge
Springwater—Honeoye Inlet
Sterling—Sterling Creek Marsh
Stuyvesant—Lewis A. Sawyer Preserve
Tonowanda—Tonowanda Creek Wildlife Management Area
Tyrone—Lamoka Lake Wetland
Warwick—Little Cedar Pond
Westchester—Marshlands Conservancy
Westerlo—Bear Swamp

New York City

Bronx—Van Cortlandt Park, Tibett's Brook, Upper Van Cortlandt Lake, Van Cortlandt Lake, Seton Falls Park, Bronx Park, Bronx River Corridor
Brooklyn—Prospect Park, Pagoda Pond, the Ambergill, Marine Park, Spring Creek, Four Sparrows Marsh, Pelham Bay Park, Hunter Island, the Lagoon, Thomas Pell Wildlife Refuge
Manhattan—Central Park, the Loch, Inwood Hill Park
Queens—Idewild Park, Udalls Park Preserve, Udalls Cove, the Ravine, Alley Pond Park, Cattail Pond, Lily Pond, Little Alley Park, Spring Pond, Little Alley Pond, Corona Park, Willow Lake (east shore), Flushing Meadows, Cunningham Park, Kettle Ponds

New Jersey

Basking Ridge—Somerset County Environmental Center and Great Swamp National Wildlife Refuge
Batso—Wharton State Forest
Branchville—Stokes State Forest
Canton—Mad Horse Creek Wildlife Management Area
Chatham—Great Swamp Outdoor Education Center
Chester—Black River Wildlife Management Area
Colliers Mills—Colliers Mills Wildlife Management Area
Farmingdale—Allaire State Park
Fortescue—Fortescue Wildlife Management Area
Freehold Township—Turkey Swamp Park
Hackensack—Hackensack River County Park
Hamilton Township—John A. Roebling Park
Hammonton—Wharton State Forest
Higbee Beach—Higbee Beach Wildlife Management Area
Highland Lakes—Wawayanda State Park
Holmdel—Holmdel Park Activity Center
Keyport—Henry Hudson (Bike) Trail
Lyndhurst—DeKorte State Park

Matawan—Cheesequake State Park
Mt. Holly—Rancocas State Park/New Jersey
Audubon Rancocas Nature Center
Neptune—Shark River Park
New Gretna—Bass River State Forest
Oceanville—Edwin B. Forsythe National Wildlife Refuge (including Oceanville Bog)
Pennsville—Killcohook and Supawna Meadows National Wildlife Refuges
Port Republic—Port Republic Wildlife Management Area
Princeton—Charles H. Rogers Wildlife Refuge
Raritan—Duke Island Park
Salem—Mannington Marsh
Sandy Hook—Gateway National Recreation Area
Somerset—Delaware-Raritan Canal State Park (Silver Maple Natural Trail)
Stafford Township—Manahawkin Wildlife Management Area
Stone Harbor—The Wetland Institute
Sussex—High Point State Park (Dryden Kuser Natural Area) and Reinhardt Preserve
Swainton—Beaver Swamp Wildlife Management Area
Tuckerton—Great Bay Boulevard Wildlife Management Area
Upper Freehold—Assunpink Wildlife Management Area
Vernon Township—Wallkill River Bottomlands National Wildlife Refuge
Woodbine—Belleplain State Forest
Wyckoff—James A. McFaul Environmental Center

Pennsylvania

Blooming Grove—Delaware State Forest, Blooming Grove 4-H Trail
Chester/Philadelphia—John Heinz National Wildlife Refuge
Dowington—Marsh Creek State Park
Edgemere—Delaware State Forest, Stillwater Natural Area
Elverson—French Creek State Park
Erie—Presque Isle State Park
Goshen—S. B. Elliot State Park
Guy Mills—Erie National Wildlife Refuge
Hawley—Decker Marsh, State Game Lands
Lords Valley—Smith's Swamp (State Game Lands 180)
Mauch Chunk—Hughes Swamp (Broad Mountain, State Game Lands 141)
Pond Eddy—Delaware State Forest, Buckhorn Natural Area
Promised Land—Promised Land State Park and

Palmyra State Forest (Panther Swamp and Balsam Swamp)
Pymatuning—Pymatuning State Park
Shohola Falls—Shohola Waterfowl Management Area (State Game Lands 180)
State College—Bears Meadows Natural Area
Stroudsburg—Tannersville Bog, c/o Messing Nature Center, Pine Lake Natural Area (Delaware State Forest)
Thornhurst—Tannery Bog (Lackawanna State Forest)
Tobyhanna—Tobyhanna State Park, Bender Swamp/Jim Smith Run Trail, Pond Swamp (State Game Lands 127), Bear Swamp
Twelve Mile Pond—Delaware State Forest, Pennel Run Natural Area
Union City—Boleratz Bog
White Haven—Hickory Run State Park

Delaware

Bethany Beach—Assawoman Wildlife Area
Blackbird—Blackbird State Forest
Dewey Beach—Delaware Seashore Park
Laurel—Trap Pond State Park
Lewes—Cape Henlopen State Park
Little Creek—Little Creek Wildlife Area
Milton—Prime Hook National Wildlife Refuge and Prime Hook Wildlife Area
Newark—Walter S. Carpenter, Jr., State Park
Port Penn—Port Penn Interpretive Center, c/o Fort Delaware State Park
Rehoboth Beach—Delaware Seashore State Park
Selbyville—Great Cypress Swamp
Smyrna—Bombay Hook National Wildlife Refuge
Wilmington—Brandywine Creek State Park
Woodland Beach—Woodland Beach Wildlife Area

Maryland

Assateague Island—Assateague State Park, Assateague Island National Seashore, Chincoteague National Wildlife Refuge (Virginia)
Bay View—Whitaker Swamp
Benedict—Patuxent River Marshes
Bittinger—Cunningham Swamp
Brandywine—Cedarville State Forest
Cambridge—Blackwater National Wildlife Refuge
Clinton—Suitland Bog (Clearwater Nature Center)
Indian Head—Mattawoman Creek Natural Area
LaPlata—Zekiah Swamp

Lothian—Jug Bay Wetlands Sanctuary, Anne Arundel County Department of Recreation & Parks, Patuxent River State Park

Princess Anne—Deal Island Wildlife Management Area and Chesapeake Bay National Estuarine Sanctuary (Monie Bay component)

Queen Anne—Tuckahoe State Park

Rock Hall—Eastern Neck National Wildlife Refuge

Scotland—Point Lookout State Park

Snow Hill—Pocomoke River State Park, Pocomoke State Forest

Upper Marlboro—Jug Bay Natural Area

Table C.1. Some poisonous wetland plants in the Northeast and Great Lakes states.

Common Name (Species Number)	Scientific Name	Poisonous Parts	Animals Affected
Herbs			
Jack-in-the-pulpit (62)*	Arisaema triphyllum	All	Humans
Swamp Milkweed (81)*	Asclepias incarnata	Leaves, fruits, stems	Livestock
Marsh Marigold (59)	Caltha palustris	Leaves	Livestock
Water Hemlock (91b)	Cicuta maculata	All	Humans, horses, livestock, pets
Poison Hemlock (91b)	Conium maculatum	All	Humans, horses, livestock, pets
Blue Flag (61)*	Iris versicolor	Roots, rhizomes	Humans, livestock
Cardinal Flower (72)*	Lobelia cardinalis	All	Humans, goats
Sensitive Fern (25)	Onoclea sensibilis	Leaves	Horses
May Apple	Podophyllum peltatum	All (except ripe fruit)	Humans, livestock
Bittersweet Nightshade* (Figure 8.7)	Solanum dulcamara	Leaves, fruit	Humans, horses, livestock
Swamp Buttercup*	Ranunculus septentrionalis	All	Horses, livestock
Golden Ragwort (93a)*	Senecio aureus	Leaves	Humans, horses, livestock
Skunk Cabbage (60)	Symplocarpus foetidus	All	Humans
Seaside Arrow-grass (Figure 5.1)	Triglochin maritimum	All	Livestock
Stinging Nettle (71b)	Urtica dioica	Hairs	Human, dogs
Death Camas	Zigadenus spp.	All	All
Woody Plants			
Common Elderberry (104)	Sambucus canadensis	All (except ripe fruit)	Humans, livestock
Poison Ivy (128)	Toxicodendron radicans	All	Humans
Poison Sumac (105)	Toxicodendron vernix	All	Humans
Black Cherry*	Prunus serotina	Leaves, seeds	Horses, livestock
Oaks (151–152)*	Quercus spp.	Acorns, young leaves	Horses, cattle

Note: Species number refers to species illustrated and described in Chapter 10 of this book. Species followed by an asterisk (*) are wetland species of genera listed as poisonous by Cornell University's Poisonous Plants Information Database (http://www.ansci.cornell.edu/plants/).

Glossary

adventitious root—root originating from above ground, often in response to flooding conditions or heavy sedimentation.

aerenchyma—plant tissue with many large air spaces, common in the roots and stems of many wetland and aquatic species; facilitates air movement to roots for survival in anaerobic soils or substrates.

aerobic—conditions where free oxygen is present; organisms that require such conditions for life.

alluvial—river processes that cause the deposition of sediment (alluvium) on floodplains, deltas, or beds.

alternate (leaves)—arranged singly along the stem, alternating from one side to the other.

anaerobic—conditions lacking free oxygen, typically caused by prolonged inundation or saturation; bacteria that require such conditions for life.

angled (stem)—having distinct edges; three-angled stems are triangular in cross section, while four-angled stems are square.

annual (plant)—living for only one year; propagates from seeds.

anoxic—lacking free oxygen.

anther—distal end of a stamen, where pollen is produced.

anthropogenic—relating to the impact of humans on the environment.

Appalachian Plateau—major physiographic region, following the Appalachian mountains in the eastern United States from New York to northern Alabama.

appressed—packed closely together, as in an appressed inflorescence.

aquatic—living in water.

aquic moisture regime—condition in which the soil is saturated by groundwater (including the capillary fringe) or by flooding long enough to create a seasonal reducing environment that is virtually free of dissolved oxygen.

aromatic—fragrant or sweet-smelling.

artificial—created or altered by humans, as in artificial wetland and artificially drained.

ascending—rising or pointing upward.

awn—bristle-shaped appendage common in some grasses.

axil—angle formed by a branch or leaf with the stem.

basal (leaves)—arising directly from the roots; may ascend along the stem as sheaths and give the appearance of alternately arranged leaves, as in cattails (*Typha* spp.).

berry—fleshy or pulpy fruit.

biomass—organic matter produced by living organisms (plants and animals).

biotic—relating to plants and animals.

blade—flattened leaf.

bog—type of wetland forming on acidic peats (peatland), typically formed by the accumulation of peat moss (*Sphagnum* spp.) in a nutrient-poor environment and colonized by ericaceous shrubs like leatherleaf or evergreen trees like black spruce.

brackish—somewhat salty, such as water of estuaries resulting from mixing of sea water with fresh water from rivers.

bract—leaflike appendage that subtends a flower or is part of an inflorescence.

bristle—long, stiff, hairlike structure.

bud—unexpanded flower, leaf, or shoot.

buttressed (trunks)—enlarged main stem of tree, much wider at base than normal and narrowing up the trunk for some distance (typically more than two feet), then assuming standard diameter above; a response of some trees like bald cypress (*Taxodium distichum*) and swamp black gum (*Nyssa sylvatica biflora*) to prolonged inundation.

capsule—dry fruit composed of two or more cells or chambers.

Carolina bay—type of wetland formed in a somewhat circular or egg-shaped (elliptical) depression with a northwest to southeast orientation, found along the coastal plain from southern New Jersey to Florida.

catkin—scaly spike of minute flowers lacking petals, as in willows, oaks, and birches.

channeled—having distinct grooves or ridges.

chroma—property of color indicating its purity or strength, typically used to describe soil colors; low-chroma colors are associated with hydric soils, organic-enriched topsoil, and soils derived from gray- or other dark-colored parent materials.

clasping (leaves)—closely surrounding the stem and attached directly without a stalk.

clay—soil particles less than 0.002 mm.

coastal plain—major physiographic region extending from the ocean inland, characterized by low topography (outer or lower coastal plain) and by low, rolling plains (inner or upper coastal plain), extending from Long Island to Florida.

compound (leaves)—divided into two or more separate blades (leaflets).

concretion—soft or hard mass of precipitated elements (e.g., iron and manganese) in the soil.

coniferous—cone bearing, such as pines, hemlock, and larch.

deciduous—not persistent, dropping off the plant when no longer functioning, as in the case of leaves that are shed annually in the fall.

Delmarva bay—type of wetland formed in a somewhat circular depression on the coastal plain; another name for a Carolina bay occurring on the Delmarva Peninsula.

detritus—organic particles formed from dead stems and leaves broken down by various organisms.

disk—tubular flower forming the central head of asters, goldenrods, and other members of the family Compositae.

dissected—deeply divided into threadlike sections, as in dissected leaves.

dominant species—organisms that are the most abundant species or among the most abundant species in a given area, usually characteristic of the area.

drained—condition in which surface or groundwater is removed either naturally or artificially (by humans); "effectively drained" refers to a condition in which drainage is such that the hydrology is no longer similar to that of a wetland (e.g., the soil is not saturated at the depth or for the duration and frequency required to support hydrophytes and typical wetland functions).

ecotypes—genetically distinct populations within a single species, each adapted to local conditions.

eluvial—resulting from removal of materials (e.g., iron or organic matter) from a soil layer.

emergent—in general terms, refers to plants growing upright out of the soil or water; in specific terms, refers to nonwoody (herbaceous) plants assuming this posture.

entire (leaves)—having smooth margins, lacking teeth.

epipedon—top mineral layer of a soil, formed at the surface.

ericaceous—related to the heath family (Ericaceae) or to areas dominated by these broad-leaved evergreen shrubs.

estuary—usually semienclosed, tidal aquatic-wetland system where sea water is measurably diluted with fresh water.

evaporation—water loss from exposed surfaces (e.g., soil or a body of water) by the conversion of liquid water to water vapor, brought about by increased temperatures and accentuated by wind action.

evapotranspiration—water loss through the combination of evaporation and transpiration by plants.

evergreen—persistent, as in the case of leaves that remain on a plant through winter and usually for more than one growing season.

facultative—adapted to or found in a wide range of conditions, from wet to dry. In plants, the term refers to species that do not grow exclusively in wetlands, but occur in both wetlands and nonwetlands (uplands); facultative wetland species have a higher affinity for wetlands than facultative species (nearly equally abundant in wetlands and nonwetlands) or facultative upland species (more typical of uplands).

fen—type of wetland growing on variably mineral-rich peats, typically with significant groundwater inflow, and dominated by sedges and mineral-loving species; characteristic of boreal and glaciated regions; more or less the antithesis of a bog.

filament—bottom of a stamen that supports the anther.

flatwood—forest growing on a broad, flat area of the coastal plain; includes both forested wetlands and upland forests with a few species of pines, hardwoods, or mixed stands and often with imperfectly drained soils.

fleshy—soft, thickened tissue; succulent.

flooding—in general terms, an event that results in, or the condition of, surface water being present on the land; in soil science, refers to surface water derived from flowing waters, such as streams overflowing their banks, tidal inundation, or runoff from adjacent slopes.

floodplain—level landform that is subject to periodic flooding at varying frequencies; e.g., annual floodplain (flooded once a year on average) or 100-year floodplain (flooded once every 100 years, or has a probability of 1 in 100 of being flooded in a single year).

flyway—traditional migratory route traveled by birds during fall and spring, as in the Atlantic flyway along the North Atlantic coast.

fresh (water)—having no measurable trace of salinity; less than 0.5 parts per thousand of ocean-derived salts.

frond—leaf of a fern.

glade—a grassy opening in the woods or forest.

gleyed (soil)—developed under anaerobic, reducing conditions that typically result in grayish, bluish, or greenish subsoils characteristic of many hydric mineral soils.

glume—thin bract at the base of a grass spikelet.

grain—fruit of certain grasses.

graminoids—grasslike plants, including grasses (family Poaceae), sedges (Cyperaceae), and rushes (Juncaceae).

groundwater—water underground in completely saturated soils, sediments, and rocks.

groundwater discharge—release of groundwater to the earth's surface via springs and seeps, and to bodies of water (e.g., streams, rivers, lakes, and oceans) via bank seepage or upward flow into lake, river, or ocean bottoms.

groundwater recharge—downward (gravity-induced) movement of water into the soil and underlying sediments and rocks from rain, snowmelt, or surface water.

growing season—time of year when plants are growing; term is often applied to crops, as in the growing season for corn; for wild species, however, it includes the period of root growth as well as aboveground growth; in soil science and for wetland delineations, generally defined as the period when soil temperatures are 5°C or 41°F measured at a depth of 20 inches below the soil.

halophytes—salt-loving plants; plants adapted for life in salt water or saline soils.

head—dense cluster of sessile or nearly sessile flowers, characteristic of the aster family (Compositae), but also found in other plants such as buttonbush and bur-reeds.

heath—low shrub thicket dominated by members of the family Ericaceae.

herbaceous—nonwoody; plants with soft stems that are easily crushed.

high marsh—zone of tidal marsh that is irregularly flooded (less than once a day).

histic—relating to heavy accumulation of organic matter, as in "histic epipedon," a shallow peat or muck layer (8–16 inches thick) on top of mineral materials, diagonistic of the wettest hydric mineral soils.

histosols—organic soils; soils with more than 16 inches of organic material at the surface, or with shallower layers overlying bedrock.

horizon—distinct layer of soil, more or less parallel with the soil surface, having properties such as color, texture and permeability; most soils have three horizons, while some soils have a fourth horizon: A-horizon (topsoil, layer of accumulation of organic matter), B-horizon (subsoil, below-surface layer where clay, iron, or organic matter accumulates), C-horizon (undisturbed, unaltered parent material), and sometimes E-horizon (layer between the A- and B-horizons where materials are leached out, giving it a characteristic grayish color).

husk—hard covering or shell of certain seeds, as in the husks of walnuts.

hydric soils—soils that are saturated, flooded, or ponded long enough during the growing season to develop anaerobic conditions in the upper part; soils that favor the growth and reproduction of hydrophytes.

hydrogen-sulfide odor—smell of rotten eggs given off by some highly reduced soils.

hydrograph—graph showing fluctuations in the water table and surface-water levels over time.

hydrologic cycle—pattern of circulation of water among land, bodies of water, and the atmosphere.

hydrology—scientific study of properties, distribution, and circulation of water in the earth.

hydroperiod—duration, depth, and frequency of flooding or near-surface soil saturation over time.

hydrophyte—plant growing in water or on a substrate that is at least periodically deficient in oxygen (anaerobic) due to excessive water content; may represent the entire population of a species or only a subset of specially adapted individuals; plants typically adapted to wetland and aquatic habitats.

hypertrophied—enlarged or expanded, as in "hypertrophied lenticels," which form in response to prolonged inundation and are believed to aid in oxygen uptake, or in "hypertrophied stems" with expanded bases.

infiltration—downward movement of water into the soil and underlying materials; often referred to as percolation.

inflorescence—flowering part of a plant.

inflow—condition in which water enters a wetland from a stream or another wetland, but there is no exit; a wetland with an inlet and no outlet.

intertidal—zone that is periodically flooded and exposed by the tides.

invertebrates—animals without backbones, such as worms, crabs, and insects.

irregular (flower)—having similar parts (e.g., petals) of differing size or shape.

irregularly (flooded)—inundated less than once a day by the tides.

jointed (stem)—having obvious nodes.

lance-shaped (leaves)—appearing in the shape of the head of a lance, several times longer than wide, broadest just above the base, and tapering to a tip.

lateral—borne on the sides.

lemma—lower of two bracts enclosing the flower of a grass.

lenticel—corky spot or line, sometimes raised, on the bark of many trees and shrubs; the point of gas exchange between the internal tissues of the plant and the atmosphere; hypertrophied lenticels are swollen or expanded lenticels that develop just below the water's surface in certain wetland trees and shrubs.

ligule—membranous or hairy structure at the junction of the leaf blade and the leaf sheath in grasses and a few sedges.

linear—narrow and elongate, several to many times longer than wide.

lips—upper and lower parts of irregular flowers, as in certain tubular flowers.

lobe—indented part of a leaf or flower, not divided into a separate part but still connected.

low marsh—zone of tidal marsh that is regularly flooded (once or twice daily).

kettle pond or hole—glacially formed pond or depression resulting from the melting of ice blocks left by the retreating glacier about 10,000 years ago.

marsh—type of wetland dominated by emergent (herbaceous) plants growing in shallow water for all or most of the growing season, often characterized by one or a few species.

matrix—the dominant color of a particular soil horizon.

meadow, wet—type of wetland dominated by herbaceous plants growing in soils saturated at or near the surface for extended periods during the growing season; may be inundated for brief periods, usually in early spring and for longer periods in winter, especially depressional areas; characterized by few or many species.

microbial—pertaining to the work of microorganisms (too small to see with the naked eye).

midrib or midvein—middle rib or vein of a leaf.

mineral soil—soil composed mostly of combinations of sand, silt and clay.

mollusc—snail (univalve) or clam or mussel (bivalve) in the phyllum Mollusca.

morphology—structure and form of an organism or the soil.

mottled zone—layer in the soil marked by mottling due to seasonal wetness.

mottling—in soils, refers to spots or blotches of nondominant color interspersed on the matrix or dominant color; in hydric soils, mottles are mainly due to oxidation-reduction processes and are called "redoximorphic features."

muck—highly decomposed organic matter.

mud flat—periodically exposed muddy bottom of a body of water; in tidal areas of the Northeast, most mud flats are flooded and exposed twice daily.

muskeg—type of wetland found in arctic and boreal regions on permanently saturated soils.

nerve—prominent vein of a leaf.

nontidal—not subject to tidal flooding.

nutlet—small, dry, hard fruit (technically called an achene).

obligate—depending solely on a particular habitat or set of environmental conditions, as in obligate wetland plant (grows only in wetlands).

oblong (leaves)—longer than wide, with nearly parallel sides.

opposite (leaves)—arranged in pairs (opposite each other) along the stem.

organic soil—soil composed mostly of the remains of plants; peats and mucks.

outflow—water exiting.

outwash deposits—clay, sand, and gravel deposited by glacial meltwater streams, sometimes forming extensive plains beyond the ice front of a glacier.

oval (leaves)—broadly egg-shaped, widest at middle and tapering to ends.

oxidation—process of combining with oxygen, as in the decomposition or breakdown of organic matter when exposed to air.

oxidation-reduction—chemical process in which one or more electrons are transferred from one molecule to another; oxidation involves removing an electron, while reduction involves adding an electron.

palea—upper of two bracts enclosing the flower of a grass.

paludification—process of swamping, in which

peat mosses (*Sphagnum* spp.) grow upslope, covering uplands with organic material in areas with cool and moist climates; an important wetland-forming process common in boreal and arctic regions, where it shapes the landscape.

palustrine (wetlands)—beyond the influence of tidal brackish waters and typically dominated by persistent vegetation (e.g., trees, shrubs, and robust emergents) that remain standing into the next growing season; most inland wetlands fall into this classification.

panicle—many-branched flowering inflorescence.

panne—depression in salt and brackish marshes that temporarily holds water.

parotoid gland—in toads, the swollen, saclike gland behind the eyes.

peat—relatively undecomposed organic matter; most of the plants are still recognizable after gentle rubbing between thumb and forefinger.

peatland—area where peat dominates, including bogs and fens.

perched (water table)—lying above the regional water table and separated from it by unsaturated material; typically forms due to the presence of an impermeable layer of clay, compacted silts (former lake bed), dense basal till, or other restricting material.

perennial—plant living for many years, usually supported by underground rhizomes, tubers, or bulbs.

permeability—soil property that describes the ease with which water moves downward through the profile, measured in inches per hour; low or slow permeability is less than 6 inches per hour, and high or rapid permeability greater than 6 inches per hour.

phase, soil—subdivision of a soil series based on features such as slope, surface texture, stoniness, thickness, artificial drainage, and flooding.

Piedmont—major physiographic region of rolling hills west of the coastal plain.

Pine Barrens—region of southern New Jersey dominated by pitch-pine forests.

pistil—seed-bearing structure of a flower.

pith—soft, fleshy, or spongy center of a stem.

plano-convex—flattened but somewhat curved.

plastron—bony undershell of a turtle or tortoise.

playa—type of wetland found in the Southwest, characterized by shallow depressions with greatly fluctuating water levels and ranging from lake to marsh to exposed clay bottoms during the wet-to-dry cycle.

pneumatophore—specialized extension of the root system that projects above the ground surface; in wetland species, helps promote root aeration in periodically flooded soils.

pocosin—type of wetland found on the coastal plain (from southeastern Virginia into Florida) on interstream divides; characterized by shallow organic soils and ericaceous shrubs and pond pines (*Pinus serotina*); a Native American word for "swamp-on-a-hill."

pod—dry fruit capsule.

podzolization—process by which aluminum and iron are leached from the E-horizon and precipitated in the B-horizon.

ponding—condition in which surface water is present in a closed depression; the water can be removed only by percolation, evaporation, or transpiration.

pothole—type of wetland formed in a distinct depression typically found in the upper Midwest (prairie pothole wetlands), in glacially formed shallow basins, and other depressional wetlands elsewhere, such as on the Delmarva Peninsula.

prickly—bearing small spines.

profile—vertical section of the soil extending through all its horizons into the parent material.

prostrate—lying flat on the ground.

pupil—contracile part of the eye, usually round, but vertical in poisonous snakes of the viper family.

raceme—spikelike inflorescence bearing stalked flowers.

rachis—main axis of a spike, branching inflorescence, compound leaf, or fern frond.

rank—row of an organ, such as leaves (2-ranked or 3-ranked) along a stem.

ray—outer, often petal-like, flower of a composite head of asters.

recurved—curved downward.

redox concentrations—accumulations of iron and manganese concretions, usually as soft masses, oxidized rhizospheres, or nodules; evidence of periodic oxidization in an alternating wet-and-dry environment.

redox depletions—accumulations of low-chroma (often grayish, bluish, or greenish) colors where iron, manganese, or clay has been reduced and stripped from the soil horizon; evidence of at least a periodically reducing environment.

reduction—process of changing an element from a higher oxidation state to a lower one, as in the reduction of ferric iron (Fe^{3+}) to ferrous iron (Fe^{2+}), which is typically brought about by anaerobic conditions associated with prolonged saturation.

regular (flower)—having similar parts (e.g., petals) with the same size and shape; radially symmetrical.

regularly flooded—subject to daily tidal flooding; flooded once or twice a day by the tides.

rhizome—underground part of a stem, usually horizontal and rooting at nodes, producing erect stems.

saline—salty.

samara—winged dry fruit bearing one seed, as in maples and ashes.

sand—soil particles 0.05 to 2.0 mm.

saturation—condition in which all or nearly all the pores of soil are filled with water, including the water table and the capillary fringe; in terms of wetland hydrology (saturated wetland), water table at or near the surface for extended periods, with surface water uncommon.

savanna—open grassland with scattered trees and shrubs.

scale—modified leaf, or a thin, flattened structure, like the scale covering nutlets in sedges.

scutes—horny or bony plates forming the shell or plastron of a turtle.

seasonally flooded—inundated for extended periods during the growing season of most years.

seasonally saturated—soil is saturated at or near the surface for extended periods, usually from late winter into early spring.

seep—point of groundwater discharge where the earth's surface is saturated but flowing water is not typically observed; discharge may be seasonal or permanent.

semipermanently flooded—covered by surface water throughout the growing season in most years.

sepal—outermost part of a flower, typically the green leaflike or thin bracts subtending the petals.

sessile—lacking stalks, as in case of some flowers and sessile leaves.

sheath—tubular envelope surrounding the stem, as in leaf sheaths of grasses.

shrub—erect woody plant, usually with multiple trunks and typically less than 20 feet tall at maturity.

silt—soil particles 0.002 to 0.05 mm.

simple (leaves)—not divided into separate parts; leaf blade is continuous, but may be lobed.

sinkhole—hole or depression in the earth's surface caused by the dissolution of underlying limestone; may connect to underground caverns.

slough—another name for a marsh, sometimes referring to a small creek through a marsh or a wetland in a creek or broad drainageway.

soil—natural bodies of mineral or organic materials that support or are capable of supporting the growth of free-standing plants, including emergent species, (e.g., trees, shrubs, and erect herbaceous plants) but not submergent plants.

soil series—group of soils with similar characteristics in the soil profile, except for the texture of the surface layer.

sori—clusters of fruit dots of ferns.

spadix—fleshy spike.

spathe—leafy or hoodlike bract or pair of bracts enclosing an inflorescence, as in skunk cabbage and other members of the arum family.

spike—simple, unbranched inflorescence, composed of a central axis with sessile or nearly sessile flowers.

spine—sharp-pointed outgrowth of a stem.

sporangia—spore cases of ferns and related plants.

spore—reproductive organ of ferns and related plants.

spring—point of groundwater discharge to the earth's surface where flowing water is observed as a pool or stream.

spur—hollow tubular extension of a flower, usually bearing nectar.

stamen—pollen-bearing part of a flower.

stigma—part of a pistil that receives and germinates pollen.

stipules—pair of appendages at the base of a leaf stalk or on each side of its attachment to the stem.

stolon—slender aboveground stem, lying flat on the ground and producing new plants at nodes.

stratified drift—material deposited by glaciers in an organized arrangement.

style—part of a pistil connecting the stigma to the seed-bearing ovary.

substrate—consolidated and unconsolidated (non-soil) materials that do not support free-standing plants; includes rocks, salt flats, and bottoms of bodies of water.

succulent—fleshy, water-filled tissue, or a plant with this characteristic.

surface water—water above ground.

swamp—type of wetland dominated by woody vegetation (trees or shrubs); in general usage, refers to any vegetated area that is wet for extended periods on a frequent, recurring basis.

sword-shaped (leaves)—appearing bayonet-shaped, flattened and tapering to a sharp-pointed tip.

temporarily flooded—inundated for brief periods during the growing season, usually in early spring and after heavy rains.

throughflow—condition in which water enters a

wetland, passes through it, and exits via a stream or groundwater flow to another wetland; a wetland with an inlet and outlet.

tidal—subject to the influence of ocean-driven tides.

till—unsorted and often compacted, deposited material left when glacial ice melted; unstratified clay, sand, gravel, rocks, and boulders.

trailing—running across the ground surface.

transpiration—water loss by plants to the atmosphere through stomata (minute pores) on leaves and stems.

tree—woody plant characterized by a main trunk; usually grows taller than 20 feet at maturity.

triangular (stem)—three-sided or triangular in cross section.

tuber—short, thickened, usually underground stem having buds ("eyes") and storing food.

twining—climbing by wrapping around another plant or object.

tympanum—ear drum.

umbel—branched inflorescence, with flowering stalks arising from a single point.

valve—piece of an open capsule.

vascular—having vessels or ducts for moving internal fluids.

veins—threads of vascular tissue, as in a leaf.

vernal pool—shallow, often seasonal pond (typically surrounded by upland woods) that serves as a breeding ground for salamanders and woodland frogs and is usually devoid of fish.

vertebrates—backboned animals such as birds and mammals.

vine—plant that climbs other plants or objects.

wetland—see Chapter 1.

watershed—land and associated surface-water drainage network above a stream, river, lake, or other body of water that contributes water to the stream, river, lake, etc.

water table—level of the soil that is wholly saturated with water (i.e., where free water is present); typically varies seasonally.

whorl—three or more organs arranged in a circle around a stem, as in whorled leaves.

woody—having stems with a protective covering of bark; trees, shrubs, and some vines.

zonation—in vegetation, refers to a distinct banding pattern of plant communities along an environmental gradient.

References

Other references are listed at the end of each chapter.

Alpaugh, G. L. 1995. *New Jersey Forestry and Wetlands Best Management Practices Manual.* New Jersey Department of Environmental Protection, Trenton, NJ.

Ammann, A. P., and A. Lindley Stone. 1991. *Method for the Comparative Evaluation of Nontidal Wetlands in New Hampshire.* New Hampshire Department of Environmental Services, Concord, NH. NHDES-WRD-1991-3.

Amos, W. H. 1967. *The Life of the Pond.* McGraw-Hill Book Co., New York.

Archer, J. H., D. L. Connors, K. Laurence, S. C. Columbia, and R. Bowen. 1994. *The Public Trust Doctrine and the Management of America's Coasts.* University of Massachusetts Press, Amherst, MA.

Barnes, B., and J. P. Costanzo. 2004. Alaskan wood frogs. http://www.earthsky.com

Blinn, C. R., R. Dahlman, L. Hislop, and M. A. Thompson. 2004. *Temporary Stream and Wetland Crossing Options for Forest Management.* USDA Forest Service, North Central Research Station, St. Paul, MN. General Technical Report NC-202.

Boyd, H. P. 1991. *A Field Guide to the Pine Barrens of New Jersey.* Plexus Publishing, Inc., Medford, NJ.

Boyle, W. J., Jr. 1986. *A Guide to Bird Finding in New Jersey.* Rutgers University Press, New Brunswick, NJ.

Brooks, R. P., D. E. Arnold, and E. D. Bellis. 1987. *Wildlife and Plant Communities of Selected Wetlands: Pocono Region of Pennsylvania.* U.S. Fish and Wildlife Service, Washington, DC. National Wetlands Research Center, Open File Report 87-02.

Brooks, R. P., and M. J. Croonquist. 1990. Wetland, habitat, and trophic response guilds for wildlife species in Pennsylvania. *Journal of the Pennsylvania Academy of Science* 64:93–102.

Brown, M. P., 1992. *New Jersey Parks, Forests, and Natural Areas: A Guide.* Rutgers University Press, New Brunswick, NJ.

Carter, V. 1996. Wetland hydrology, water quality, and associated functions. In *National Water Summary on Wetland Resources,* compilers J. D. Fretwell, J. S. Williams, and P. J. Redman. U.S. Geological Survey, Reston, VA. Water-Supply Paper 2425: 35–48.

Coon, N. 1979. *Using Plants for Healing.* Rodale Press, Emmaus, PA.

Council on Environmental Quality. 1975. *The Delaware River Basin: An Environmental Assessment of Three Centuries of Change.* U.S. Government Printing Office, Washington, DC.

Cowardin, L. M., V. Carter, F. C. Golet, and E. T. LaRoe. 1979. *Classification of Wetlands and Deepwater Habitats of the United States.* U.S. Fish and Wildlife Service, Washington, D.C. FWS/OBS-79/31.

Cox, D. D. 1959. *Some Postglacial Forests in Central and Eastern New York State As Determined by the Method of Pollen Analysis.* New York State Museum and Science Service, University of the State of New York, Albany, NY. Bulletin No. 377.

Craig, L. J. 1991. Small mammal communities of mature and logged Atlantic white cedar swamps. M.S. thesis, Rutgers University, Camden, NJ.

Croonquist, M. J., and R. P. Brooks. 1991. Use of avian and mammalian guilds as indicators of cumulative impacts in riparian-wetland areas. *Environmental Management* 15(5):701–714.

Crum, H. 1988. *A Focus on Peatlands and Peat Mosses.* University of Michigan Press, Ann Arbor, MI.

Cullen, J. B. 1996. *Best Management Practices for Erosion Control on Harvesting Operations in New Hampshire.* University of New Hampshire Cooperative Extension, Forestry and Wildlife Program, Durham, NH.

Dann, K., and G. Miller. 1982. *30 Walks in New Jersey.* Rutgers University Press, New Brunswick, NJ.

Dansereau, P., and F. Segadas-Vianna. 1952. Ecological study of the peat bogs of eastern North America. *Canadian Journal of Botany* 30: 490–520.

Day, C. 2002. The pendulum has swung. New Jersey Field Office, U.S. Fish and Wildlife Service, Pleasantville, NJ. *Field Notes* (Spring 2002): 1–2.

DeGraff, R. M., and D. A. Richard. 1987. *Forest Wildlife of Massachusetts*. University of Massachusetts, Cooperative Extension, Amherst. Publication C-182.

DeGraff, R. M., and G. M. Witman. 1979. *Trees, Shrubs, and Vines for Attracting Birds*. University of Massachusetts Press, Amherst, MA.

Elliott, C. G. 1912. *Engineering for Land Drainage: A Manual for the Reclamation of Lands Injured by Water*. John Wiley & Sons, New York.

Federal Interagency Stream Restoration Working Group. 1998. *Stream Corridor Restoration: Principles, Processes, and Practices*. U.S. Government, Departments of Agriculture, Commerce, Defense, Housing and Urban Development, and Interior; U.S. Environmental Protection Agency; Federal Emergency Management Agency; Tennessee Valley Authority.

Fernald, M. L. 1970. *Gray's Manual of Botany*. Van Nostrand Reinhold Company, New York.

Forman, R.T.T. (editor). 1979. *Pine Barrens: Ecosystem and Landscape*. Academic Press, New York.

Fretwell, J. D., J. S. Williams, and P. J. Redman (compilers). 1997. *National Water Summary on Wetland Resources*. U.S. Geological Survey, Reston, VA. Water-Supply Paper 2425.

Fugro East, Inc. 1995. *A Method for the Assessment of Wetland Function*. Northborough, MA.

Fusillo, T. V. 1981. *Impact of Suburban Residential Development on Water Resources in the Area of Winslow Township, Camden County, New Jersey*. U.S. Geological Survey, Water Resources Division, Trenton, NJ. Water Resources Investigations 8-27.

Geis, J. W., and J. L. Kee. 1977. *Coastal Wetlands along Lake Ontario and St. Lawrence River in Jefferson County, New York*. State University of New York, College of Environmental Science and Forestry, Institute of Environmental Program Affairs, Syracuse, NY.

Gibbs, J. P. 1993. Importance of small wetlands for the persistence of local populations of wetland-associated animals. *Wetlands* 13(1): 25–31.

Gilmore, M. R. 1919. *Uses of Plants by the Indians of the Missouri River Region*. Bureau of American Ethnology. Thirty-third Annual Report.

Ginsberg, H. S. (editor). 1993. *Ecology and Environmental Management of Lyme Disease*. Rutgers University Press, New Brunswick, NJ.

Gosner, K. L. 1979. *A Field Guide to the Atlantic Seashore*. Houghton Mifflin Company, Boston, MA.

Government of Canada and U.S. Environmental Protection Agency. 2002. *The Great Lakes: An Environmental Atlas and Resource Book*. Environment Canada, Ontario Region, Downsview, Ontario, and U.S. EPA, Great Lakes National Program Office, Chicago, IL.

Green, J. 1971. *The Biology of Estuarine Animals*. University of Washington Press, Seattle, WA.

Harshberger, J. W. 1970. *The Vegetation of the New Jersey Pine-Barrens*. Dover Publications, New York. Reprint of 1916 publication.

Herdendorf, C. E., C. N. Raphael, and E. Jaworski. 1986. *The Ecology of Lake St. Clair Wetlands: A Community Profile*. U.S. Fish and Wildlife Service, Washington, DC. Biol. Rep. 85 (7.7).

Homebuilders Association (New York). 1994. Redelineation of wetlands possible under '87 manual. *Homebuilders Association Newsletter*, January 1994.

Jeffrey, J. A. 1921. *Text-book of Land Drainage*. Macmillan Company, New York.

Joabsson, A., and T. R. Christensen. 2002. Wetlands and methane emission. In *Encyclopedia of Soil Science*, ed. R. Lal, pp. 1429–1432. Marcel Dekker Inc., New York.

Johnson, T. 1995. Study casts doubt on flood plain buyout option. *Star-Ledger*, Newark, NJ, October 17, 1995, pp. 1 and 6.

Jorgensen, N. 1977. *A Guide to New England's Landscape*. Globe Pequot Press, Chester, CT.

Josselyn, J. 1972. *New-Englands Rarities Discovered*. Massachusetts Historical Society, Boston, MA.

Keller, C.M.E., C. S. Robbins, and J. S. Hatfield. 1993. Avian communities in riparian forests of different widths in Maryland and Delaware. *Wetlands* 13: 137–144.

Kenney, L. P. 1994. *Wicked Big Puddles: A Guide to the Study and Certification of Vernal Pools*. Vernal Pool Association, Reading Memorial High School, Reading, MA.

Kier Associates. 1994. *Fisheries, Wetlands and Jobs: The Value of Wetlands to America's Fisheries*. Prepared for the Campaign to Save California Wetlands, Sausalito, CA.

Kilgo, J. C., R. A. Sargent, B. R. Chapman, and K. V. Miller. 1998. Effects of stand width and adjacent habitat on breeding bird communities in bottomland hardwoods. *J. Wildl. Manag.* 62: 72–83.

Kingsley, J. S. (editor). 1884. *The Standard Natural History*. Volume II. Crustacea and Insects. S. E. Cassino and Company, Boston, MA.

Klots, E. B. 1966. *The New Field Book of Freshwater Life*. G. P. Putnam's Sons, New York, NY.

Kraft, H. C. 1986. *The Lenape: Archeology, History, and Ethnography*. New Jersey Historical Society, Newark.

Leck, C. F. 1984. *The Status and Distribution of New Jersey's Birds*. Rutgers University Press, New Brunswick, NJ.

Lobeck, A. K. 1932. *Atlas of American Geology*. C. S. Hammond & Company, Maplewood, NJ.

Lugo, A. E., S. Brown, and M. M. Brinson. 1990. Concepts in wetland ecology. In *Forested Wetlands*, ed. A. E. Lugo, M. Brinson, and S. Brown, pp. 53–85. Elsevier, Amsterdam.

Luttenberg, D., D. Lev, and M. Feller. 1993. *Native Species Planting Guide for New York City and Vicinity*. Natural Resources Group, Parks and Recreation, City of New York.

Lyford, W. H. 1964. *Water Table Fluctuations in Periodically Wet Soils of Central New England*. Harvard University. Harvard Forest, Petersham, MA. Paper No. 8.

Mack, J. J. 2001. *Ohio Rapid Assessment Method for Wetlands, v. 5.0: User's Manual and Forms*. Ohio Environmental Protection Agency, Division of Surface Water, 401/Wetland Ecology Unit, Columbus, OH. Technical report WET/2001-1. http://epa.state.oh.us/dsw/401/oram50sf_s.pdf.

MacKenzie, C. L., Jr. 1992. *The Fisheries of Raritan Bay*. Rutgers University Press, New Brunswick, NJ.

Magee, D. W. 1981. *Freshwater Wetlands: A Guide to Common Indicator Plants of the Northeast*. University of Massachusetts Press, Amherst, MA.

Maryland Department of the Environment. 1997. *Maryland Department of the Environment Landscape Level Wetland Functional Assessment Method: Additional Guidance for Use, Regionalization, and Revisions*. Baltimore, MD. March 26, 1997, memorandum.

McColligan, E. T., Jr., and M. L. Kraus. No date. Exotic wetland plants of New Jersey. New Jersey Department of Environmental Protection, Division of Coastal Resources, Trenton. Unpublished mimeo.

Michaud, J. 1992. Rhode Island butterflies and their host plants: #2—Bogs and their coppers. *Rhode Island Wild Plant Society Newsletter* 9:2–4.

Miller, R. E., and B. E. Gunsalus. 1997. *Wetland Rapid Assessment Procedure (WRAP)*. South Florida Water Management District, Natural Resource Management Division, West Palm Beach, FL. Tech. Pub. REG-001.

Millspaugh, C. F. 1974. *American Medicinal Plants*. Dover Publications, New York.

Minnesota Department of Natural Resources. 2004. *Minnesota Rapid Assessment Method for Evaluating Wetland Functions (MnRAM 3.0)*. St. Paul, MN. http://www.bwsr.state.mn.us/wetlands/mnram

Minnesota Department of Natural Resources. 1987. *The Minnesota Peat Program Summary Report 1981–86*. St. Paul, MN.

Mirick, P. G. 1994. Empire builders: The life and history of America's amazing rodent. *Massachusetts Wildlife* 44(4):6–21.

Moore, P. D., and D. J. Bellamy. 1974. *Peatlands*. Springer-Verlag, New York.

Morgan, A. H. 1930. *Field Book of Ponds and Streams*. G. P. Putnam's Sons, New York, NY.

Motzkin, G. 1992. Vegetation classification and environmental characteristics of calcareous fens of western New England and adjacent New York State. Draft report submitted to the Massachusetts Natural Heritage and Endangered Species Program and the Eastern Heritage Task Force, the Nature Conservancy.

Motzkin, G. 1991. *Atlantic White Cedar Wetlands of Massachusetts*. Massachusetts Agricultural Experiment Station, Amherst, MA. Research Bulletin No. 731.

New York State Forestry. 2000. *Best Management Practices for Water Quality. BMP Field Guide*. New York State Department of Environmental Conservation, Division of Lands and Forests, Albany, NY.

Nicholas, G. P. 1991. Putting wetlands into perspective. *Man in the Northeast* 42:29–38.

Nicholas, G. P. 1991. Places and spaces: Changing patterns of wetland use in southern New England. *Man in the Northeast* 42:75–98.

Niering, W. A. 1966. *The Life of the Marsh*. McGraw-Hill Book Co., New York.

Novitski, R. P. 1982. *Hydrology of Wisconsin Wetlands*. U.S. Geological Survey, Reston, VA. Information Circular 40.

O'Connor, A. 2003. South Florida freezes are linked to draining of wetlands. *New York Times*, November 25, 2003.

Office of Technology Assessment, U.S. Congress. 1984. *Wetlands: Their Use and Regulation*.

Superintendent of Documents, U.S. Government Printing Office, Washington, DC. OTA-O-206.

Orson, R. A., R. S. Warren, and W. A. Niering. 1987. Development of a tidal marsh in a New England river valley. *Estuaries* 10:20–27.

Orson, R. A., R. L. Simpson, and R. E. Good. 1990. Rates of sediment accumulation in a tidal freshwater marsh. *Journal of Sedimentary Petrology* 60:859–869.

Owens, O. D. 1993. *Living Waters: How to Save Your Local Stream*. Rutgers University Press, New Brunswick, NJ.

Palmer, E. L., and H. S. Fowler. 1975. *Fieldbook of Natural History*. McGraw-Hill Book Co., New York.

Paustian, K. 2002. Organic matter and global C cycle. In *Encyclopedia of Soil Science*, ed. R. Lal, pp. 895–898. Marcel Dekker, Inc., New York.

Pinelands Commission. 1980. *Comprehensive Management Plan for the Pinelands Natural Reserve (National Parks and Recreation Act, 1978) and Pinelands Area (New Jersey Pinelands Protection Act, 1979)*. New Lisbon, NJ.

Reed, P. B., Jr. 1988. *National List of Plant Species That Occur in Wetlands: Northeast Region*. U.S. Fish and Wildlife Service, Washington, D.C. Biological Report 88(26.1).

Reed, P. B., Jr. (compiler). 1997. *Revision of the National List of Plant Species That Occur in Wetlands*. U. S. Fish and Wildlife Service, Washington, D. C.

Reschke, C. 1990. *Ecological Communities of New York State*. New York Department of Environmental Conservation, Natural Heritage Program, Latham, NY.

Robinson, A. 1995. Small and seasonal does not mean insignificant: Why it's worth standing up for tiny and temporary wetlands. *Journal of Soil and Water Conservation* 50(6):586–590.

Rummer, B. 2004. Managing water quality in wetlands with forestry BMPs. *Water, Air, and Soil Pollution: Focus* 4: 55–66.

Scala, G. G. 1995. Preservation efforts sealed in area land deal with state. *Lacey Beacon*, Manahawkin, New Jersey, July 27, 1995.

Shaler, N. S. 1890. *General Account of the Freshwater Morasses of the United States, with a Description of the Dismal Swamp District of Virginia and North Carolina*. U.S. Geological Survey, Washington, DC. 10th Annual Report, 1888–1889.

Shaw, S. P., and C. G. Fredine. 1956. *Wetlands of the United States: Their Extent and Their Value to Waterfowl and Other Wildlife*. U.S. Fish and Wildlife Service, Washington, DC. Circular 39.

Simmons, R. C., A. J. Gold, and P. M. Groffman. 1992. Nitrate dynamics in riparian forests: groundwater studies. *J. Environ. Qual.* 21: 659–665.

Simon, B. G. 1991. Prehistoric land use and changing paleoecological conditions at Titicut Swamp in southeastern Massachusetts. *Man in the Northeast* 42:63–74.

Soil Survey Division Staff. 1993. *Soil Survey Manual*. U.S. Department of Agriculture, Washington, DC. Handbook No. 18.

Stevens, W. K. 1995. Restored wetlands could erase threat of Mississippi floods. *New York Times*, August 8, 1995, pp. C1 and C4.

Stewart, R. E., Jr. 1996. Wetlands as bird habitat. In *National Water Summary on Wetland Resources*, compilers J. D. Fretwell, J. S. Williams, and P. J. Redman. U.S. Geological Survey, Reston, VA. Water-Supply Paper 2425: 49–56.

Thien, S. J. 1979. A flow diagram for teaching texture-by-feel analysis. *Journal of Agronomic Education* 8:54–55.

Thomson, K. S., W. H. Weed III, and A. G. Taruski. 1971. *Saltwater Fishes of Connecticut*. State Geological and Natural History Survey of Connecticut, Hartford, CT. Bulletin 105.

Tiner, R. W. 1985. *Wetlands of New Jersey*. U.S. Fish and Wildlife Service, Region 5, Newton Corner, MA.

Tiner, R. W. 1987. *A Field Guide to Coastal Wetland Plants of the Northeastern United States*. University of Massachusetts Press, Amherst, MA.

Tiner, R. W. 1987. *Preliminary National Wetlands Inventory Report on Vermont's Wetland Acreage*. U.S. Fish and Wildlife Service, Region 5, Hadley (formerly Newton Corner), MA.

Tiner, R. W. 1988. *Field Guide to Nontidal Wetland Identification*. U.S. Fish and Wildlife Service, Region 5, Newton Corner, MA, and Maryland Department of Natural Resources, Annapolis, MD. (Reprinted by IWEER, P. O. Box 288, Leverett, MA 01054.)

Tiner, R. W. 1990. *Pennsylvania's Wetlands: Current Status and Recent Trends*. U.S. Fish and Wildlife Service, Newton Corner, MA.

Tiner, R. W. 1992. *Preliminary National Wetlands Inventory Report on Massachusetts' Wetland Acreage*. U. S. Fish and Wildlife Service, Region 5, Hadley (formerly Newton Corner), MA.

Tiner, R. W. 1993. *Field Guide to Coastal Wetland*

Plants of the Southeastern United States. University of Massachusetts Press, Amherst, MA.

Tiner, R. W. 1994. *A Guide to Identifying Wetlands of the Southeast.* Institute for Wetland & Environmental Education & Research, Leverett, MA.

Tiner, R. W. 1994. *Maine Wetlands and Their Boundaries.* State of Maine, Department of Economic and Community Development, Office of Community Development, Augusta, ME.

Tiner, R. W. 1999. *Wetland Monitoring Guidelines.* Operational draft. U.S. Fish and Wildlife Service, Northeast Region, Hadley, MA.

Tiner, R. W., and D. G. Burke. 1995. *Wetlands of Maryland.* U.S. Fish and Wildlife Service, Region 5, Hadley, MA and Maryland Department of Natural Resources, Annapolis, MD. Cooperative publication.

Tiner, R., R. Lichvar, R. Franzen, C. Rhodes, and W. Sipple. 1995. *Supplement to the List of Plant Species That Occur in Wetlands: Northeast (Region 1).* U.S. Fish and Wildlife Service, Washington, DC. Supplement to Biological Report 88 (26.1).

Tiner, R. W., J. Q. Swords, and B. J. McClain. 2002. *Wetland Status and Trends for the Hackensack Meadowlands.* U.S. Fish and Wildlife Service, Northeast Region, Hadley, MA.

Tiner, R. W., and P.L.M. Veneman. 1995. *Hydric Soils of New England.* University of Massachusetts Extension, Amherst, MA. Bulletin C-183R.

Trettin, C. C., and M. F. Jurgensen. 2003. Carbon cycling in wetland forest soils. In *The Potential of U.S. Forest Soils to Sequester Carbon and Mitigate the Greenhouse Effect,* ed. J. M. Kimble, L. S. Heath, R. A. Birdsey, and R. Lal, pp. 311–331. CRC Press, Boca Raton, FL.

Twilley, R. R., and R. Chen. 1998. A water budget hydrology model for a basin mangrove forest in Rookery Bay, Florida. *Marine and Freshwater Research* 49(9): 309–323.

U.S. Army Corps of Engineers. 2004. *Waterways Experiment Station, Vicksburg, MS.* http://www.wes.army.mil/el/emrrp/emris/emrishelp6

U.S. Army Corps of Engineers. 1995. *Section 404 of the Clean Water Act and Wetlands.* Regulatory Branch, Washington, DC. Special Statistical Report.

U.S. Army Corps of Engineers. 1995. *The Highway Methodology Workbook Supplement: Wetland Functions and Values: A Descriptive Approach.* New England Division, Concord, MA. NEDEP-360-1-30a. http://www.nae.usace.army.mil/reg/hwsplmnt.pdf

USDA Soil Conservation Service. 1982. *National List of Scientific Plant Names.* U.S. Department of Agriculture, Washington, DC. SCS-TP-159.

U.S. Fish and Wildlife Service. 1990. *Regional Wetlands Concept Plan: Emergency Wetlands Resources Act.* U.S. Fish and Wildlife Service, Northeast Region, Hadley, MA.

U.S. General Accounting Office. 2004. *Waters and Wetlands. Corps of Engineers Needs to Evaluate Its District Office Practices in Determining Jurisdiction.* Report to the Chairman, Subcommittee on Energy Policy, Natural Resources and Regulatory Affairs, Committee on Government Reform, House of Representatives, Washington, DC. GAO-04-297. (http://www.gao.gov/cgi-bin/getrpt?GAO-04-297)

Van Diver, B. B. 1985. *Roadside Geology. New York.* Mountain Press Publishing Company, Missoula, MT.

Viosca, P. 1928. Louisiana wetlands and the value of their wildlife and fisheries resources. *Ecology* 9:216–229.

Voesenek, L.A.C.J., C.W.P.M. Blom, and R. H. Pouwels. 1989. Root and shoot development of Rumex species under waterlogged conditions. *Canadian Journal of Botany* 67:1865–1869.

Walton, T. E., and R. Patrick (editors). 1973. *The Delaware Estuary System: Environmental Impacts and Socio-economic Effects.* Delaware River Estuarine Marsh Survey. The Academy of Natural Sciences, Philadelphia, The University of Delaware, Newark, and Rutgers University, New Brunswick, NJ.

Wander, W. 1980. Breeding birds of southern New Jersey cedar swamps. *Records of New Jersey Birds,* 6(4):51–65. Occasional Paper No. 138. NJ Audubon.

Ward, H. B., and G. C. Whipple. 1918. *Fresh-Water Biology.* John Wiley & Sons, Inc., New York.

Warming, E. 1909. *Oecology of Plants: An Introduction to the Study of Plant-Communities.* Clarendon Press, Oxford, England. (Updated English version of a 1896 text).

Wayland, R. H., III. 1995. The Clinton administration's perspective on wetlands protection. *Journal of Soil and Water Conservation* 50(6):581–584.

Weatherbee, P. B., and G. E. Crow. 1992. Natural plant communities of Berkshire County, Massachusetts. *Rhodora* 94:171–209.

Weir, W. W. 1920. *Productive Soils*. J. B. Lippincott Co., Philadelphia.

Wharton, C. H., W. M. Kitchens, E. C. Pendleton, and T. W. Sipe. 1982. *The Ecology of Bottomland Hardwood Swamps of the Southeast: A Community Profile*. U.S. Fish and Wildlife Service, Washington, D.C. FWS/OBS-81/37.

Whitworth, W. R., P. L. Berrien, and W. T. Keller. 1968. *Freshwater Fishes of Connecticut*. State Geological and Natural History Survey of Connecticut, Hartford, CT. Bulletin 101.

Widmer, K. 1964. *The Geology and Geography of New Jersey*. D. Van Nostrand Company, Princeton, NJ.

Wiest, R. L. 1998. *A Landowner's Guide to Building Forest Access Roads*. U.S. Forest Service, Northeast Area, Radnor, PA. NA-TP-06-98.

Wisconsin Department of Natural Resources. 1992. *Rapid Assessment Methodology for Evaluating Wetland Functional Values*. Madison, WI.

Wright, J. O. 1907. *Swamp and Overflowed Lands in the United States*. U.S. Department of Agriculture, Office of Experiment Stations, Washington, DC. Circular 76.

Zhang, L., and W. J. Mitsch. 2002. Water budgets of the two Olentangy River experimental wetlands in 2001. In *Annual Report*, ed. W. J. Mitsch and L. Zhang, pp. 23–28. Olentangy River Wetland Research Park at the Ohio State University, Columbus, OH.

Index

About the Author

Ralph Tiner is a wetland ecologist with over twenty-five years of experience in wetland identification, classification, mapping, and delineation. A nationally recognized expert in wetland delineation, he has written more than sixty publications and serves as an adjunct professor in the Department of Plant and Soil Sciences at the University of Massachusetts, Amherst.

CONVERSION TABLE

English Units	Metric Equivalents
1/25 inch	1 millimeter
1/5 inch	5 millimeters
1/4 inch	6 millimeters
1/2 inch	12 millimeters
1 inch	2.5 centimeters
1 foot	30 centimeters
3.3 feet	1 meter
10 feet	3 meters

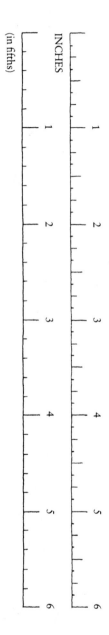